# Design of Industrial Structures

# Design of Industrial Structures

## Reinforced Cement Concrete and Steel

Ashoke Kumar Dasgupta

CRC Press
Taylor & Francis Group
Boca Raton London New York

CRC Press is an imprint of the
Taylor & Francis Group, an **informa** business

First edition published 2022
by CRC Press
6000 Broken Sound Parkway NW, Suite 300, Boca Raton, FL 33487-2742

and by CRC Press
2 Park Square, Milton Park, Abingdon, Oxon, OX14 4RN

© 2022 Ashoke Dasgupta

CRC Press is an imprint of Taylor & Francis Group, LLC

Reasonable efforts have been made to publish reliable data and information, but the author and publisher cannot assume responsibility for the validity of all materials or the consequences of their use. The authors and publishers have attempted to trace the copyright holders of all material reproduced in this publication and apologize to copyright holders if permission to publish in this form has not been obtained. If any copyright material has not been acknowledged please write and let us know so we may rectify in any future reprint.

Except as permitted under U.S. Copyright Law, no part of this book may be reprinted, reproduced, transmitted, or utilized in any form by any electronic, mechanical, or other means, now known or hereafter invented, including photocopying, microfilming, and recording, or in any information storage or retrieval system, without written permission from the publishers.

For permission to photocopy or use material electronically from this work, access www.copyright.com or contact the Copyright Clearance Center, Inc. (CCC), 222 Rosewood Drive, Danvers, MA 01923, 978-750-8400. For works that are not available on CCC please contact mpkbookspermissions@tandf.co.uk

*Trademark notice*: Product or corporate names may be trademarks or registered trademarks and are used only for identification and explanation without intent to infringe.

---

**Library of Congress Cataloging-in-Publication Data**

---

Names: Dasgupta, Ashoke Kumar, author.
Title: Design of industrial structures : reinforced cement concrete and steel / Ashoke Kumar Dasgupta.
Description: First edition. | Boca Raton : CRC Press, 2022. | Includes bibliographical references and index.
Identifiers: LCCN 2021026047 (print) | LCCN 2021026048 (ebook) | ISBN 9781032078380 (hbk.) | ISBN 9781032078397 (pbk.) | ISBN 9781003211754 (ebk.)
Subjects: LCSH: Building, Iron and steel. | Industrial buildings--Design and construction. | Buildings, Reinforced concrete--Design and construction.
Classification: LCC TH1611 .D335 2022 (print) | LCC TH1611 (ebook) | DDC 693/.71--dc23
LC record available at https://lccn.loc.gov/2021026047
LC ebook record available at https://lccn.loc.gov/2021026048

---

ISBN: 978-1-032-07838-0 (hbk)
ISBN: 978-1-032-07839-7 (pbk)
ISBN: 978-1-003-21175-4 (ebk)

DOI: 10.1201/9781003211754

Typeset in Times
by Deanta Global Publishing Services, Chennai, India

# *Dedication*

*To my parents*

# Contents

Preface ..................................................................................................... xv
Acknowledgment ................................................................................. xvii
Author .................................................................................................. xix

**Chapter 1**  Contracts and Specification ................................................ 1
  1.1  Type of Contracts ..................................................................... 1
  1.2  Tender Specifications for EPC and Item Rate Contracts .......... 3
  1.3  Tender Evaluation ..................................................................... 5
  1.4  Technical Specifications ........................................................... 7
  1.5  Reference Codes of Practice and Standards ........................... 11
      1.5.1  American Standards ..................................................... 11
      1.5.2  British Standards .......................................................... 12
      1.5.3  Indian Standards .......................................................... 12

**Chapter 2**  Site Survey and Soil Investigation ................................... 15
  2.1  Reconnaissance and Topographical Survey ............................ 15
  2.2  Site Preparation and Enabling Work ...................................... 16
  2.3  Soil Investigation Work Including Applications for Field
       Tests ........................................................................................ 17
  2.4  Ground Improvement Works .................................................. 23
      2.4.1  General ......................................................................... 23
      2.4.2  Types of Soil Requiring Ground Improvement ........... 23
      2.4.3  Methods of Ground Improvement ............................... 23
      2.4.4  Preloading and Prefabricated Vertical Drains ............. 24
            2.4.4.1  Preloading and Sand Drains ......................... 25
      2.4.5  Dynamic Compaction .................................................. 25
            2.4.5.1  Dynamic Compaction by Falling Drop
                     Hammers ....................................................... 25
            2.4.5.2  Dynamic Compaction by Vibro-
                     Compaction and Vibro-Replacement ........... 27
      2.4.6  Cement Deep Mixing Piling ........................................ 28
      2.4.7  Compaction Piling ....................................................... 28

**Chapter 3**  Materials and Uses ........................................................... 31
  3.1  General .................................................................................... 31
  3.2  Cement .................................................................................... 32
  3.3  Water ....................................................................................... 33
  3.4  Aggregate ................................................................................ 33

|     |     |     |     |
| --- | --- | --- | --- |
| | 3.5 | Admixture | 34 |
| | 3.6 | Reinforcing Steel | 35 |
| | 3.7 | Formwork | 36 |
| | 3.8 | Concrete Mixes | 38 |
| | | 3.8.1 Nominal Mix Concrete | 39 |
| | | 3.8.2 Design Mix Concrete | 39 |
| | 3.9 | Method of Concreting | 40 |
| | | 3.9.1 Placement of Concrete | 42 |
| | | 3.9.2 Vibration for Compaction | 42 |
| | | 3.9.3 Concreting under Special Conditions | 42 |
| | | 3.9.4 Method of Construction | 44 |
| | 3.10 | Curing | 46 |
| | 3.11 | Quality Control | 46 |
| | 3.12 | Joints and Materials | 47 |
| | | 3.12.1 Construction Joints | 48 |
| | | 3.12.2 Expansion Joints | 51 |
| | | 3.12.3 Separation or Isolation Joints | 53 |
| | | 3.12.4 Contraction Joint | 53 |
| | 3.13 | Embedment Details | 54 |
| | 3.14 | Structural Steel Materials | 55 |
| | 3.15 | Bolts and Weld | 56 |
| **Chapter 4** | Design Basis Report | | 57 |
| | 4.1 | Introduction | 57 |
| | 4.2 | Items to Be Covered | 57 |
| | | 4.2.1 Description of the Building and Structure | 58 |
| | | 4.2.2 Units and Language | 58 |
| | | 4.2.3 Site Information and Development | 58 |
| | | 4.2.4 Soil Characteristics and Type of Foundation | 58 |
| | | 4.2.5 Codes and Standards to be Followed for Loads and Design | 58 |
| | | 4.2.6 Load and Load Combinations | 58 |
| | |     4.2.6.1 Load | 58 |
| | |     4.2.6.2 Combination of Loads | 60 |
| | | 4.2.7 Materials to Be Used | 60 |
| | | 4.2.8 Design Concept of the Building Structure | 60 |
| | | 4.2.9 Protection against Corrosion (Subsoil and in the Air) | 60 |
| | | 4.2.10 Architectural Works | 61 |
| | 4.3 | Design Criteria of an Auxiliary Plant Building: RCC Structure without a Crane (International Standards-Based) | 61 |
| | | 4.3.1 Description of the Building | 61 |
| | | 4.3.2 Units and Language | 61 |
| | | 4.3.3 Site Information | 61 |

|  |  | 4.3.3.1 | Meteorological Data | 61 |
|---|---|---|---|---|
|  | 4.3.4 | Soil Characteristics and Foundation System | | 62 |
|  |  | 4.3.4.1 | Design Pile Capacity | 62 |
|  | 4.3.5 | Codes and Standards | | 62 |
|  | 4.3.6 | Load and Load Combinations | | 62 |
|  |  | 4.3.6.1 | Dead Load | 62 |
|  |  | 4.3.6.2 | Equipment Load | 62 |
|  |  | 4.3.6.3 | Live Load | 63 |
|  |  | 4.3.6.4 | Wind Load | 63 |
|  |  | 4.3.6.5 | Seismic Load | 63 |
|  |  | 4.3.6.6 | Temperature Load | 63 |
|  |  | 4.3.6.7 | Load Combinations | 63 |
|  | 4.3.7 | Materials | | 64 |
|  |  | 4.3.7.1 | Grade of Concrete | 64 |
|  |  | 4.3.7.2 | Reinforcement Steel | 65 |
|  | 4.3.8 | Design Concept | | 65 |
|  | 4.3.9 | Corrosion Protection | | 65 |
|  |  | 4.3.9.1 | Concrete Cover | 65 |
|  | 4.3.10 | Architecture Finishes | | 65 |
| 4.4 | Design Criteria of a Steam Turbine Generator Building: Steel Structure with an EOT Crane (Indian Standards-Based) | | | 65 |
|  | 4.4.1 | Description of the Building | | 66 |
|  | 4.4.2 | Units and Language | | 67 |
|  | 4.4.3 | Site Information | | 67 |
|  | 4.4.4 | Soil Characteristics and Foundation System | | 67 |
|  | 4.4.5 | Codes and Standards | | 68 |
|  | 4.4.6 | Load and Load Combinations | | 68 |
|  |  | 4.4.6.1 | Dead Load | 68 |
|  |  | 4.4.6.2 | Live Load | 69 |
|  |  | 4.4.6.3 | Wind Load | 69 |
|  |  | 4.4.6.4 | Seismic Load | 69 |
|  |  | 4.4.6.5 | Temperature Load | 69 |
|  |  | 4.4.6.6 | Load Combinations | 70 |
|  | 4.4.7 | Materials | | 71 |
|  |  | 4.4.7.1 | Structural Steel | 71 |
|  |  | 4.4.7.2 | Concrete Works | 71 |
|  |  | 4.4.7.3 | Reinforcement Bars | 72 |
|  | 4.4.8 | Design Concept | | 72 |
|  | 4.4.9 | Corrosion Protection | | 72 |
|  |  | 4.4.9.1 | Steel Structures | 72 |
|  |  | 4.4.9.2 | Concrete Cover | 72 |
|  | 4.4.10 | Architecture Finishes | | 72 |
| 4.5 | Conclusions | | | 73 |

| Chapter 5 | Conceptual Design | 75 |
|---|---|---|
| | 5.1 Introduction | 75 |
| | 5.2 Selection of the Foundation Type | 75 |
| | 5.3 Type of Structure and Material of Construction | 76 |
| | 5.4 Hybrid Structures | 79 |
| | 5.5 Cost Optimization | 80 |

| Chapter 6 | Procurement of Materials | 81 |
|---|---|---|
| | 6.1 General | 81 |
| | 6.2 Structural Steelwork | 81 |
| | 6.3 Concrete Structure | 86 |

| Chapter 7 | Design Office Procedures for Engineering and Drawing | 89 |
|---|---|---|
| | 7.1 Timeline Planning | 89 |
| | 7.2 Civil Input Data | 89 |
| | 7.3 Structural Framing, Loading, and Analysis | 91 |
| |     7.3.1 Structural Framing System | 91 |
| |     7.3.2 Method of Analysis | 93 |
| | 7.4 Drawings for Reinforced Concrete Works and Bar Bending Schedules | 94 |
| | 7.5 Drawings for Structural Steelwork | 95 |
| | 7.6 Interdisciplinary Coordination | 96 |
| | 7.7 Quantity Estimation | 96 |
| | 7.8 Design Coordination with the Construction Site | 97 |
| | 7.9 As-Built Drawings | 97 |

| Chapter 8 | Steel Structures | 99 |
|---|---|---|
| | 8.1 General | 99 |
| | 8.2 Purlins and Side Girts | 99 |
| | 8.3 Roof Trusses | 100 |
| | 8.4 Lattice Girders | 101 |
| | 8.5 Roof Bracings | 101 |
| | 8.6 Beam and Girders | 101 |
| | 8.7 Crane Girders | 102 |
| | 8.8 Column Bracing and Tie Members | 104 |
| | 8.9 Column Base Plates and Anchor Bolts | 105 |
| | 8.10 Miscellaneous Structures: Gable End Structures, Stairs, and Platforms | 106 |
| | 8.11 Pipe and Cable Racks | 106 |
| | 8.12 Connections | 108 |
| |     8.12.1 Bolted Connections | 109 |
| |         8.12.1.1 Bearing Type Connections | 109 |

# Contents

|  |  | 8.12.1.2 | Friction Grip or Non-Slip or Slip-Resistant Type | 109 |
|---|---|---|---|---|
|  | 8.12.2 | Weld Connections | | 110 |
|  |  | 8.12.2.1 | Methods of Welding | 110 |
|  |  | 8.12.2.2 | Connection to RCC Members | 117 |
| 8.13 | Coal Bunker | | | 118 |
| 8.14 | Design Examples | | | 125 |

## Chapter 9  Reinforced Concrete and Associated Work ..... 223

- 9.1 General ..... 223
- 9.2 Roof Slabs ..... 224
- 9.3 Floor Slabs ..... 224
- 9.4 Ground Floor Slabs ..... 225
- 9.5 Walls ..... 225
- 9.6 Beams ..... 226
- 9.7 Columns ..... 226
- 9.8 Foundations Including Pile and Caisson ..... 226
  - 9.8.1 Foundation Types and Selection Criteria ..... 227
  - 9.8.2 Loading Pattern ..... 227
  - 9.8.3 Type of Foundations ..... 229
    - 9.8.3.1 Foundations Resting on Soil ..... 229
    - 9.8.3.2 Foundations Resting on Pile or Caisson ..... 230
    - 9.8.3.3 Well Foundations ..... 231
  - 9.8.4 Selection of Foundation Types ..... 231
  - 9.8.5 Settlement ..... 232
  - 9.8.6 Shallow Depth Foundations ..... 233
    - 9.8.6.1 Isolated Footing ..... 233
    - 9.8.6.2 Combined Footing and Strip Footing ..... 237
    - 9.8.6.3 Mat or Raft Foundation ..... 238
  - 9.8.7 Deep-Seated Foundations Including Caisson or Well Foundations ..... 239
    - 9.8.7.1 Pile Foundations ..... 239
    - 9.8.7.2 Pile Load Tests ..... 245
    - 9.8.7.3 Pile Types ..... 245
  - 9.8.8 Caisson or Well Foundation ..... 249
- 9.9 Steel Tank Foundations ..... 249
  - 9.9.1 Sand Pad ..... 250
  - 9.9.2 RCC Ring Wall Foundation ..... 251
  - 9.9.3 RCC Mat Resting on Pile ..... 251
- 9.10 Machine Foundation ..... 252
  - 9.10.1 Types of Machines and Foundations ..... 253
  - 9.10.2 Codes and Standards ..... 256
  - 9.10.3 Materials Used ..... 256
  - 9.10.4 Vibration Isolation ..... 256

|  |  |  |
|---|---|---|
|  | 9.10.5 | General Guidelines ................................................... 257 |
| 9.11 | Miscellaneous Structures ....................................................... 257 |
|  | 9.11.1 | RCC Reservoir .......................................................... 257 |
|  | 9.11.2 | Pipe and Cable Trenches ........................................ 258 |
|  | 9.11.3 | RCC Culverts at Road Crossings ........................... 259 |
| 9.12 | Retaining Walls ...................................................................... 259 |
| 9.13 | Sheet Pile and Diaphragm Walls ........................................... 260 |
| 9.14 | Embankment Lining and Shore Protection Work ................ 261 |
|  | 9.14.1 | Embankment Lining ................................................ 261 |
|  | 9.14.2 | Shore Protection Work ........................................... 262 |
| 9.15 | Silos ......................................................................................... 262 |
| 9.16 | Stacks ...................................................................................... 270 |
| 9.17 | Design Examples .................................................................... 278 |

## Chapter 10  Prestressed Precast Concrete .............................................................. 299

   10.1  Introduction ............................................................................ 299
   10.2  Member Design ...................................................................... 299

## Chapter 11  Plant Area Roads and Drainage ........................................................... 315

   11.1  Roads and Pavements ............................................................ 315
      11.1.1  Flexible Pavement .................................................. 315
      11.1.2  Rigid Pavement ....................................................... 315
      11.1.3  Interlocking Paver Blocks ...................................... 316
   11.2  Drainage .................................................................................. 318
   11.3  Design Steps for Stormwater Drains with Buried Pipelines ....... 319
      11.3.1  Reference Documents for Design Information ........ 319
      11.3.2  Preparation of Catchment Plan and the
               Network for Design ................................................. 319
      11.3.3  Pipe Size Calculation .............................................. 319

## Chapter 12  Design of Industrial Structures ........................................................... 321

   12.1  Design of a RCC Four-Story Building for
         a Packaging Plant ................................................................... 321
      12.1.1  Introduction ............................................................. 321
      12.1.2  References ............................................................... 322
      12.1.3  Material .................................................................... 322
      12.1.4  Load on the Building .............................................. 323
      12.1.5  Structural Analysis .................................................. 327
      12.1.6  Design of Columns ................................................. 328
      12.1.7  Design of Foundations ............................................ 330
      12.1.8  Design of Grade beams ........................................... 333
      12.1.9  Design of Floor & Roof Beams .............................. 335
      12.1.10 Design of Slabs & Stair Case ................................. 337

12.1.11 Computer Model ......................................................... 341
12.1.12 Drawings ................................................................... 343
12.2 Design of a Factory Shed with a Crane Facility for a
Machine Shop ......................................................................... 354
    12.2.1 Description of Building Structure ............................ 354
    12.2.2 Materials and Codes .................................................. 354
    12.2.3 Reference Codes and Documents ............................ 354
    12.2.4 Method of Ananlysis and Design ............................. 354
    12.2.5 Sketch ......................................................................... 355
    12.2.6 Loading ...................................................................... 356
    12.2.7 Analysis of Frame ..................................................... 358
    12.2.8 Design of Primary Members ..................................... 359
        12.2.8.1 Purlin with hollow rectangular section
                (TATA section) ......................................... 359
        12.2.8.2 Truss: Top chord with hollow
                rectangular section (TATA section) .......... 360
        12.2.8.3 Truss: Bottom chord with hollow
                rectangular section (TATA section) .......... 362
        12.2.8.4 Truss: Diagonals with hollow
                rectangular section (TATA section) .......... 365
        12.2.8.5 Truss: Vericals with hollow rectangular
                section (TATA section) ............................. 367
        12.2.8.6 Column: Twin legs with hollow
                rectangular section (TATA section) .......... 369
        12.2.8.7 Column bracing: Single leg hollow
                rectangular section (TATA section) .......... 371
        12.2.8.8 Rafter bracing: Single leg hollow
                rectangular section (TATA section) .......... 373
        12.2.8.9 Design of RCC Grade beam ..................... 375
        12.2.8.10 Design of RCC footing ............................. 376
        12.2.8.11 Column Base and Anchor bolt ................. 379
        12.2.8.12 Design of RCC pedestal ........................... 380
        12.2.8.13 Design of Steeel Crane Girder .................. 381
    12.2.9 Design Drawings ....................................................... 384
12.3 Design of Pipe and Cable Racks in a Power Plant ............... 391
    12.3.1 Single-Tier Pipe Rack for a Raw Water Pipeline ..... 391
        12.3.1.1 Description ................................................ 391
        12.3.1.2 Reference .................................................. 391
        12.3.1.3 Material ..................................................... 391
        12.3.1.4 Dimensioning ............................................ 391
        12.3.1.5 Load ........................................................... 392
        12.3.1.6 Analysis ..................................................... 393
        12.3.1.7 Design of member sections ...................... 394
    12.3.2 Two tier Pipe rack for Ash Pipes and Cable racks ....... 400
        12.3.2.1 Description ................................................ 400
        12.3.2.2 Reference .................................................. 400

|  |  |  |  |
|---|---|---|---|
| | 12.3.2.3 | Material | 400 |
| | 12.3.2.4 | Sketch | 401 |
| | 12.3.2.5 | Load | 402 |
| | 12.3.2.6 | Member forces in Cross beams | 405 |
| | 12.3.2.7 | Member forces in Lattice Girder | 406 |
| | 12.3.2.8 | Member forces in Plan bracing | 407 |
| | 12.3.2.9 | Member forces in Trestle | 408 |
| | 12.3.2.10 | Member forces in Braced Columns — Bottom part of trestle | 408 |
| | 12.3.2.11 | Member forces in Column bracings | 409 |
| | 12.3.2.12 | Summary of member forces and designed section | 409 |
| | 12.3.2.13 | Member design according to IS 800 : 2007 [by working stress method] | 410 |
| 12.3.3 | Single tier Pipe and Cable racks crossing road | | 441 |
| | 12.3.3.1 | Description | 441 |
| | 12.3.3.2 | Reference | 441 |
| | 12.3.3.3 | Material | 442 |
| | 12.3.3.4 | Sketch | 442 |
| | 12.3.3.5 | Design parameters | 443 |
| | 12.3.3.6 | Loading per meter length of Bridge | 444 |
| | 12.3.3.7 | Analysis | 445 |
| | 12.3.3.8 | Member design | 451 |

12.4 Design of RCC Pipe Sleepers with Cable Racks for an Ash Handling System ... 476
    12.4.1 Description ... 476
    12.4.2 Reference ... 476
    12.4.3 Design parameters ... 476
    12.4.4 Sketch ... 477
    12.4.5 Design of sleeper ... 478
    12.4.6 Cross beams ... 483

12.5 Design of Riverbed Protection Works for an Underwater Discharge Pipeline ... 485
    12.5.1 Introduction ... 485
    12.5.2 Reference documents ... 485
    12.5.3 Design parameters ... 485
    12.5.4 Material ... 485
    12.5.5 Conceptual Sketch ... 486
    12.5.6 Riprap and Bedding [Indian Standard Code Guideline - IS 14262] ... 487
    12.5.7 Riprap/Bedding Stone Sizing [US Army Corps of Engrs Hydraulic Design EM _1110-2-1601] ... 490

Index ... 493

# Preface

The primary objective of writing this book is to guide young engineers entering the practical field of work and professional engineers working in design offices on industrial projects and residential houses. This book is also intended for engineers working on construction sites or urban areas aspiring to design a structure or building but without access to computer software and guides available in city offices. Step-by-step long hand calculations are shown in worked examples.

The essence of the book is to bridge the gap between university curriculum and application in the practical field. The language is simple and supplemented with illustrations and sketches to aid in a clearer understanding.

Chapters have been added on contracts, specifications, soil surveys, and design criteria for designers to understand the environment and objectives of their design work. Industrial structural design is a multidisciplinary task unlike designing a dwelling house. Procedures are given to guide the designer on how to proceed with the construction in phases on the site, negotiating changes in equipment, and design development.

Safety, quality, and economic requirements of structural design are highlighted, and national and international codes and standards are referenced to ensure good engineering practice.

An introduction is given to the latest methods of analysis and design, and the use of advanced construction materials is provided. A brief description of analysis software and drafting tools is also given for the design engineer's reference.

This book is a helpful resource for professional civil and structural engineers, designers, graduate students, and associated designers.

**Ashoke Kumar Dasgupta**
*BE (Civil), FIE*

# Acknowledgment

To my family, friends, and colleagues, as well as my publisher who pursued me to write this book.

Special thanks to my brother-in-law, Mr. Subhendu B. Roy, for editing the language, and my son, Dr. Suman Dasgupta, for his invaluable help.

# Author

**Ashoke Kumar Dasgupta** has worked in reputed consulting engineering companies, Development Consultants Pvt Ltd (India) and The Kuljian Corporation (USA), at their design offices and construction sites in India, the United States, Bangladesh, Middle Eastern countries (Turkey, Saudi Arabia, Syria, Egypt, UAE, Qatar, and Kuwait), and Far East counties (China, Japan, Indonesia, and Vietnam) from 1974 to 2019.

His expertise lies in structural, civil, and foundation engineering for power plants and industrial structures.

# 1 Contracts and Specification

## 1.1 TYPE OF CONTRACTS

Let us try to understand the different types of contracts for industrial projects through an example scenario. Suppose you want to paint your house before the rainy season arrives, but you do not have prior experience of undertaking such an endeavor. Some of the things you may want to know to determine your budget and schedule would be the type of paint needed for outdoor and indoor protection, the cost of labor and materials, time of completion, and so on. So you visit your neighbor, who had their house painted recently and shares with you their experience and knowledge on these items. Based on this preliminary information, you contact three reputed painting companies, and their representatives inspect your house separately. All parties agree to supply material and labor to paint the interior and exterior of your house as per your selection of color and brand of paint from a reputed manufacturer, and indicate their prices and time schedule within the timeline proposed by you. The following week, they present formal offers including a schedule describing individual items of work and their rates. The schedule includes all items of work in detail, for example, type of paint with color shade and number of coats, including a primer coat and surface preparation to be applied on the ceiling and walls for the living room, kitchen, toilet, balcony, doors and windows including grills, stair handrails, and so on, as well as a weatherproof coat on external surfaces, repairing of damp surfaces, and wall plaster in spots where required. This document also includes the supply of labor, materials, scaffolding, and so on. The schedule of items includes items of work, estimated areas in square feet, and individual rates of items. The summary of the cost and time schedule is furnished at the end.

There are three offers for your evaluation. After carefully reviewing all the offers, you discuss with all parties separately and negotiate in order to get their best prices without any deviation from the scope of work, quality, and time of completion.

Although the negotiated prices from all the bidders were slightly higher than your estimated budget, you select the one most suitable. Possible reasons for rejecting the other two parties could be that one quoted the highest and the other asked for a longer time of completion than the one you selected.

After further negotiation with the selected bidder, the value of the contract price was reduced after adding a condition that the supply of paint will be provided by you, according to the list of procurement furnished by the contractor after the work order was given to him. However, the total value of the job is now close to your budgetary estimate.

DOI: 10.1201/9781003211754-1

It was also agreed that the area of work may vary by ±15% and his or her unit rate will remain the same and valid within the agreed time period. Additional work executed within the premises, which is not in the schedule of items, is to be considered as extra work at an additional price for material and labor rates prevailing in the market plus 20% profit over it. The payment to the contractor will be given in parts, say 60% of the total in three phases as progress payment and the rest after final measurements and completion of all the work as per work specification and to the satisfaction of the owner. Work specification means the paint manufacturer's recommendation for application and method of work, which was supplied to the selected contractor as a part of the contract.

This contract procedure is called an item rate contract.

Another friend of yours, who did not have adequate time and manpower, but needed to paint his house, went for a different option. He wanted the job to be executed by a reputed contractor, who would advise him on the type of painting based on a serviceability requirement satisfying the owner's choice of color coat and appearance. The contract agreement stipulated the type and brand of paint materials, number of coats, and method of surface preparation and application will be decided by the contractor based upon his or her experience and in conformity with standard codes of practice and ensuring serviceability over a period of five years. The contractor is responsible for repairing any kind of damage due to the inferior quality of materials or incorrect method of application within one year after completion: 5% of the total payment will be held up during this warranty period of one year. The contractor will quote a lump sum price including detailed specification of material, method of work, and payment schedule in his or her offer. The contractor will supply all the materials, bear the cost of labor, equipment, scaffolding, and all associated work, assess defective areas of damaged surfaces, select appropriate treatment, undertake repairs before painting, and hand over the site after completion. Like you, your friend had also invited quotes from reputed companies after a prequalification study of local contractors who had adequate experience and financial capabilities and finally selected the best suitable offer giving merit to the time of completion after commercial negotiation.

This type of contract is called an EPC (engineering, procurement, and construction) contract.

The following are the type of contracts commonly used for the execution of industrial projects:

a) Item rate contract
b) EPC contract
c) Turn-key contract
d) BOT (build, operate, and transfer)

In an item rate contract, there is a schedule of items that shows the description of works and estimated quantity against each item. The payment for the work is made as per quoted rates. The quantum of work can vary up to an agreed percentage of total prices. Here, the owner has the flexibility to execute the work as per the design

# Contracts and Specification

and drawing issued by them. The owner's consultant prepares and supplies drawings and specifications. The contractor's responsibility is to supply materials and construct the building and structure in conformity with the drawings and specifications supplied within a scheduled time.

An EPC contract is currently most popular because of reduced involvement and manpower deployment by the owner. Here the contractor is bounded by specific requirements, for example, the number of buildings, dimensions, and type of materials. The price is quoted in a lump sum and the contractor has to deliver it in complete form as per the approved drawing. Any changes in prices of materials, fuel, and so on within the scheduled period of work have to be borne by the contractor. The engineering design and preparation of construction drawings are done by the contractor and submitted to the owner/consultant for review and approval or comments, if any. After incorporations of review comments, the drawing documents are finally approved and released for construction. Payment is made in steps as per agreed terms and conditions in the contract.

A turn-key contract is similar to an EPC, but here the contractor is solely responsible for delivering the finished product as per the agreed terms and conditions, Approval of the contractor's construction drawings is generally not a requirement.

In a BOT contract, the contractor builds the plant as per their design but meeting the terminal output as per the contract agreement, running the plant for a specified period and selling output to recover their investment, and finally handing over the plants and buildings to the owner for running its working life.

For each of these cases, tenders are invited for the job and intended bidders quote for the work. All such tenders complete with price and technical deviations, if any, are submitted in a sealed form to the owner or the owner's consultant within a stipulated date and time. These tender documents are then evaluated by the consultant/approved authority and successful bidders are awarded the job.

The lowest-priced bid is not always chosen for the award unless the technical capabilities and past experiences meet the selection criteria.

In each type of contract, the contract document includes the scope of work, time schedule, general and commercial terms, technical specifications, and tender drawings.

The bidder is instructed to visit the site, assess the nature of the job, soil conditions, and relevant information before delivering the quote.

Time is of the essence in all types of contracts.

## 1.2 TENDER SPECIFICATIONS FOR EPC AND ITEM RATE CONTRACTS

The tender specification document is prepared according to standardized documents and methods followed by the owner or the consulting company. The consultancy firm or the owner's engineer prepare the document and issue it for tendering purposes. For large projects, tender documents are issued to selected vendors only. These vendors are prequalified based on their experience in similar jobs done in the past five years.

The tender document contains the following:

a) Notice inviting tenders (NIT)
b) Conditions of tendering
c) Tender form
d) General condition of contract (GCC)
e) Supplementary condition of contracts
f) Bank guarantee form
g) Contract agreement form
h) General specifications
i) Technical specifications
j) Schedule of items
k) Drawings
l) Documents showing the experience of the tenderer including completion certificates or status of running projects
m) List of construction equipment and qualified workers available for the contract
n) Documents showing the current financial status and earnest money deposit
o) Other documents that need to be submitted as per the tender document

The tenderer has to submit the above documents duly filled in and signed for evaluation by the owner or consultant. These signed tender documents with specifications and the schedule of items form a part of the contract.

Notice inviting tender describes the name of the work, earnest money deposit required with the tender, and time of completion. Information for purchase and submission place and dates are furnished in the NIT.

General conditions of contract (GCC) include the following:

a) Definitions and interpretations of the terms in the contract; location and access to the site by road/rail and nearest airports; Water and Power supply; land to be provided for construction; storage of material, fabrication yard, concrete mix batching plant, workshops, field offices, and land for residential accommodation, etc.
b) General instruction like the scope of work, time of completion, validity, security deposit, and other details to be submitted with the tender; rates in figure and words; corrections and erasures, etc.
c) General obligations like force majeure, compensation for delay, subletting of work, list of subcontractors, power of entry, patent and royalties, etc.
d) Performance and execution; coordination and inspection; dewatering during monsoon or subsoil ingress; work on holidays; responsibility for setting out the work, leveling, and alignment; drawings to be supplied by the owner or contractor; yard lighting for construction, watchtower, etc.
e) Material to be supplied by owner and contractors, tests for quality, period of liabilities, guarantees, etc.

# Contracts and Specification

f) Schedule of rates and payments; procedure for measurements; billings and running bills for work in progress; final billing and completion certificate, etc.
g) Taxes and insurance
h) Safety regulations and safety codes
i) Proformas for bank guarantees, agreements, etc.

Supplementary conditions to contracts are generally given by the owner. They include how the progress reports, schedule of works, and quality control are to be maintained if materials like steel, cement, and reinforcements bars are to be supplied by the owner (if applicable), conditions of supply, and so on.

General specifications consist of the scope of work in detail, general climatic details and characteristics of subsoil strata, workmanship, time of completion, schedule of handing over of the structure or building in a phased manner, etc.

Technical specifications cover all the requirements for the supply, construction, and maintenance for each type of work such as earthwork, cement concrete, structural steelwork, masonry and finishes, doors and windows, and sanitary and plumbing as necessary for the project work. National building codes and standards form a part of this specification. In case there is no technical specification furnished by the owner in the tender documents, codes and standards for the working country will be followed for design and construction.

For some projects, the tender is submitted in two parts. Part I is technical, which contains techno-commercial parts including the deviation statement, if any. The price schedule is submitted in Part II.

The schedule of items and prices are quoted item-wise as per the forms given in the tender documents.

Typical examples of forms generally used for the building and construction of industrial plants are shown in Tables 1.1 and 1.2.

a) For item rate contracts (Table 1.1)
b) For EPC contracts (**for civil works only) (Table 1.2)
   [Note: (**) In EPC work, price includes all works including equipment, mechanical, electrical, civil and arch, and control and instrumentation work; here the tables are a part of a schedule showing civil and arch work only.]

Methods of measurement for various items are applicable for item rate contracts. This is as per applicable building codes and standards unless noted otherwise in the technical specification.

## 1.3 TENDER EVALUATION

Tender evaluation is done by consultants and then submitted to the owner for approval, who then awards the job to the successful bidder after commercial negotiations.

For two-part tenders, the techno-commercial part is evaluated first and then the bidders are asked to settle deviations, if any. After discussions, all the bidders are

## TABLE 1.1
### Schedule of Rates for Building and Architectural Work

|  |  |  |  | Rates in Rupees | |
|---|---|---|---|---|---|
| Item No. | Description | Approx. Quantity | Unit | In Figures | In Words |
| 1 | Earthwork in excavation | | m³ | | |
| 2 | Earthwork in backfilling | | m³ | | |
| 3 | Plain cement concrete/lean concrete (M10) | | m³ | | |
| 4 | Reinforced cement concrete for foundation (M25) | | m³ | | |
| 5 | Reinforced cement concrete for superstructure (M20) | | m³ | | |
| 6 | Formwork for foundation and substructure below plinth | | m² | | |
| 7 | Formwork for superstructures above plinth | | m² | | |
| 8 | Reinforcement steel for items 4 and 5 | | M Ton | | |
| 9 | Structural steelwork including fabrication, erection, and painting | | M Ton | | |
| 10 | Misc. steel embedment in concrete | | M Ton | | |
| 11 | Architectural items including masonry work, door, windows, finishes, floorings, and painting, all complete | | Lump sum | | |
| | | | Total: Rs | | |

brought in so that the deviation statements are withdrawn. Otherwise, the deviations items will be subjected to commercially loading during the price evaluation process.

The price evaluation part is done in a spreadsheet where all the bidders are put in columns and items of works are in rows, so that quoted rates and prices are seen in a row for comparison. An example of a worksheet is shown in Table 1.3 for better understanding.

Names of tenderers are put in alphabetical order. After evaluations, the recommendation is shown in a report, which is called a tender evaluation report.

A typical guideline for preparing a tender evaluation report (techno-commercial) is given as follows:

a) Cover sheet. This will describe the name of the project, owner, consultant, project document numbers, and document title.
b) List of contents, here items c) to g) as stated below.
c) Introduction.

## TABLE 1.2
## Price Schedule for Building Work

| Item No | Description | Cost in Rupees | |
|---|---|---|---|
| | | In Figures | In Words |
| 1 | Topographical survey and geotechnical investigations | | |
| 2 | Site preparation including area grading and landfilling | | |
| 3 | Outdoor facilities: roads, drainage, pavements, trenches, culverts, road crossings, and buried services | | |
| 4 | Piling work | | |
| 5 | Foundations for buildings and equipment | | |
| 6 | Buildings – civil, structural, and Architectural work | | |
| | Building – A | | |
| | Building – B | | |
| | Building – C | | |
| | Building – D | | |
| 7 | Outdoor pipe and cable rack | | |
| 8 | Other work if any | | |
| Total | | | |

d) List of tenderers. Here the names of tenderers/bidders shall be furnished in alphabetical order and in a tabular form (Table 1.4).

e) Discussion on tenderer. Completeness of tender document; competency of the tenderer, e.g., experience in jobs of similar nature in the past five years; time of completion; deviation statements; financial Resources and other commercial terms; validity period of the offered tender or any other special conditions like available equipment, concrete mix batching plant, etc. are to be addressed. Each tender document needs to be carefully scrutinized and discussion notes shall be made separately for each tenderer.

f) Summary of conclusion. Here all the above points are to be summarized and the names of bidders shall be advised in order of merit. Competency, list of equipment, and time of completion shall be major points in the merit list.

g) A comparison statement may be attached in the annex of the report (Table 1.5).

## 1.4 TECHNICAL SPECIFICATIONS

Technical specifications detail the various methods of construction and quality control measure available.

Where technical specifications for a project are not available, engineers should refer to the government specifications used in the Public Works Department (PWD)/Central Public Works Department (CPWD)/Indian Road Congress (IRC)/The National Building Code of India (NBC).

### TABLE 1.3
### Price Evaluation Sheet

| SL No | Item Description | Quantity | Unit | ABC Construction Company (Name of Tenderer) | | DEF Construction Company (Name of Tenderer) | | Remarks |
|---|---|---|---|---|---|---|---|---|
| | | | | Rate RS | Price RS | Rate RS | Price RS | |
| 1 | Earthwork | 3000 | m³ | 100 | 300,000 | 130 | 390,000 | |
| 2 | RCC (M25) | 500 | m³ | 4800 | 2,400,000 | 4950 | 2,475,000 | |
| 3 | Other items (not shown for clarity) | | | | | | | |
| | Total Price in Rs | | | | 2,700,000 | | 2,865,000 | |

# Contracts and Specification

**TABLE 1.4**
**Names of Tenderer**

| SL No | Names of Tenderer | Tenderer's Covering Letter | Notice Inviting Tender |
|---|---|---|---|
| 1 | M/s ABC Construction Company | | Xxx /yyy/ dt August 6, 2019 |
| 2 | M/s DEF Construction Company | | Xxx /yyy/ dt August 6 |
| 3 | M/s GH Construction Company | | Xxx /yyy/ dt August 6 |
| 4 | M/s JK Construction Company | | Xxx /yyy/ dt August 6 |

Technical specifications are part of the contract document, and the contractor must follow this during the execution of the work. If there is any deviation found necessary due to site conditions or non-availability of any particular material, the contractor has to bring it to the attention of the engineer-in-charge and submit alternate proposals for the approval of the owner or consultant.

Technical specifications for EPC projects are precise and specific and do not give many alternatives. For example, in a structural steel specification, there may be more than one method of tightening field bolts, say the turn of nuts method, torque wrench tightening, etc. In an EPC contract, there should be one method specified, for example, the torque wrench method with the diameter-wise torque listed.

For a general plant building, the following technical specifications are necessary:

   I. Technical specifications for earthwork
  II. Technical specifications for piling
 III. Technical specifications for cement concrete work
  IV. Technical specifications for the fabrication of structural steelwork
   V. Technical specifications for the erection of structural steelwork
  VI. Technical Specifications for metal deck formwork
 VII. Technical specifications for building work and services
   i. Properties, storage, and handling of common building materials
   ii. Masonry works
   iii. Finish to masonry and concrete
   iv. Floor finishes
   v. Door, window, ventilators, louvers, etc.
   vi. Glazing works
   vii. Roll-up shutters
   viii. Painting and protection
   ix. Metal cladding for roof and sides
   x. Roof waterproofing
   xi. Water supply
   xii. Sanitary and plumbing
   xiii. Fencing and gates
   xiv. Anti-termite treatment

**TABLE 1.5**
**Comparison Statement**

| Name of Tenderers | Completeness of Documents | Competency | Time of Completion | Deviations | Financial Capabilities | Equipment | Validity period |
|---|---|---|---|---|---|---|---|
| | | | | | | | |

For special works like chimneys, silos, and pump houses, separate technical specifications, such as slip-form work, acid and temperature resistant lining, steel lining for flue ducts, and gates and screens, are added in addition to the common list of items mentioned previously.

## 1.5 REFERENCE CODES OF PRACTICE AND STANDARDS

We have already explained the definitions of general specification of contracts and technical specification of contracts. Now let us discuss the codes of practices and technical specifications.

To begin with, we must remember that all building codes and standards published by national standards institutions (Indian/American/British) should be used as guidance for work in accordance with technical specifications laid out by the owner/consultants in the contract document. In case a conflict arises between two documents, national codes and standards will take precedence. However, these issues are to be mutually settled between the owner/consultant and the contractor's engineers.

Where separate technical specifications are not available, work is carried out in accordance with standard codes of practices. The essence of this discussion is to advise engineering professionals that one should understand the purpose of the codal clauses and use them in a proper sense in design and construction work.

A structure/building shall be built following guidance and safety limits given in the codes and specifications. However, an engineer should apply judgment based on his or her experience; applications of clauses will not relieve the constructor/engineer from his or her responsibility in designing, supplying, and constructing a work that will be safe and workable during its specified design life.

The following codes are used for the design and construction of civil and structural work.

### 1.5.1 AMERICAN STANDARDS

ASCE 7 Minimum Design Loads for Buildings and Other Structures
Uniform Building Code 1997, (UBC 1997)
Uniform Plumbing Code, (UPC)
Construction Industry Research and Information Association (CIRIA) C577, Guide to Construction of Reinforced Concrete in the Arabian Peninsula.
ACI 318, American Concrete Institute, Building Code Requirements for Structural Concrete and Commentary
ACI 543R, Design, Manufacture, and Installation of Concrete Piles
ACI 350, Code of Practice for Environmental Engineering Concrete Structures
CRSI, Concrete Reinforcing Steel Institute, Manual of Standard Practice
AISC, American Institute of Steel Construction, Manual of Steel Construction
ACI 530.1, American Concrete Institute, Specification for Masonry Structures
AWS D1.1, American Welding Society Structural Welding Code for Steel
ASTM, American Society for Testing and Materials Standards (as applicable)

NFPA 850, National Fire Protection Association, Recommended Practice of Fire Protection for Electric Generating Plants
SSPC, Steel Structures Painting Council for Protective Coatings
AWWA, American Water Works Association
AASHTO, American Association of State Highway and Transportation Officials

### 1.5.2 BRITISH STANDARDS

Euro Code 8 for Design Provisions for Earthquake Resistance Structures
British Standard, BS 6399 Part 1, Part 2, and Part 3 for Building Loads and Wind Forces
British Standard, BS 8110 Part 1 and Part 2 for Structural Use of Concrete
British Standard, BS 449-2 Specification for the Use of Structural Steel in Building
British Standard, BS 4-1 for Structural Steel Sections
British Standard, BS 4449 for Concrete Reinforcing Steel Bars
BS 8007 Design of Concrete Structures for Retaining Aqueous Liquids

### 1.5.3 INDIAN STANDARDS

- IS:800 General Construction in Steel – Code of Practice
- IS:456 Plain and Reinforced Concrete – Code of Practice
- SP 16: Design aid for Reinforced Concrete to IS:456
- IS:875 Code of Practice for Design Loads (Other Than Earthquake) for Buildings and Structures
  Part 1 Dead Loads – Unit weights of building materials and stored materials
  Part 2 Imposed Loads
  Part 3 Wind Loads
  Part 4 Snow Loads
  Part 5 Special Loads and Combinations
- IS:1893 Criteria for Earthquake Resistant Design of Structures
  Part 1General Provisions and Buildings
  Part 2 Liquid Retaining Tanks – Elevated and ground supported
  Part 3 Bridges and Retaining Walls
  Part 4 Industrial Structures Including Stack-Like Structures
  Part 5 Dams and Embankments
- IS:2911 Code of Practice for Design and Construction of Pile Foundation
  Part 1/Section 1 Driven cast-in-situ concrete piles
  Part 1/Section 2 Bored cast-in-situ piles
  Part 1/Section 3 Driven precast concrete piles
  Part 1/Section 1 Bored precast concrete piles
- IS:3370 Concrete Structures for Storage of Liquids – Code of Practice
  Part 1 General Requirements
  Part 2 Reinforced Concrete Structures

Part 3 Prestressed Concrete Structure
Part 4 Design Table
- IS:2502 Code of Practice for Bending and Fixing of Bars for Concrete Reinforcement
- SP 34: Handbook of Concrete Reinforcement and Detailing
- IS:2974 Code of Practice for Design and Construction of Machine Foundations (all parts)
- IS:2062 Steel for General Structural purposes – Specification
- IS:806 Code of Practice for Use of Steel Tubes in General Building Construction
- IS:4000 High strength bolts in Steel Structures – Code of Practice
- IS:4923 Hollow Steel Section for Structural Use – Specification
- IS:10430 Criteria for Design of Lined Canals and Guidance for Selection of Type of Lining
- IS:1343 Prestressed Concrete – Code of Practice
- IS:4998 Criteria for Design of Reinforced Concrete Chimneys
- Relevant Codes and Standards Followed by Indian Road Congress (IRC)
- IRC: 5 Standard Specifications and Code of Practice for Road Bridges;
- Section I: General Features of Design
- IRC: 6 Standard Specifications and Code of Practice for Road Bridges;
- Section II: Loads and Stresses
- IS:6403 Code of Practice for Determination of Allowable Bearing Pressure and Shallow Foundation
- IS:8009(Part I) Code of Practice for Calculation of Settlement of Foundation Subject to Symmetrical Vertical Loads for Shallow Foundation
- IS:2212 Code of Practice for Brickwork
- IS:2250 Code of Practice for Preparation and Use of Masonry Mortar
- IS:1661 Code of Practice for Cement and Cement-lime Plaster Finish on Walls and Ceilings
- IS:2191 Wooden Flush Door Shutter (Cellular and Hollow Core Type)
- IS:2202 Wooden Flush Door Shutters (Solid Core Type)
- IS:1038 Steel Doors, Windows, and Ventilators
- IS:4351Steel Door Frames
- IS:12118 Specification for Two Parts Polysulfide Based (Parts I and II) Sealants
- IS:6248 Metal Rolling Shutters and Rolling Grills
- IS:3548 Code of Practice for Glazing in Building
- IS:2114 Code of Practice for Laying In-Situ Terrazzo Floor
- IS:427 Specification for Distemper, dry color as required
- IS:428 Specification for Distemper Oil Emulsion, color as required
- IS:2338 Code of Practice for Finishing of Wood and Wood
- (I and II) based materials
- IS:2395 Code of Practice for Painting Concrete, Masonry, and Plaster Surface
- IS:2932 Specification for Enamel, Synthetic, Exterior, Type-I

- IS:5410 Specification for Cement Paint, color as required
- IS:2441 Code of Practice for Fixing Ceiling Coverings
- IS:1346 Code of Practice for Waterproofing of Roofs with Bitumen Felts
- IS:2065 Code for Practice for Water Supply in Buildings
- IS:1172 Code of Basic Requirements for Water Supply, Drainage, and Sanitation
- IS:1230 Cast Iron Rainwater Pipes and Fittings
- IS:1742 Code of Practice for Building Drainage
- IS:5329 Code of Practice for Sanitary Pipe Work above Ground for Buildings
- IS:2470 Code of Practice for Designs and Construction of Septic Tank for Small and Large Installations
- IS:3889 Centrifugally Cast (Spun) Iron Spigot and Socket Soil Waste and Ventilating Pipes, Fittings, and Accessories
- IS:1729 Sand Cast Iron Spigot and Socket Soil, Waste and Ventilating Pipes, and Accessories
- IS:651 Salt Glazed Stoneware Pipes and Fittings
- IS:774 Flushing Cisterns for Water Closets and Urinals (valveless siphon type)
- IS:2527 Code of Practice for Fixing Rainwater Gutters and Downpipes for Roof Drainage
- IS: 3414 Code of Practice for Design and Installation of Joins in Buildings

# 2 Site Survey and Soil Investigation

## 2.1 RECONNAISSANCE AND TOPOGRAPHICAL SURVEY

A reconnaissance survey means visual surveying of the plot, preliminary investigation work, and gathering data, which helps to get a feel of the land and an understanding of the feasibility of the project. The engineer or surveyor should walk the plot or land or accessibly traverse it by car. This survey also includes visiting drivable approach roads to the site from nearby highways or interstate roads. The engineer should observe surface conditions, boundaries, trees, streams, nullahs, or ditches; terminal points for storm and sanitary sewerage; and the ground slope and other obstructions that may interrupt carrying out a topographical survey and exploratory geotechnical investigation. Trial pits may be dug to examine the topsoil stratum and the existence of fill material, if any. Photographs should be taken to represent the interpretative site picture. The engineer should also collect information for sources of drinking water and construction water and electricity. In this survey, the engineer also gathers information such as locally available materials, meteorological data from authorities, and maximum flood levels for the past 50 years.

The engineer makes a preliminary report on this visit and advises about a topographical survey and exploratory boreholes at suitable locations for subsoil and groundwater contamination tests as necessary to complete the preliminary field report. Photographs of borehole cores are taken and included in the report. This report will be considered as first-stage information.

All these documents will be included in the tender document as preliminary information and guidance to the tenderer for his or her proper assessment of the worksite, thus saving time and cost.

While a reconnaissance survey is essentially a human observation, a topographical survey is for actual measurements of land and its surface features with precise instruments and subsequently preparing a map showing all permanent features existing within the area. Magnetic north is shown in the drawing.

The surveying contractor sets suitable reference monuments or benchmarking pillars of permanent construction located on the corner of a grid system. Elevations and coordinates are marked on a metal plate embedded on top of pillars. These elevations are calculated with respect to mean sea level or official data previously established in the vicinity of the plot or existing plant benchmark. The survey drawing indicates grid points, contour intervals, boundary corner coordinates, roads, drains, and all permanent features existing within the plant area. Existing roads, drainage, railway tracks, overhead power lines, buildings or houses, walls, fences, embankments, streetlights, water tanks, septic tanks, boundary corner points, manholes and underground facilities, trees, and so on are shown with a legend in the drawing, as applicable.

After the awarding of the contract, the contractor undertakes the work for the final design and construction work. The information given in the tender is preliminary; hence it is the contractor's responsibility to carry out a topographical survey and geotechnical investigation work to assess the actual site conditions.

A bathymetric survey is necessary for waterfront structures and structures on the riverbed.

All these works should comply with the codes and standards of the country.

## 2.2 SITE PREPARATION AND ENABLING WORK

This is also named as early civil work, which is necessary to proceed with the general construction of buildings and structures.

This part of civil work includes:

a) Clearing the area of vegetation and roots for area grading and leveling.
b) Cutting trees and the demolition of existing buildings or structures, if any.
c) Making embankment and land filling where necessary to raise the grade level above the flood level.
d) Cutting land and filling ground to reach the final grade level, as per the contour level given in the topographical survey.
e) Ground improvement work, if necessary, as per the soil report and recommendations. A special study is carried out for the method of ground improvement by specialized agencies.
f) Construction of roads and drainage to keep the plot dry during construction.
g) Protective fencing or boundary walls.
h) Establishing site office, workshop, store, fabrication yard, and an open yard for stacking equipment, pipes, structures, and precast pile shafts, where necessary.

The work for clearing and grabbing includes the stripping of surface materials like grasses, vegetation, roots, weeds, debris from demolition, trash, seashells, slush, organic or contaminated soils, and so on. The area shall be cleared for final area grading, i.e., cutting or filling (with selected soil) to raise the level up to the plant grade level. Stripped-off surface material (other than topsoils of approved quality) and debris from the demolition works should be stacked and later disposed of at a designated area.

Trees up to 300 mm in trunk size need to be cut and uprooted. For trees of more than 300 mm girth, approval should be obtained from the appropriate authorities.

Before starting excavation work, initial levels of the ground should be taken in a grid pattern. These levels are referred to for final grading work and payments.

The excavations in ordinary soils, including hard soil, are generally done by mechanical excavators. The rough excavation shall be carried out up to a depth of about 150 to 200 mm before reaching the final level. The final level shall be reached by careful dressing to avoid extra excavation. The excavation of rocks is done by controlled blasting. The use of explosives and other methods of blasting should be done following codes of safe blasting.

If there are underground cables, pipes, or other service lines in the area that need to be excavated, they should be removed or protected before commencing actual work.

For low-lying areas, riverside plots, or reclaimed land, which require filling of large areas, they should be protected with a peripheral embankment for a flood protection bund before the filling work starts. The sources and type of filling material and method of compaction or consolidation should be determined by the geotechnical engineer based on laboratory test results of the virgin soil.

As a guideline, cohesive soils may be compacted to 90% of Proctor dry density at optimum moisture content and sand or non-cohesive soils should be compacted to a minimum of 75% of relative density unless stated otherwise in the contract specifications.

The area filling work is done in layers of about 150–300 mm depending on the type of soil and equipment used for leveling. The original ground should be rolled and compacted by an 8–10 Mton roller in multiple passes (at least six passes) before placing the fill materials.

Safety rules should be followed for all these works.

More details on ground-improving work, roads, drains, embankments, and so on will be discussed in the following chapters.

## 2.3  SOIL INVESTIGATION WORK INCLUDING APPLICATIONS FOR FIELD TESTS

Soil investigation work is necessary for the assessment of the subsoil layer for construction and design. A construction engineer selects the method of excavation and the protection system of the excavated pit or area based on the type of soil and level of the groundwater table. The foundation engineer needs the engineering properties of the soil to determine the type of foundations and estimated settlement to ensure safety and serviceability during the plant's design life.

The investigation work is carried out following the scope of work and test details given in the soil investigation specification, which is prepared by the owner or consultants based on the type of plant to be constructed, estimated loading on the founding layer, and knowledge of the nature of the subsoil. The field test data and laboratory test reports of soil and water samples collected at different depths are analyzed by geotechnical experts. Their findings are presented in a report with recommendations for foundation design and construction underground.

The geotechnical investigation work includes deep boreholes, seismic velocity tests, chemical analysis of soil and groundwater, a soil liquefaction study, and various field and laboratory tests that are necessary to determine the engineering properties of the soil for design and construction. This geotechnical report includes recommendations on the foundation system and ground improvements as necessary. The work is done by agencies specialized in soil mechanics and foundation engineering.

The geotechnical investigation report gives us a fair idea of the soil stratum and its engineering properties. The engineer should note that the information received from this report is based on the investigation and test samples collected from boreholes and other test pits. Since test locations are spaced apart, it gives representative data and may vary with the actual results after the area is dug for excavation. In case bad patches or loose soil are found at the foundation level, the same has to be replaced by compacted sand fill or mass concrete depending on the loading and importance of the foundation.

A geotechnical report includes a description of tests, test location plan, field data, and laboratory results. It is summarized with a recommendation report. As a good engineering practice, the foundation engineer should see the bore log description and test data before reading the recommendation part of the report. It will help the engineer to make his or her own judgment if there are any discrepancies in the final recommendations. In such cases, the foundation designer should take the help of the geotechnical engineer for clarification.

Some of the major engineering properties of soil and the type of tests necessary to get these data are listed for guidance (Table 2.1).

Let us discuss the application of these engineering properties of soil in design and construction.

In the design of the foundations, the primary requirement is to ensure that the properties of the soil below the foundations should behave consistently during the service life of the structure. It means the superstructure or foundation will not tilt or settle down beyond the permissible limit under the stipulated design load.

To ensure this requirement, the engineer must determine the following properties of the soil stratum:

a) Safe bearing capacity
b) Settlement at design load
c) Shear strength
d) Liquefaction potential

The safe bearing capacity is the allowable pressure or load intensity that can be placed on the soil stratum at the foundation level without introducing any damage to the foundations or superstructure.

The downward movement or settlement of the foundations takes place for various reasons, such as the elastic compression of granular soil, the consolidation of cohesive soil, changes in pore water pressure due to the lowering of groundwater, underground vibration caused by an earthquake or equipment generated force, movement of the soil below the foundations, and increased effective pressure induced by compressible fill overground. The settlement of foundations is an unavoidable phenomenon that is time dependent. The foundation engineer must compute the magnitude of the settlement of the foundation system at design load and check the computed settlement is within the allowable limit or settlement criteria. The settlement criteria of different structures vary according to the functional requirement of buildings/structures.

## TABLE 2.1
## Soil Properties and Test Designation

| Sl No. | Soil Properties | Test Designation |
|---|---|---|
| 1 | Bulk density and dry density | Laboratory test |
| 2 | Natural moisture content | Laboratory test |
| 3 | Grain size classification (sieve analysis and by hydrometer) | Laboratory test |
| 4 | Atterberg limits (liquid limit, plastic limits, plasticity index, shrinkage limit, liquidity index, relative consistency) | Laboratory test |
| 5 | Unconfined compression test | Laboratory test |
| 6 | Shear test by triaxial compression | Laboratory test |
| 7 | Cohesion (c) and angle of friction ($\Phi$) | Laboratory tests: direct shear test (consolidated drained); triaxial compression test |
| 8 | Consolidation test | Laboratory test |
| 9 | Specific gravity of rock | Laboratory test |
| 10 | Crushing strength of rock (in soaked and unsoaked condition) | Laboratory test |
| 11 | Chemical analysis of soil and subsoil water (to determine pH value, sulfate, chloride, carbonate, and organic content) | Laboratory test |
| 12 | Standard Proctor density test to determine the relationship of water content versus dry density under light compaction | Laboratory test |
| 13 | Standard penetration test (SPT value at various depth = N) for the direct measure of consistency and relative density | Borehole test |
| 14 | Undisturbed soil sample below ground level | Open trial pit |
| 15 | Relative density | SPT; Dutch cone test/static cone penetration test |
| 16 | Soil resistivity | Electrical resistivity test |
| 17 | Dynamic parameters (P-wave velocity; modulus of elasticity (E), Poisson's ratio; shear modulus coefficients of elastic uniform compression (Cu0, elastic uniform shear (C$\tau$), elastic non-uniform compression (C$\Phi$), and elastic non-uniform shear (C$\psi$); dynamic shear modulus (G); damping) | Seismic tests – cross hole and down hole |
| 18 | Swelling characteristics (swelling pressure and free swell index) | Laboratory test |
| 19 | Permeability | Field permeability test |
| 20 | Groundwater level | Borehole (piezometer) |
| 21 | Subgrade modulus | Plate load test |
| 22 | Shear modulus (static) (shear stress/strain, $\mu$) | Vane shear test |
| 23 | Shear strength | SPT, Dutch cone test, Menard pressure meter test, Vane shear test |
| 24 | Rock quality designation (RQD) and rock mass rating (RMR) | Borehole and taking core samples or open excavation for shallow depth |
| 25 | California bearing ratio (CBR) test for road and pavement | In situ penetration test |

The value of safe bearing capacity can be determined in the following steps:

1) To compute the ultimate bearing capacity of the soil structure below foundation level using Terzaghi's equation for bearing capacity. The properties of soil are the average value of underlain layers up to a depth of 1.5 to two times the width of the foundation. Allowable or safe bearing pressure (p) will be equal to the ultimate value divided by the appropriate factor of safety (three to five).
2) To determine the settlement of soil strata, apply pressure equal to the safe bearing pressure (p) at the foundation level. Determine the pressure distribution (two vertical: one horizontal) to soil layers below the foundation up to the depth at which the pressure increase from the foundation is less than 10% of the existing effective overburden pressure at that depth. The depth at which the pressure increase is less than 10% will provide the total thickness (H) of soil to be evaluated in the settlement computation. Calculate the total settlement (S).

For example, the permissible total settlement of a building is 25 mm. The bearing pressure (p) obtained from the Terzaghi equation is 250 kN/m² and computed settlement (S) of the soil stratum below the foundations is 50 mm at applied load (p) i.e., 250 kN/m².

In that case, the design safe bearing pressure ($p_{design}$) shall be 125 kN/m², in order to meet the settlement criteria, here 25 mm.

Like bearing capacity, shear strength is another important property of the soil structure. Soil cannot resist any tensile force like concrete or steel. So uplift is prevented by a mass of backfill on top of the foundation and frictional resistance. Pile bearing capacity in granular soil depends on the internal friction of soil.

Figure 2.1 represents a typical case of a spread footing with uplift load (T) and a pile cap resting on a group of pile bearing pedestal load in compression (W).

Liquefaction study of the site is a specialized subject and not discussed here. It is done by the geotechnical engineer and furnished in the recommendations of the soil report.

Some important mathematical equations are given below for reference:

### A. TERZAGHI EQUATION FOR SAFE BEARING CAPACITY CALCULATION:

Continuous footings:

$$Q_{ult} = c\, N_c + \gamma D_f N_q + 0.5 \gamma B N_\gamma$$

Square and circular footings:

$$Q_{ult} = 1.3\, N_c + \gamma D_f N_q + 0.6 \gamma B N_\gamma,$$

Where,
$Q_{ult}$ = Ultimate bearing capacity, kN/m²
$c$ = Cohesion of soil, kN/m²
$D_f$ = Depth of foundation from the ground surface to the bottom of the footing

# Site Survey and Soil Investigation

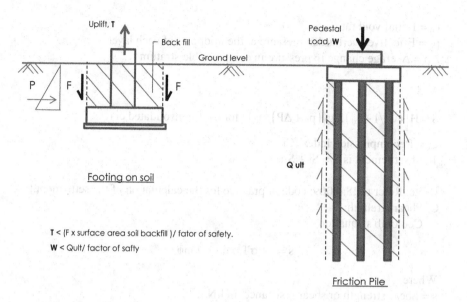

**FIGURE 2.1** Frictional resistance due to the shear strength of the soil.

$\gamma$ = Unit weight of soil, kN/m³ if above water level (submerged weight, if below water level)

B = Width of footing (for rectangular footing B = the smaller side, for circular footing B = diameter)

Nc, Nq, N$\gamma$ = Bearing capacity factor, see textbooks and codes (*) for these values. These values are dependent on $\Phi$, where

$\Phi$ = Angle of shearing resistance or internal friction in degrees.

[Note: (*) Refer to IS: 6403: code of practice for the determination of bearing capacity of shallow foundations.]

## B. SETTLEMENT CALCULATIONS:

Immediate settlement:

$$Si = \sum \Delta H . \Delta \sigma . (1-v^2)/Eu$$

$\Delta H$ = thickness of the layer
$\Delta \sigma$ = Average increase in stress for this particular layer
$v$ = Poisson's ratio of soil
$E_u$ = Constrained modulus

Consolidation settlement:

$Sii = H\left[ (C_c/1+e_0) \log((p_0 + \Delta P)/p_0) \right]$ for normally consolidated clay

$C_c$ = Compression index.

$e_0$ = Initial void ratio
$p_0$ = Effective overburden pressure at the midpoint of each layer
$\Delta p$ = Average change in pressure in compressible stratum

And

$$S = H\left[(C_s/1+e_0)\log((p_0+\Delta P)/p_0)\right] \text{ for over-consolidated clay}$$

$C_s$ = Recompression index (Cr)
Total settlement is S = Si + Sii.

[Note: Refer to IS: 8009: code of practice for the calculation of the settlement.]
C. Shear strength:
   Coulomb's equation

$$s = c + \sigma'f = c + \sigma'\tan\phi$$

Where,
s = Shear strength or shear resistance, in kN/m²
c = Cohesion, kN/m²
σ' = inter-granular pressure acting perpendicular to the shear plane, kN/m²;
   = ( σ – u ) in this case, σ = total pressure and u = pore water pressure.
f = coefficient of friction
Φ = Angle of shearing resistance or internal friction in degrees.

Bearing capacity and settlement are the prime factors in the design of the foundations (Figure 2.2).

It is noted from the above equations that the values of soil properties, for example, cohesion, unit weight, Poisson's ratio, modulus of elasticity, angle of internal friction, compression index, and void ratio, are needed for the computation of bearing pressure, shear strength, and settlement. There are other properties like permeability, seismic wave velocity for the determination of seismic coefficients, subgrade

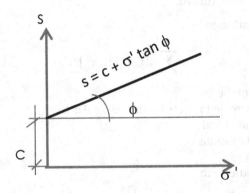

**FIGURE 2.2** Shear strength of soil.

# Site Survey and Soil Investigation

modulus, electrical resistivity of soil, and so on, which are also necessary for the design of a foundation system for a plant or building.

A geotechnical engineering report provides all these data based on field tests and laboratory works for the use of foundation engineers.

## 2.4 GROUND IMPROVEMENT WORKS

### 2.4.1 General

Ground improvement is a site-specific requirement depending on the engineering properties of soil as determined by geotechnical investigation work. Ground improvement works involve large volumes of earthwork, to be done by specialized agencies using appropriate systems, equipment, and skilled personnel, which involves time and cost.

### 2.4.2 Types of Soil Requiring Ground Improvement

Weak soils having low shear strength and high porosity are susceptible to large consolidation settlement and are recommended for ground improvement solutions. Without proper ground improvement, pile foundations will also not be effective, because the load-bearing capacity of piles through these weak layers is not significant. The presence of weak soil at the top layer does not give resistance to the lateral movement of piles. Hence, we get lower capacity piles without ground improvement.

Ground improvement is also necessary for soil strata that are susceptible to liquefaction in seismic conditions. This is a phenomenon that occurs in some types of classified fine-grained soil and leads to an abrupt failure of all the foundations. Such a liquefaction stage occurs when the ground acceleration forces ("g" in all directions) break the shear resistance or inter-particle frictional strength of soil due to cyclic shear forces generated by the soil mass in seismic conditions. Remember that seismic forces are not external forces, it is a force generated by the movement of mass. However, such a reduction of inter-granular friction creates a reduction in volume in loose soil, and thus pore pressure increases. In fine-grain soil, water cannot dissipate quickly, so the soil structure is brought to a zero-friction stage. This is said to be in a liquefied state. At this state, the soil bearing capacity drops to zero, thus resulting in a catastrophic failure over the entire zone under liquefaction.

In both the cases stated here, ground improvement works are necessary to improve bearing capacity and control settlement.

### 2.4.3 Methods of Ground Improvement

There are various methods of soil improvement work.
For example:

a) The replacement of soil by large area excavations and then area filling with good soil

b) Use of geosynthetic material in the form of cloth or mesh to improve the strength of the soil
c) Chemical stabilization

These are solutions for ground improvement work at a shallow depth.

For deep underground improvement, the following methods are used:

a) Consolidation by preloading
b) Sand drains or vertical drains of prefabricated drainpipes with preload
c) Dynamic compaction
d) Vibro-compaction and vibro-replacement
e) Compaction pile
f) CDM pile

There are other methods of ground improvement work. All these methods of ground improvement work have major commercial implications on the project cost. Hence, engineers should select the appropriate method of ground improvement while considering soil, estimated design load, and allowable settlement criteria.

In this chapter, we will discuss the following types of ground improvement procedures, which have been used and found effective in many countries.

a) Preloading and prefabricated vertical drains
b) Dynamic compaction
c) CDM piling
d) Compaction piling

### 2.4.4 Preloading and Prefabricated Vertical Drains

Figure 2.3 shows a typical arrangement of ground improvements by consolidation settlement of soft soil. This is done by applying a preloading surcharge on ground/

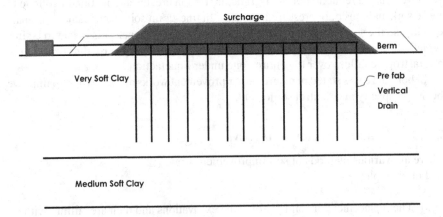

**FIGURE 2.3** Ground improvement by preloading and the PVD method.

# Site Survey and Soil Investigation

fill and inserting prefabricated vertical drain (PVD) drainpipes connected to vacuum pumps aboveground.

In this process, ground improvements will be done by the consolidation of the soft soil under surcharge load. PVDs will be used for enhancing the drainage and thus reducing the time of consolidation. Vacuum pumping will further expedite the drainage. A suitably designed pumping system reduces the height of the preloading surcharge.

The spacing of PVD pipes, preloading material, and the height of surcharge is determined by theoretical calculations using international codes and standards and the theory of soil mechanics. The ultimate vertical settlement is calculated considering the estimated design load on the soil stratum during the plant's service life.

A sand mat layer and geomembrane are placed over the ground below the stack of sand used for preloading. Drains around the preloading area are provided to keep the area dry. These PVD pipes are flexible tubes made of PVC/HDPE material. PVD pipes are installed by static pressing or the vibration method.

The entire work is performed under constant monitoring and analyzing results.

Post-soil investigation is done after the work is complete to ensure the degree of consolidation and the resulting improvement of soil.

### 2.4.4.1 Preloading and Sand Drains

Similar to Section 2.4.4, but with vertical sand drains of 150–750 mm diameter round holes filled with sand. Additional surcharge over ground is provided as preloading for consolidation and to expedite ground improvement.

The use of geotextile, non-woven fiber material in vertical wick drains is also a cost-effective solution.

## 2.4.5 DYNAMIC COMPACTION

Dynamic compaction is a kind of mechanical stabilization that includes various methods like rolling the soil with mechanical rollers, tamping ground, or preloaded fill by a heavyweight drop hammer or plate hammer and vibro-compaction and vibro-replacement.

### 2.4.5.1 Dynamic Compaction by Falling Drop Hammers

Compaction of soil by tamping has been a widespread practice for many years. Such work is done manually using a lightweight (5 kg) drop hammer. The compaction is done on each layer and the effect of tamping is good for a shallow depth.

Hence, the method of lifting heavyweight hammers by crane and dropping them from a great height, in a kind of free fall, has become an effective and well-established method of compaction for different types of soil ranging from clayey silt to granular fill.

These heavyweight drop hammers (5–30 $t/m^2$ approx.) are made of thick mild steel plates stacked together by large diameter bolts and nuts. The free-fall height is controlled by a mechanical crane. This hammer repeatedly falls on the ground or fills that need to be compacted. Number of drops, weight of hammer in each pass,

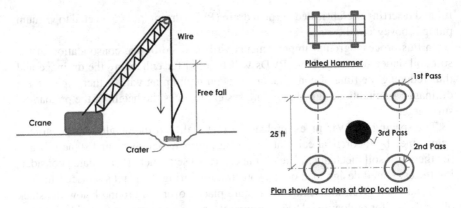

**FIGURE 2.4** Ground compaction by a dropping hammer.

**TABLE 2.2**
**Sample Data from a Completed Project in India**

| Pass | Weight of Plate Hammer (MTon) | Height of Free Fall (M) | Total No. of Drops |
|---|---|---|---|
| First pass | 15.5 | >20 to 30 | 14 |
| Second pass | 15.5 | 20 | 12 |
| Third pass | 15.5 | 20 | 10 |
| Iron pass | 4.5 | 11 | 2 |

spacing of drop location, and so on are based on design calculations conforming to engineering properties of soil strata (Figure 2.4) (Table 2.2).

There have been many published studies on dynamic compaction. The final design is established by using empirical formula and trial compaction tests.

In this process, the soil layer at the top is forced to settle by the free-fall weight of the hammer. It creates a crater, and the vibration effect reaches deep into the ground. The different locations of hammer drops are spaced in a square grid at spacings as per the design requirements. The hammer drops are applied in a number of passes and the hammer weights are also changed.

The soil layers are subject to high stress causing a reduction of voids and an instant increase in pore pressure. After the dissipation of pore pressure, the properties of the soil structure change due to an increase in density and shear stress, which then leads to an increase in soil bearing capacity. It is seen that the pore pressure generated by dynamic compaction dissipates faster than it would have by other methods of consolidation. This depends on the coefficient of permeability of the soil being compacted. In the case of coarse grain granular soil, the pore pressure dissipates instantaneously, so this method is found to be very effective.

# Site Survey and Soil Investigation

In fine-grain soil, this increased pore water pressure at lower levels sometimes builds up to a limit that equals the weight of the overburden fill. At that stage, the soil structure becomes unstable similar to liquefaction. But this effect is localized and is eliminated by applying a greater number of hammering passes.

The effectiveness of this technique is dependent on the percentage of fines and particularly the presence of clay. In general, this method is good for non-cohesive soil of fines content not greater than 15% to 20% and soils of low plasticity.

### 2.4.5.2 Dynamic Compaction by Vibro-Compaction and Vibro-Replacement

Vibro-compaction and vibro-replacement methods are found effective for stabilizing soil below large diameters, for example, oil storage tanks and large storage areas, i.e., a coal storage area in a power plant. Cement deep mixing (CDM) pile is also a useful solution for improving large area storage foundation plinth over soft to very soft soil stratum.

#### 2.4.5.2.1 The Vibro-Compaction Process

In this process, a vibro-float is inserted into the ground. It goes down by its own weight assisted by water jetting. After the design depth is reached, the bottom jets are closed. Water is flushed from the top jets to begin the compaction process. The vibro-float is then gradually withdrawn in stages, forming a densely compacted column with feeding of backfill material stacked around the hole. The inter-granular pressure is temporarily destroyed by vibration and soil particles rearrange themselves in the densest possible state under gravitational force (Figure 2.5).

#### 2.4.5.2.2 The Vibro-Replacement Method

Stone columns are formed by the vibro-replacement method. Large diameter holes are drilled into the ground by a wet or dry process depending on the type of soil and

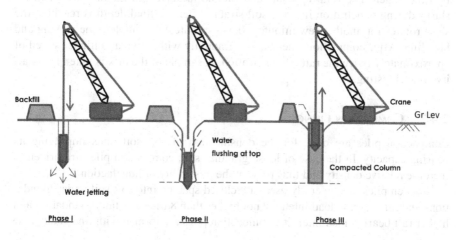

**FIGURE 2.5** The vibro-compaction process in stages.

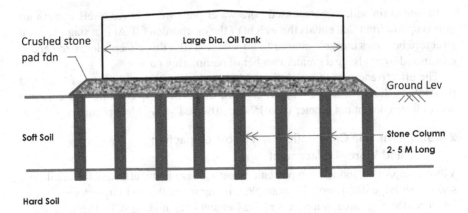

**FIGURE 2.6** Stone columns by vibro-replacement.

method of construction set by the contractor. After penetration of the vibrator up to the design depth, coarse granular fill material is fed from the surface up to the base in stages of compaction by vibration. The diameter of the hole varies from 500–700 mm and the length can be 2–8 m (Figure 2.6).

### 2.4.6 Cement Deep Mixing Piling

CDM pile is used to improve engineering properties of weak soil up to a sufficient depth in order to increase the soil bearing capacity. In this process, soft clayey soils are mixed with cement slurry by mechanically driven augur and mixing puddles. The auger is driven into the ground at a slow speed (0.5–1.5 m per minute). Cement slurry is fed into the soil through the tip of the advancing auger shaft, which is a hollow tube with a mixing puddle fixed outside. The slurry is pumped into the auger. The rotating shaft and puddle mixer blend the soil with cement slurry during penetration into the soil strata. After the final depth is reached, the shaft rotates for another few minutes (0.5–2 minutes) to complete the mixing and blending. After completion, the shaft is gradually withdrawn at a higher speed of approximately twice the rate of penetration. A sample of the mix is taken for quality control testing.

### 2.4.7 Compaction Piling

Compaction piles are done for the densification of soft soil, thus improving its bearing capacity. In the case of loose granular soil, compaction piles are driven to increase relative density and thus prevent the possibility of liquefaction.

Wooden piles are generally used at a closed spacing into soft soil under foundations and raft bearing load intensity not higher than 8 t/m². For tank foundation and higher raft bearing load intensity, compaction piles are formed with fine and coarse aggregate. Steel casings are driven into the ground to the design length. The tubular casing is then gradually withdrawn following subsequent filling by a mix of fine and coarse aggregate. The aggregate fills the voids during the withdrawal process and forms the shape of the compaction pile.

Diameter, spacing, and depth of such compaction piles are determined as per the geotechnical design for pile bearing capacity.

# 3 Materials and Uses

## 3.1 GENERAL

All materials used for design and construction should conform to national standard codes and standards prevailing in the host country unless there is a more stringent requirement specified in the contract specification. The designer shall indicate the grade and type of material to be supplied and used for construction in the construction drawing. If certain materials are found unavailable in the local market or are not being produced by approved and reputed manufacturers, this should be brought to the notice of the engineer for a recommendation of alternative materials or various new sources.

The designer should try to use locally available materials, if found to be suitable, considering cost-effective solutions. Although in some cases, it may not be practicable and the engineer may have to import materials from distant sources. For example, if the design drawing indicates clay bricks for masonry construction but they are not available in the region, an alternate solution like cement concrete blocks may be used.

The selection of appropriate and cost-effective material is also an art of design. For sites in medium to high seismic zones, it is recommended to use a lightweight steel structure covered with metal side cladding. The roof can be made of a sandwiched panel of insulated sheet metal or a thin concrete slab laid on a permanent metal deck formwork.

For structures in severe wind conditions, the cladding materials and external finishes should be chosen carefully so that they do not require frequent maintenance. For example, the use of glass windows and large panels on exterior faces should be minimized.

Material for structures in a corrosive environment and groundwater contaminated zones should be selected to prevent the effect of corrosion. Sulfate-resistant cement and corrosion-resistant bars are used to delay the effect of ground contamination.

Engineers should keep themselves updated with the latest technology and the development of materials. Currently, the use of high-strength steel in structural steel frameworks, high-strength concrete, and prestressed precast concrete units in building structures in combination with cast-in concrete and structural steelwork have become popular for cost-effective reasons. Fiber-reinforced glass and PVC products in small structures, stairs, platforms, and plumbing materials, and so on have also become popular. Heat-resistant colored glass wall panels are not only aesthetically pleasing but also reduce heating and air conditioning costs.

Ecofriendly materials like compressed wood panels as a partition wall reduces the self-weight of the building and thus the foundation cost.

Let us discuss some common materials and their properties in the following sections.

## 3.2 CEMENT

Cement is a basic material to prepare cement concrete and mortar for construction.

It is a powdery material formed by crushing, milling, proportioning, and calcining limestone (CaO), sand ($SiO_2$), aluminum oxide ($Al_2O_3$), iron oxide, and clay. It is mixed with water to form a binding agent for mortar (cement: sand) and concrete mix (cement: sand: coarse aggregate). After mixing with water, chemical reactions begin, and after the initial setting time (about 60 minutes), it starts to form hardened concrete. The setting and hardening (crystallization) process continues for months. Laboratory tests established that it reaches 100% of its compressive strength (characteristic strength of concrete mix) under moist-cured conditions within 28 days after mixing with water.

The water-cement paste not only acts as a binding material but also fills up the pores between fine and coarse aggregates to make a dense concrete product. In a dense concrete mass, the ingress of corrosive air-water is less, hence it is more durable. Durability is one of the major concerns when resisting a corrosion attack and ensures the strength of the section over the design life.

The following types of cement are used for construction:

Indian Standards

- a) Ordinary Portland cement (OPC): IS: 269
- b) Portland pozzolana cement: IS: 1489
- c) Portland slag cement: IS: 455
- d) Hydrophobic cement: IS:8043
- e) Rapid hardening Portland cement: IS: 8041
- f) High-strength ordinary Portland cement: IS: 8112
- g) Super sulfated cement: IS: 6909

American Standard

- a) Portland cement: ASTM C150
- b) Blended hydraulic cements: ASTM C595
- c) Expensive hydraulic cement: ASTM C 845
- d) Hydraulic cement: ASTM C 1157
- e) Fly ash and natural pozzolan: ASTM C 618
- f) Ground granulated blast furnace slag: ASTM C989
- g) Silica fume: ASTM C 1240

The type of cement chosen depends on the exposure category. Sulfate-resisting cement should be used for concrete exposed to an injurious concentration of sulfate. Classification of exposure categories, requirements of cement type, and grade of concrete are defined in the respective codes (IS: 456/ACI 318 M).

# Materials and Uses

For example, the exposure class is considered severe when the presence of water-soluble sulfate in soil ($SO_4$) is found to be more than 0.2% (by weight) and very severe when $SO_4$ content exceeds 2% (by weight).

The engineer should also consider protection against corrosion depending on the degree of exposure to external sources of moisture and chloride. Concrete in direct contact with salt, chemicals, saltwater, seawater, and so on are subject to such contaminations. Permissible limits of chlorides are furnished in codes of practice. The Indian code suggests the maximum permissible limit of chlorides (as Cl) is 2000 mg/liter for plain concrete work and 1000 mg/liter for reinforced concrete work.

## 3.3 WATER

Water should be clean, soft, and potable – free from organic materials, oil, acid, salt, and visible floating matters that are harmful to the concrete mix. For example, soft water collected from ponds and settled to clear suspended impurities is better than hard water obtained from a deep tube well. However, the availability of the purest sources, water analysis reports, and a purifying process will ensure the quality of water.

Water used for cement concrete work should meet the following limitations of impurities.

Water for mixing in concrete:

a) Chloride ions                                <500 ppm
b) Sulfate $SO_3$                                  <500 ppm
c) Alkali carbonates and bicarbonates    <750 ppm
d) Other dissolve salts                        <2000 ppm

Total dissolve salts a + b + c + d <3000 ppm
The pH value of water should generally not be less than 6.

## 3.4 AGGREGATE

Aggregates should comply with the requirements of local codes of practice (IS: 383/ASTM C33 for normal weight and ASTM C330 for lightweight).

Aggregates should be hard and dense, natural or crushed gravel, or crushed rock – quartzite, sandstone, gravel, granite, basalt, and limestone.

Aggregate between 4.75 mm and 150 mm is called coarse aggregate. Aggregate smaller than 4.75 mm and below (within the grading limits set in codes of practice) is called fine aggregate.

Aggregates should be free from earth, shale, and decomposed matters, and other impurities likely to affect the durability of concrete.

The quality and proposed source of aggregate should be thoroughly investigated by the contractor and samples of fine and coarse aggregates should be tested in compliance with the technical specifications of the contract prior to construction.

Aggregates should not contain any materials that may react with alkalis in the aggregate itself or in the cement or mixing water in contact with the finished concrete and mortar.

The acid soluble chloride content as chloride ions in aggregate as a percent by mass (BS 812 Part 117) should not exceed the following limits:

Coarse aggregate    0.03%
Fine aggregate      0.06%

Water absorption of aggregate should not exceed 3%.

The recommended size of coarse aggregate is shown in Table 3.1.

Fine aggregates are natural sand. Natural sands are available in fine and coarse grain at sources. In such cases, these two types may be combined to meet the grading requirement.

Crushed stone sand may also be used with the approval of an engineer. Due to the scarcity of natural sand or river sand in some places, manufactured sand (M sand), which is prepared in combination with crushed stone and natural sand, may be used.

## 3.5 ADMIXTURE

Admixtures for concrete should be used with the approval of the owner/consultant's engineer. The admixture should conform to the relevant codes of practice (IS: 9103/ASTM C494).

The use of admixture is important for concrete quality. Errors in the mix proportion or impurities in concrete admixtures result in multiple shrinkage cracks after drying. Sometimes, it does not harden enough even after 28 days.

The admixture should be procured from reputed manufacturers. The mix proportion and its application should be undertaken as per the manufacturer's recommendations. After mixing with concrete, a sample test should be performed in compliance with the testing requirements.

Admixture improves the properties of concrete including setting time and design strength. The quantity of cement can be lowered in a design mix using admixture without sacrificing the characteristic strength. Hence, the possibility of drying with shrinkage cracks is also reduced. Admixture of a plasticizer kind improves the

**TABLE 3.1**
**Recommended Aggregate sizing for Concrete items**

| Concrete Item and Location | Max. Size in mm |
|---|---|
| Very narrow space | 12 |
| Reinforced concrete except foundations | 20 |
| Foundations and plain cement concrete | 40 |
| Mass concrete | 80 |
| Mass concrete in very large structures | 150 |

# Materials and Uses

flowability without changing the water:cement ratio. Free-flow concrete with admixture is used for concreting in a highly reinforced congested structure and to reach inaccessible zones like grouting under a base plate.

Trial mixes should be carried out with the proposed admixture in a concrete mix, considering the method of transportation, time required for placement, finishing, and curing.

Admixture containing chlorides should not be permitted.

The following types of admixtures are generally used:

a) Air-entraining admixture (ASTM C 260)
b) Water-reducing admixture (ASTM C 494, Type A)
c) Water-reducing and retarding admixture (ASTM C 494, Type D)
d) Water-reducing high-range admixture, superplasticizer (ASTM C 494, Type F or G, ASTM C 1017, Type 1 or 2)
e) Accelerating admixture (ASTM C 494, Type C)

A special admixture as a corrosion inhibitor is also used, where necessary.

The use of GGBS (ground granulated blast furnace slag) with Portland cement to form concrete is very common in the Middle East and other countries. It enhances durability, increases the ability to provide protection against a sulfate attack, and also provides protection in steel reinforcement against chloride ingress. GGBS is blended with cement in the mixture during concrete production. GGBS should conform to the requirements of BS 6699 or ASTM C 989 (grade 120).

## 3.6 REINFORCING STEEL

Reinforcement steel should be produced as per the relevant code of practice and all bars should be of tested quality.

The following types of bars are used in reinforced concrete structures.

**INDIAN STANDARDS:**

a) Mild steel bars (IS: 432) – yield stress 240 N/mm$^2$
b) High yield strength deformed bars (IS: 1786) – yield stress 415 N/mm$^2$ and 500 N/mm$^2$
c) Corrosion-resistant steel bars (CRSI) – ASTM; yield stress above 500 N/mm$^2$

**AMERICAN STANDARDS:**

a) Carbon steel: ASTM A615 M
b) Low alloy steel: ASTM A 706M

[For design yield stresses refer to ACI 318 M – the building code requirement for structural concrete.]

Fabrication and placement of re-bars in concrete work should be done by skilled workers. The strength of a concrete member greatly depends on this work.

Reinforcement work includes stacking bars separately diameter-wise; cleaning, cutting to the required length, bent to shape, and placing them correctly in positions as per the drawing. Bars intended to be in contact at crossing points are securely tied together at all points with 20 gauge annealed soft iron wires. The fabrication of bars should be done as per guidance furnished in IS: 2502 or ACI 315. All bars should be cold bent unless required otherwise. Bars should not be bent or straightened in a manner that would damage the material. Bends should be made around a pin having a diameter as stated below:

a) 16 mm and below:   4 × diameter of bars
b) 20 mm to 25 mm:    6 × diameter of bars
c) 28 mm to 36 mm:    8 × diameter of bars
d) 40 mm:             10 × diameter of bars

The contractor should prepare the bar bending schedule showing the shape, number, and diameter of each bar in drawings or in standard sheets to identify bars by name (bar mark) for fabrication and placement at the site.

Splices of reinforcement should be done as per the design contact length with IS: 456 or ACI 318. This contact length is dependent on the grade of concrete and its allowable bond stress and diameter of bars. All contact lap splices should be securely tied with iron wire. When two bars of different diameters are to be lap jointed at a point, the lap length should be governed by the smaller diameter. For example, if two bars, i.e., 16 mm and 12 mm in diameter, are to be jointed at a location with a specified lap length of 50 × diameter of the bar, the overlapped length will be 50 × 12 = 600 mm.

Welded splice should be avoided as far as possible.

As a good engineering practice, tension bars should not be kinked or joggled as far as possible. The embedded length of the bars should be inserted into the compression zone at a member joint.

## 3.7 FORMWORK

Formwork for the substructure and superstructure can be classified separately because the surface of the superstructure, i.e., aboveground concrete works, is visible and calls for better finishing. As a result, materials and workmanship for superstructures need special attention. The types of supporting systems and safety requirements for elevated structural formwork are more stringent than forms of ordinary substructures, hence unit rates of substructures are generally less.

Forms for footings, trenches, pits, walls, columns, and so on that are below ground are made of mild steel plate boards formed by thin sheets and welded ribs, wooden planks made to shape, and plywood boards with wooden frames. All these boards are made to standard sizes so that these can be used as modular units for different shapes of concrete structures. Bituminous paints are applied on the contact face of the plyboard to make it usable for a number of occasions. Plastic sheets are also fixed on the concrete receiving face to keep the surface free from cement slurry.

There have been remarkable improvements in superstructure formwork.

There are special types of made-to-order wood and plyboard work for architectural ornamental works and curved and special shapes.

In general, superstructure formwork is made of steel or coated (paint or plastic) plywood panels. The contact face is covered with plastic sheets. These panel boards are framed with nailed wood runners, which carry the weight of green concrete. Steel tubes are used as props. These tubes are joined with removable clamps and bolts. All the joints are standard-type joints so as to facilitate fixing and removal after completion. The steels props are placed in a square grid pattern and tied with horizontal ties and runners. Clamp joints can bear axial forces. The diameter and spacing of tubular structured supporting props are designed to carry the weight of fresh concrete and the construction load (should not be less than 300 kg/m$^2$). Timber props (Sal wood) and bamboo are also used as prop members for small structures.

For special structures like deep underground excavations and waterfront structures, steel sheet piles are used. These sheet piles are made to form diaphragm walls braced with steel members – the joists or channel section.

All such formworks are designed to retain earth pressure and construction loads.

There are permanent types of formwork too. Slabs over large halls at a high altitude are cast on steel metal deck forms. These are made with cold-formed steel (0.8–1.2 mm thick) formed to shape like corrugated roof sheets but with deep valleys (up to 175 mm). These valleys are spaces 140 to 200 mm apart. Sheets are laid across the roof/floor beams so that valleys span over the supporting beams. The available length of sheets is up to 6 m; hence, one long sheet can work like a continuous member spanning over the supports with deflection less than the supported member. The sheets are stitched together side by side and fixed with a supporting beam by metal fasteners. Bottom re-bars are placed along valleys so that it forms a rib of a slab cast on a metal deck. Top re-bars are made by fabric wire mesh or ordinary re-bars. The thickness of cast-in slabs with a metal deck is generally provided up to 200 mm (max). The surfaces of these sheets are phosphate coated in a factory. The use of such deck sheets is popular in the construction of bridges and industrial buildings. These metal deck forms have eliminated vertical props and thus allow faster construction. More details on allowable load and sizes of deck sheets are supplied by manufacturers in their product catalogs.

For the construction of chimneys, silos, and elevator shafts in high-rise buildings and conveyor supporting towers, mechanical formwork is used. This is called the slip form method of construction. In this method, a collapsible type of steel shuttering system is used. The shuttering unit is assembled on the ground, bracing the first lift of the concrete shell or structure. This slip form shuttering unit is slowly mechanically jacked upward at a controlled rate until the required elevation is reached. There is a working deck in and around the shutter units, which also moves up. The starter section of the shell, which is cast on a foundation base, has reinforcement bars on both sides projecting up and a set of jack rods placed at intervals in between re-bars along the center of the wall. The slip form unit and working deck are made of a steel structure that is a closed and stable structure like a chassis of a car or vessel. This entire unit is clamped on the jack rod like a floating deck supported on jack rods.

These jack rods are of a high-strength steel and allow the slip form structure to climb upwards driven by an oily pump machine installed on the ground. The power cable and supply pipes go up with the deck. Workers undertake reinforcement placing, binding, and pour the concrete mix supplied by the concrete pump hose as the slip form shutter moves up.

The supply of all materials and working personal are provided by cranes and a winch-operated lift erected inside the center of the shell. The concrete mix is designed with proper admixtures so that the slip form shutter slides smoothly up, leaving the hardened concrete below. The hardened shell structure serves as a supporting tower until the total height is reached. The rate of concreting and jacking of the slip form is controlled as per the calculated strength of cast-in concrete.

This is work undertaken by specialist agencies. More details will be available in the manufacturer's catalogs and the relevant ACI codes of practices.

## 3.8 CONCRETE MIXES

Concrete mixes are designated in grades as shown in Table 3.2.

In the designation, M refers to the concrete mix and the number is the specified characteristic compressive strength of a 15 cm cube at 28 days in $N/mm^2$.

Grades of concrete lower than M15 should not be used in reinforced concrete.

In American standards, the strength of concrete is designated by cylinder strength. A strength test is the average of the strengths of at least two $150 \times 300$ mm cylinders or at least three $100 \times 200$ mm cylinders made from the same sample of concrete and tested at 28 days or the test age designated for determination of $f'c$.

Minimum cylinder compressive strength = 0.8 compressive strength specified for 15 cm cubes.

The maximum chloride content of reinforced concrete should not exceed 0.15% by weight of cement.

The maximum chloride content of unreinforced concrete should not exceed 0.6% by weight of cement.

**TABLE 3.2**
**Grades of Concrete**

| Grade Designation | Characteristic Compressive Strength at 28 days (N/mm²) |
|---|---|
| M10 | 10 |
| M15 | 15 |
| M20 | 20 |
| M25 | 25 |
| M30 | 30 |
| M35 | 35 |
| M40 | 40 |

# Materials and Uses

The maximum sulfate content of concrete calculated from the ingredients should not exceed 4% by weight of cement.

There are two ways of preparing a concrete mix of desired strength:

a) Nominal mix concrete or ordinary concrete
b) Design mix concrete or control concrete

## 3.8.1 Nominal Mix Concrete

In this type, the portioning of the concrete should be done by weight of cement, fine and coarse aggregates, and water as per proportions for nominal mix concrete furnished in IS: 456. However, for small works, the proportions may be determined by volume. Table 3.3 may be used as guidance.

If the sand is moist, the volume of sand can be increased by a +20% maximum for bulking allowance.

As per customary practice in small jobs, the cement is measured with a standard box or container (230 × 230 × 310 mm size, containing 16.5 l of cooking oil). One bag full of cement fills about half of a container. The volume of one bag of cement (50 kg) = $0.0347$ cm$^3$. Referring to the size of this container/cement volume as unit volume, wooden boxes are made of the same volume for the handling of sand and coarse aggregate. All these boxes and containers are fitted with handles for processing the materials and pouring them into the mixture hopper mouth. The coarse aggregate should be of 20 mm (nominal size) size and uniform graded. The popular term is 20 mm and down, which means that there are 20 mm and smaller sizes of stones mixed up to form a uniform graded material.

## 3.8.2 Design Mix Concrete

The Design mix concrete or control concrete should be used for all types of major construction works in order to achieve the required grade of concrete at optimum uses of cement. In this method, the proportioning of cement, water, and fine and coarse aggregates are determined by computation and subsequent confirmation by laboratory tests.

This design mix is developed in a laboratory using materials available on-site. Standard concrete test samples (cubes or cylinders) are taken with that design mix

**TABLE 3.3**
**Nominal Mix Concrete (by volume)**

| Grade of Concrete | Cement | Sand | Coarse Aggregate | Water |
|---|---|---|---|---|
| M10 | 1 | 3 | 6 | 0.98 |
| M15 | 1 | 2 | 4 | 0.92 |
| M20 | 1 | 1.5 | 3 | 0.86 |
| M25 | 1 | 1 | 2 | 0.78 |

and then tested to find out the actual strength of that particular design mix. This is a process of trial and error to finally set the design mix proportion to meet the target design strength.

The computation of design and trial mixes is done in accordance with

> IS: 456, IS: 10262, and SP 23 (Indian standards) or ACI 211.1 (American standards), as applicable.

The results of the initial laboratory mix designs are used to develop the mix design for full-scale trials using the material, plant, and production control methods proposed to be followed on the job site.

Full-scale trial mixes should be carried out using the batching plant to be used for the work. As a minimum, the contractor should carry out the following tests on each batch (a batch of concrete means a full truckload) of fresh concrete:

a) Sump tests (a minimum of two)
b) Measurements of concrete temperature
c) Air content test
d) Fresh density test
e) Initial and final setting time
f) Bleeding test

The evaluation of test results should be based on criteria specified in IS: 456 or ACI 318, as applicable. The strength test should be the average of two cubes made from the same sample of concrete.

In mix proportion calculations, total water content should include the amount of water carried on the aggregate. The water content in the aggregate is determined periodically by testing and the amount of free water on the aggregate is subtracted from the water allowed in the mix.

The contractor should design the mix considering workability and allowable sump range, so that concrete remains workable during transportation, placing, and compaction.

## 3.9  METHOD OF CONCRETING

There are three ways of getting concrete mix ready for casting on formwork:

a) Site-mixed concrete
b) Ready-mix concrete
c) Batching plant

In **site-mix concrete**, the ingredients are fed into the hopper mouth of rotary-type concrete mixture machines in sequence. The hopper drum starts rotating at a slow speed, and sand and coarse aggregate are fed first. After a few turns, it gets mixed up. Cement in dry form followed by water is added to the rotating mass. The hopper drum is now allowed to rotate for a maximum of two minutes until a homogeneous concrete mix is formed. The hoper then turns vertically to unload the mixed-up concrete in the opposite direction. The concrete mix is discharged on a paved area at the base of the mixture machine, from where it is conveyed to the casting floor manually by head load or mechanical lift. It is a continuous process, and the machine rotates continuously producing a fresh mix until the casting is over.

The formwork is cleaned and watered before pouring fresh concrete. The concrete mixture machine should be installed close to the work site so that the time of conveying is within the initial setting time. If the casting floor is at a higher altitude and the target volume of cast-in concrete is more, the concrete mix is lifted by a winch-operated bucket lift. The number of mixture machines and skilled and unskilled laborers are to be selected according to the target volume of casting. In normal weather conditions, 30 cm$^3$ of floor slab-beam concrete should be cast at a fourth-floor elevation within six to eight hours in a working day.

**Ready-mix concrete** is very popular in construction within cities and work sites, where storing aggregates and placing the concrete mixture is difficult due to a lack of space. Because of this, it is made in a factory – the quality of the mix should be better than site-mix concrete depending on manual workers. The strength of ready-mix concrete is made to a higher grade and durability by adding admixture. The conveying time is estimated based on the distance of the worksite from the factory, the probability of delay from traffic congestion during transportation, and so on. Accordingly, admixture for retarding the initial settling time is added. The workability/slump of this product is made suitable for concrete pumps to deliver at the desired elevation. If the time between mixing and placing exceeds 45 minutes, the concrete should be transported in equipment that provides on-route agitation.

The supply of ready-mix concrete should conform to ASTM C94 or equivalent codes of practices. The ready-mix concrete producer should use only materials that comply with approved specifications and the mix design for the required strength and durability.

***Water should not be added to the ready-mixed concrete at the site.***

Ready-mix concrete should demonstrate the entire process of quality control – handling and placement of concrete and facilities and the production of high quality and durability – to the satisfaction of the owner/engineer.

**Batching plants** are generally installed on-site for large project construction. They produce fresh concrete as per the design mix to meet the requirements of strength and durability. The aggregates, admixture, cement, and water are fed into the large mixing plant by separate conveyors through an automatic weighing system. After mixing for the design period, the final product is discharged into a delivery truck (truck mixture/agitator) for transport to the casting site.

There are two types of batching plant – dry type and wet mix type.

In the dry type, all aggregates, including cement and admixture, are mixed in a batching plant mixture and then discharged into the delivery truck, which is a mixture truck. Water is added into the truck at a measured quantity after the aggregate is fed in. The truck rotates 70 to 100 times during transportation and thus the desired mix of concrete gets ready for casting.

In the wet mix type, the truck receives the complete product with adequate slump/workability at the batching plant delivery outlet. Then it is agitated in the truck during the transportation period to maintain workability. No water is to be added to the concrete mix inside the truck or later at the site.

The capacity of such batching plants is 30–120 cm$^3$ per hour approximately.

### 3.9.1 Placement of Concrete

Before placement or casting, all activities of formwork, reinforcement, fixing embedded items, electrical conduits, and so on are to be completed. The entire area and pockets at column joints and the beam web should be cleaned by a compressed air jet. Old concrete surfaces, if any, should be kept wetted for five to six hours and applied with rich cement slurry before the placement of green concrete.

Manual pouring by a small cart or bucket or pouring by chutes or a concrete pump hose pipe are allowed. The movement of all workers and equipment should be done on temporary walkways or makeshift paths of wooden planks.

No aluminum equipment should be allowed to be in contact with concrete.

The concrete mix should be placed continuously as far as possible within one hour after mixing water. The compacting and vibrating procedure should go along with pouring. Re-tampering of cast-in concrete is not allowed. The batch of concrete mix that has set and passed the initial setting time during transport should not be used for casting. The maximum free fall will not be more than 1 m and not more than 0.5 m when placed through reinforcement. For casting thick mat, the pouring should be done in a planned sequence to avoid cold joints as far as possible. The concreting for such thick foundations should be done in layers not more than 300 mm.

Concrete should be consolidated to its maximum density so that there are no air pockets or entrapped air. Beam webs, columns, and walls are compacted using a needle vibrator; slabs in a horizontal position are compacted with a long wooden mallet.

For slip form casting, the concrete is vibrated immediately ahead of the slip form.

### 3.9.2 Vibration for Compaction

Vibrators are used to compact the concrete during placement. The use of these vibratos (25–40 mm needle diameter) should be carefully controlled to avoid segregation of aggregates. Minimum use of vibrators is recommended. Vibrator sticks should not touch reinforcement bars as far as possible during work. This makes a slurry layer around re-bars pushing coarse aggregate away, and thus reduces effective shear stress of concrete in that zone.

Vibrating work should be performed complying with IS: 2505, 2506, 2514, and 4656 or ACI 309.

### 3.9.3 Concreting under Special Conditions

**Cold weather concreting.** When the ambient temperature is expected to be at 4.5°C or below during the placing and curing period, the work should

# Materials and Uses

conform to the requirements stated in IS: 456, IS:7861, or ACI 318 and ACI 306.

**Hot weather concreting**. The placement of concrete during extremely hot weather conditions should be done following the recommended practice given in IS: 456, IS: 7861, or ACI 318 and ACI 305.

The factors influencing concreting in hot weather are high air temperature, reduced relative humidity, and wind velocity. If the wind velocity is high, the concreting should be stopped temporarily. In the summer months, concreting should be done during the cooler part of the day. Generally, no concreting should be done if ambient temperature exceeds 40°C. In ambient temperatures above 30°C, the coarse aggregate should be sprayed with potable water to effect cooling by evaporation.

The maximum temperature of concrete at the time of placement should be within 5°C of the mean daily ambient air temperature but should not exceed 32°C under any conditions.

Chilled or iced cooled water is used to lower the temperature while producing concrete. Crushed ice at 0°C can also be used to control the water:cement ratio. Mixing of concrete shall be continued until the ice is completely melted. When the rate of evaporation is expected to reach 1 $kg/m^2/h$, the contractor should take the necessary precautionary measures like dampening formworks or the subgrade, erecting wind barriers, and so on. Temporary sunshades should be used for concreting in hot weather conditions.

**Concreting underwater** should be done in accordance with requirements specified in IS: 456 or ACI 318.

Under no conditions are dry mixes to be poured inside foundation pits filled with water. Wet mix concrete, same as was being used in a normal dry period added with 10% additional cement, should be poured carefully in continuous header and stretcher courses so that continuity is maintained. The whole casting should be done in a single operation without any construction joints underwater. The base of the foundations and sides should be lined with plastic sheets to arrest the loss of cementitious particles. Dewatering by a manual process or by pumping should not be done during the placement of concrete or until 24 hours after the completion of casting.

**Blinding concrete** is a thin layer of plain cement concrete placed on the soil ready to receive the foundation slab. The thickness of this layer should generally be 75 mm for all structures and 100 mm for the water-retaining structure unless shown otherwise in construction drawings. This layer should be placed immediately after the excavation is reached at the final level in order to retain the properties of soil unchanged due to the loss of moisture being exposed to air. The top of the excavation bed should be rammed and compacted and sprayed with sand before pouring the blinding concrete layer. Brick soling with gaps filled with sand is not a recommended alternative to this plain cement concrete layer. This homogenous layer prevents the loss of cementitious materials from the foundation concrete during placement. Polyethylene sheets or bitumen paints can be applied on blinding concrete to give a protective underside layer of foundation concrete when the foundation is to be cast on contaminated soil. In case the bottom of the excavated soil is seen to be slushy and very soft due to the ingress of subsoil water in the excavated pit, a layer of crushed bricks (Khowa) in the form of broken bats or stone aggregates with sand or moorum fill should be placed on the soft mud bed before placing blinding concrete. The thickness of this coarse aggregate layer should be 200 mm or more to suit site conditions.

For highly corrosive areas, membrane protection is provided on this blind concrete layer as per the drawings and instructions of the engineer.

**Self-compacting concrete** is a type of concrete made with the suitable addition of plasticizer admixture to increase flowability. It is a kind of free-flow mix design to access gaps between reinforcement congestion and joints where compaction with vibrator needles is not practically possible. The self-compacting mix is compacted by its own weight due to gravitation.

**Mass concrete** is large volume concrete work, where dimensions in plan and thickness exceed 1000 mm. This is usually seen for machine foundations, rafts of chimney, silos, large area filling with fly ash concrete, and so on. If concrete is poured continuously without precautionary measures while casting this type of massive foundation, there is the possibility of developing thermal stresses in between the concrete due to the heat of the hydration generated by chemical reactions of cement and water. This thermal stress induces detrimental cracks inside the core of the foundation. To avoid this phenomenon, pouring sequences are planned in a manner that the heat generations do not exceed the tolerable limit (20°C max). The casting is done in layers not exceeding 300 mm and at the same time overlaps passes before the initial setting time and thus avoids cold joints. To monitor this phenomenon, thermocouples are inserted inside the core at locations selected by the construction engineer at the site. Retarder admixtures are also helpful to control such generation of heat from hydration.

**Fly ash concrete** is a cost-effective and popular kind of concrete used for large area pavements and filling concrete. Portland cement and fly ash from power plants are mixed together in proportion to form the cementitious material in a design or nominal mix of concrete. Fly ash replaces a part of the cement by weight. For example, a nominal mix of a fly ash concrete [cement 0.7 + fly ash 0.35 (by weight of cement): sand 2: aggregate: 4] can be tested for strength requirements in place of an M15 grade [cement 1: sand 2: aggregate: 4] ordinary cement concrete. The quality of fly ash and mix design should meet the requirements of IS: 1489, 456, or ACI 318.

### 3.9.4 Method of Construction

The method of construction is the guiding document for construction engineers like a design basis document is for a designer.

When technical specifications of work and relevant codes of practice give a range of choices, this document will specifically indicate the materials to be used and the steps of construction that will be followed for that particular task or item in the drawing.

The construction engineer will prepare this document and submit it for the owner's approval. The advantage of this document is that, if any change notice is necessary for site conditions, it can be approved through this submission without waiting for the regular drawing change procedure. For example, if the drawing shows a particular branded cement admixture, which is not readily available in the market, the contractor can propose another suitable equivalent product for the owner's acceptance. Once the method of construction document showing the alternate product is accepted and approved by the owner, the construction can proceed without interference.

The method of construction documents are prepared in standard engineering procedure formats of the construction company.

Let us see what the items are in a method of construction document.

a) Cover sheet (with title block showing the work to be done, for example: "Placement of Concrete and Curing")
b) The scope of work
c) Abbreviation, if any
d) References – drawing, technical specifications, and quality control procedures
e) Responsibilities (site engineer and construction manager's task)
f) Resources
   Materials: ready-mix concrete (grade...), potable water for curing, Hessians, and polythene sheets
   Plant and equipment: batching plant, transit mixer trucks, concrete pumps, mobile crane, concrete vibrator, and finishing tools and laboratory equipment for testing and quality control work.
   Workforce: trained employees for field and laboratory work
g) Methodology: in this part, work procedures from supply to placement and up to curing are defined. An example of how to place blinding concrete (or lean concrete) is stated as follow:
Blinding concrete:

The ground surface will be prepared and finished at the elevation shown in the drawing. The surface will be dry, firm, and leveled to +0 mm and −25 mm from the specified level so that a minimum thickness of 100 mm as specified in the drawing shall be maintained. Groundwater, if any at the founding level, shall be lowered and maintained at least 600 mm below the surface receiving the blinding concrete; 1000-micron thick polyethylene sheet to be laid on the ground as per the drawing detail. Formwork shall be of timber plank true to line and fixed in position by pegs and side supports.

The concrete mix will be discharged directly from the transit mixer truck or pumped to suit the site conditions at the time of casting. The concrete will be spread with shovels and rakes within the formwork. The surface shall be leveled and tamped to the correct level with straight edges.

An inspection request will be sent to the owner specifying the time and location of the required inspection.

Upon receipt of owner approval and written permission on the inspection job card, blinding concrete will be cast on the ground as per the approved drawing.

The surface of cast-in concrete shall be kept moist with potable freshwater for at least 24 hours, allowing for the initial curing to start.
h) Quality control management
i) Safety requirements

## 3.10 CURING

Curing is a method of protection for cast-in-place concrete. It prevents loss of water from the concrete mix during the period of crystallization and hardening, so that chemical reactions of the cementitious material are complete at moist conditions for at least ten days, and thus helps the concrete to reach its characteristic strength, i.e., 28 days for compressive strength.

Sometimes to expedite the time of curing, steam curing processes are adopted at the factory for prestressed precast concrete products like piles and concrete pipes.

Curing should begin as soon as practicable after casting is complete.

Water is generally used as a curing material. Membrane curing is also acceptable for cases, where necessary, with a liquid membrane compound in conformity with requirements of ASTM C309.

The curing of horizontal surfaces like floors and roof slabs is done by providing a stagnant layer of water and wet Hessian on the top surface of the slab. The underside of the slab and edges are enclosed within polythene sheets, which protect against the loss of water.

Vertical surfaces, for example, columns and walls, are cured by a wet-curing system. The formwork is stripped once the concrete has gained the required strength and the exposed surface is wrapped with wet Hessian cloths. The process of wetting and rewetting continues for at least ten days.

## 3.11 QUALITY CONTROL

Structural steel fabricated items, field connection bolts, nuts, washers, and welding electrodes are produced at a factory where more stringent quality control is possible. Hence, quality control of structural steelwork is not as critical compared with reinforced cement concrete work, which mostly depends on manual work and skill. Formwork, reinforcement placement, production of concrete mix, placement and finishing, protection curing, and all works are done at field conditions.

Quality control of concrete structures is greatly dependent on construction supervision in compliance with field quality control procedures and engineering drawings.

In this section, we will discuss details of the engineering supervision of reinforcement concrete buildings and structures.

Engineering supervision or construction supervision is done to ensure that the construction work conforms with the design drawing and technical specifications. The work includes checking lines and layout, excavation, formwork, reinforcement work including the bar bending schedule, embedded parts, concreting works, curing, quality control including testing of materials and concrete mix, certification of quantity and bills for payments, and so on.

# Materials and Uses

For large projects, there are quality control monitoring systems established by big construction companies based on their experiences. The construction engineer follows the same. So, let us highlight some tests that are essential for the quality control operation (Table 3.4).

For small job sites, where proper equipment, facilities, and skilled workers are not available, an engineering supervisor has to look into following the minimum requirements:

a) Excavation foundation pits: line and level; a plastic water pipe level tube may be used as a construction water level tube.
b) Quantity of reinforcement and its placement.
c) Mix proportion of concrete. Where volumetric proportion is used and bulking of sand and grades of coarse aggregate to be checked. Coarse aggregates should be uniformly graded.
d) For hand mix concrete, the quantity of cement should be 10% extra.
e) The concrete mixing machine should be rotated for at least two minutes for each batch.
f) Water:cement ratio to be controlled.
g) The time of placement should be within the initial setting time of cement (not more than 50 minutes from mixing water in the cement to placement and consolidation).
h) Construction joints should be vertical-faced and not located at the column face.
i) Semi hardened (cold concrete that passed the initial setting time) should not be tamped to the correct level. It is bad practice, and some masons prefer leveling all slab panels at the end of casting.
j) The use of vibrators should be restricted. Vibrator nozzles should not be forced inside the congestion zone of reinforcement. Bar placing should be in multiple layers to avoid congestion at beam-column joints. Excessive use of vibrators and touching re-bars segregates aggregate and thus weakens concrete at the joint.

## 3.12  JOINTS AND MATERIALS

In all concrete works, joints are generally required for various reasons. Occasionally, they make a planned separation between buildings or structures or allow the temporary discontinuation of work during construction when all the concrete in a work can be placed continuously. These joints are provided at selected locations as shown in the drawing or as decided on the site during construction.

The types of joints generally used for concrete construction work are as follows:

a) Construction joints
b) Expansion joints
c) Separation or isolation joints
d) Contraction joints

## TABLE 3.4
## Materials and Tests for Quality Controls

| SL No | Materials | Test to be Done |
|---|---|---|
| 1 | Cement | As per relevant codal provision |
| 2 | Water | Impurity test (chloride ions, sulfate SO3, alkali carbonates and bicarbonates, other dissolve salts; total dissolved salts shall not be more than 3000 ppm) |
| 3 | Aggregates | Physical, mechanical, and chemical properties; grading, moisture content, impurities |
| 4 | Admixture | As per ASTM C494 or local codes of practice |
| 5 | Formwork | Material (waterproof, non-warping, non-shrinking); shape, lines, and dimensions, strength to support without deflection; adequacy of vertical supports, crossties, and bracings; surface finish and levels after completed |
| 6 | Reinforcing steel | Strength and material tests as per relevant code of practice; fabrication (shape, size, and dimensions) and placement as per drawing/bar bending schedule; location and number of splices as per drawing. |
| 7 | Concrete grade | Mix design at laboratory and full-scale batching plant; testing of all properties at laboratory and full-scale batching plant trials (strength tests; durability tests; slump tests, tests for air content, temperature, density, initial and final setting time, bleeding) |
| 8 | Construction joints | Location and details of joints, quality, and dimensions of water stops and splices; dimensions and alignment |
| 9 | Anchor bolts and metal plate embedment | Material, paint, or galvanized, size, lugs, levels, and location dimensions |
| 10 | Placement of concrete | Placing condition (surface of formwork, preparation of old concrete joint, if any), weather conditions, placing equipment and methods, placing time (time interval to avoid cold joints), rate of pouring, temperature monitoring, maximum free fall, vibrator uses, consolidation |
| 11 | Surface finishes | Repairing defects, screed, or float finishes or by surface finish tools for special type finishes |
| 12 | Curing | Method and duration of time (a fully weight-curing system with damp Hessian for a minimum of ten days following placement; use of curing compound should be restricted to special situations |

### 3.12.1 CONSTRUCTION JOINTS

First, let us discuss construction joints (CJ) and what should be their best location. The sketches in Figure 3.1 show sections through a typical construction joint.

Figure 3.1(a) shows us how the shearing force acts along the jointing faces; here the load is pure shear force (V kN /m width) and acting parallel to the face of the joint.

# Materials and Uses

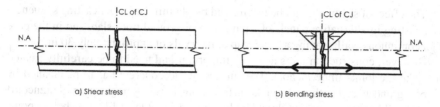

a) Shear stress    b) Bending stress

**FIGURE 3.1** Stresses in a construction joint.

In a monolithic construction, the intergranular friction of concrete mix gives the resistant force against this shear force V.

$$V = Tc \times Ac,$$

where Tc = Design shear stress of concrete in kN/mm² and
Ac = Area of concrete in mm², ignoring the shear capacity of reinforcement bars in the bottom layers.

But in the case of a construction joint, the jointing between old concrete and new concrete is not homogeneous as it would have been in a monolithic construction. Notwithstanding standard procedures, such as wetting the old surface and applying cement slurry (with or without a joint sealing compound) before pouring new concrete, the joints are not perfect like in a monolithic construction. They may be made watertight by providing water stop bars across the section, but the frictional resistance due to intergranular bonding cannot be ensured. During the process of hardening, the new concrete shrinks and leaves the old concrete surface, causing micro-level shrinkage cracks. The bonding of the cement slurry of the bonding agent cannot prevent such micro-level separation in totality. Such shrinkage cracks meander through the jointing surface and the width or crack is increased in the course of time. As a result, the value of Tc will be much less and sometimes zero when the crack propagates from top to bottom.

Figure 3.1(b) shows the bending stress across the plane section of a joint. This happens only at the mid-span of a slab panel or beam, where the bending moment is at maximum and the shear force is nil. The sectional area above the neutral axis (NA) is under compression, which means the surfaces are pressed together. Below NA, concrete has been cracked, and the tension is resisted by a reinforcement bar. At this plane, the concrete section is capable of resisting moment due to the applied load, and despite shrinkage as explained earlier, it is capable of withstanding design moment. There is no shearing force, so the intergranular resisting force is not necessary as it would have been in shear predominant locations, for example, near the supports. Hence, the joint is safe.

This is why the best location of construction joints is at the middle of the span, not at the support.

For unavoidable cases where the joints are to be provided near supports, it has to be designed by adding separate brackets or shear bars suitably designed to carry shear force at the section.

The effect of shrinkage can be minimized by planning a proper casting sequence, i.e., casting in alternative panels of a pavement or ground floor slab, and in the case of a continuous wall, leaving a panel gap at intervals of approximately 30 m.

However, construction joins are very important and need to be carefully treated to make the joints impermeable. The surface of old concrete has to be cleaned by chipping and wire brushing to make it free from loose aggregates and laitance, if any. The old concrete surface should be kept moist for at least 12 hours before pouring the next concrete unless it is cast within that time. The old concrete face should be applied with rich cement slurry paste immediately before pouring green concrete. Where bonding agents are used, the application of the same shall be as per manufacturer recommendations.

The water bar (water stop) should be provided at construction joints if shown in the drawing.

The water bars are made of PVC, rubber, or copper strip. There are various sizes and shapes of these water bars with projected ribs and a dumbbell at the center. The dimensions, shapes, and strength of water bars are available in the product catalog. Common types are made with a central bulb (dumbbell) that resists compression and tension at the construction joint. The selection of type, width, and thickness of water bars depends on the slab thickness and location. Normally the water bar is placed at the center of the slab or wall. Half of its width is embedded inside the end face of the concrete slab at the construction joint, and the half remains projected for getting embedded in the next casting. These water bars are supported by hangers made with small-diameter steel reinforcement bars. Long pieces of wooden bars hold the water bars straight along the joint.

The width of water bars should not be less than the thickness of the slab (but not more than the maximum available width in the market). The width of water bars ranges from 150 mm to 300 mm and the thickness varies from 5 to 10 mm. The main purpose of deciding width is to increase the water path, and the thickness should be adequate to resist the underwater pressure at the joint location. For joints at 4 m or more below groundwater level, the strength of the water bars should be checked. Sometimes in thick walls subject to high underground pressure, double water bars are used to keep adequate space in between (say one-third wall thickness) for concreting.

There is an external type of water sealing bar, which may be used below the base slab or outside surface of the wall. These bars are used in addition to centrally placed water bars to ensure additional protection against the ingress of high-pressured subsoil water deep underground.

A typical detail of water bar placement in a concrete slab is shown in Figure 3.2

The placement of water bars to maintain a proper alignment during concreting is an important task for the constructor. Wooden bars are used to keep them straight along a jointed face. These bars are supplied in a roll. Generally, the length should be cut in a piece in order to avoid jointing on-site. The water bars on-site are joined by lapping each other where necessary. These joints are lap joints. Ribs are cut by chiseling to prepare a flat surface and then glued at contact faces. This glued joint is then placed in a splice heater, which is made of steel plates clamped by bolts to

# Materials and Uses 51

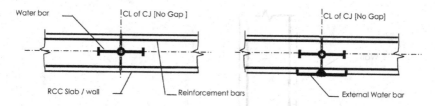

**FIGURE 3.2** Water bars in a construction joint.

press the jointed surfaces together. The joint portion is then electrically heated over a specified time to complete the vulcanization process. Factory-made joints like T and L-shaped pieces are used to facilitate fieldwork.

The use of water bars should be limited because faulty workmanship may induce more leakage at the joints.

Some details of constructions joints are shown in Figure 3.3.

Both types of joints are good for casting walls on the base slab. In type (a), the wall is directly resting over the base slab with one construction joint. In type (b), a small projection (normally 75–150 mm) is made with second-stage concrete over the base slab as a starter of the wall. This second-stage concrete work is usually done on the next day when the concrete of the base raft is set. It helps the carpenter with alignment and to support the formwork for the wall above. It is a favorite choice for many carpenters but induces one more joint between second-stage concrete and the base raft.

### 3.12.2 Expansion Joints

These joints are provided to accommodate the volume changes related to shrinkage and temperature changes. The spacing of such expansion joints has been recommended in IS: 456 and ACI 224.3R. As general guidance, the Indian code recommends 45 m maximum permissible spacing for buildings and concrete structures. American codes (PCA) recommend a spacing of expansion joints at 60 m maximum for general buildings and 36 m for water-retaining and sanitary structures. However, the maximum length of the building or structure should be decided by the amount of movement that can be tolerated by the permissible stress or strength capacity of building framing members. As a safe rule, the spacing of joints should be 30 m.

Structural members at expansion joints should be completely separated, leaving a clear gap for expansion and contraction (about 20–50 mm) as shown in the drawing.

In some expansion joints of pavement slabs, the reinforcement bars are allowed to continue over the joint in the form of dowels with sleeved capping or bond-breaking paint at one end in order to allow sliding horizontally during expansion and contraction movements but resisting vertical shear.

For exceptional cases, where the length of the building or structure exceeds recommended spacing, the building or structures have to be designed for temperature rises and member strength should be kept within the permissible limit.

Some typical examples of expansion joints are shown in Figure 3.4.

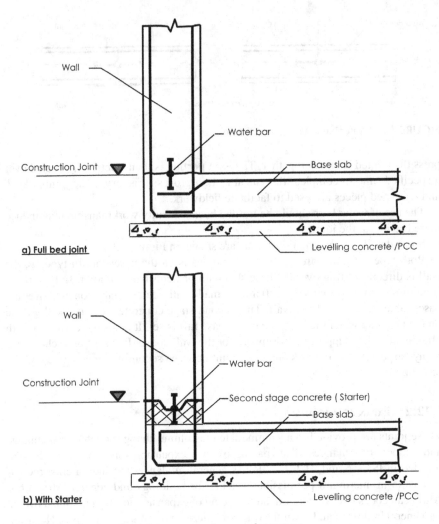

**FIGURE 3.3** Construction joints on an underground structure.

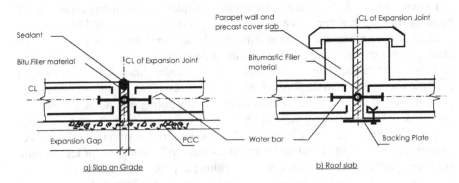

**FIGURE 3.4** Expansion Joint.

# Materials and Uses

The width of the expansion gap is generally kept within 50 mm. An estimated dimension of the gap caused by the environmental temperature change can be obtained by multiplying the coefficient of expansion of concrete (approx. $1.2 \times 10^5$ per °C) by the length of the structure and the temperature change. For example, a 60 m long building subjected to a temperature increase of 15°C would elongate about 11 mm.

The expansion joint should pass through the entire structure above the top level of the foundation. There should be pairs of columns, pedestals, and framing beams at expansion joints, leaving the defined clear gap above the foundation raft. The foundation slab should not be separated at the joint. It would be one slab holding the pair of columns or walls separated by the gap.

Generally, in basement walls, base raft expansion joints are avoided in order to avoid the possibility of joint leakage, where groundwater is at a higher elevation.

For such underground tunnels, basement structures, or long overground aqueducts where an expansion joint is needed, a copper strip with a central U-bend or V-shaped kink can be used in place of a PVC or rubber water stop.

### 3.12.3 Separation or Isolation Joints

These joints are similar to expansion joints but are provided to isolated building segments in order to gain relief from cracking because of the contraction of structural concrete. Separation joints are used to isolate vibratory structures or the machine foundation from building floors and structures to prevent the propagation of vibratory waves. In the case of different shapes and variable loading, the structures are subjected to different rates of settlement. These joints are provided to isolate them from interference and simplify structural behavior.

Figure 3.5 shows typical details of an isolation joint.

### 3.12.4 Contraction Joint

Contraction joints or crack control joints are provided on slabs and walls to make planes of weakness for cracks to form. Drying shrinkage and temperature drops induce tensile stress in concrete slabs at restrained conditions. Cracks develop when

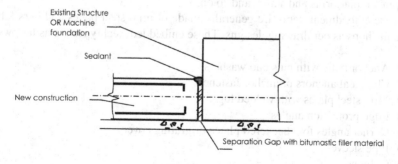

**FIGURE 3.5** Separation or isolation joint.

the developed stress exceeds allowable tensile stress in concrete. Generally, such cracks appear on the surface and at random. By providing contraction joints at a regular grid pattern and spacing, the shrinkage cracks can be controlled to occur along these defined weak lines. The cracks are thus hidden from view and give a better architectural look.

These joints are formed by a machine saw cut after the concrete initially sets. The depth of cut is restricted within the cover and later filled with joint sealant.

For structures at critical exposure conditions, the slab panels and walls are provided with designed reinforcements that resist the structure against temperature stress. But surface cracks cannot be eliminated. Hence, crack control joints are necessary for parapet, slab-on-grade, and pavements that remain exposed to the sun.

Reinforcement bars should be continued across contraction joints. Some designers prefer the reduction of reinforcement at these joints to ensure a plane of weakness.

Figure 3.6 shows typical details of crack control joints.

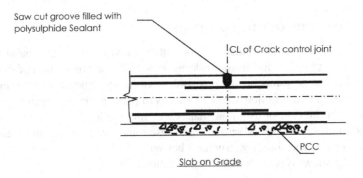

**FIGURE 3.6** Crack control joint.

## 3.13 EMBEDMENT DETAILS

Embedded steels are provided in concrete structures for connection to structural steel members, pipe and cable supports, hangers for monorails, anchor bolts for equipment, platforms and stairs, and so on.

These embedment items are generally made of mild steel plates and bars fabricated in shops as per drawing designs. These embedded steel types are as follows:

a) Anchor bolts with nuts and washers
b) Chemical anchors or anchor fasteners
c) Mild steel plates with welded lugs
d) Edge protection angles
e) Corner angles for checkered plates or grating covers
f) Pipe sleeves
g) Collates

# Materials and Uses

The embedded steel types are tested for strength and service requirements before supplying on-site. The exposed surfaces are painted as per approved corrosion-resistant paints. Embedded portions and surfaces in contact with the concrete are cleaned from rust and not painted. In some cases of high corrosive environments, all the embedded bolts, nuts and mild steel plates, angles with lugs, and so on are galvanized.

## 3.14 STRUCTURAL STEEL MATERIALS

Materials for structural steelwork should conform to local/international codes and standards.

### INDIAN STANDARDS:

| | |
|---|---|
| Rolled steel sections and plates: | IS: 2062 (yield stress, $F_y = 250$ MPa) |
| Bolts, nuts, and washers: | IS: 1367; IS: 4000 |
| Welding consumable: | IS: 814; IS: 1278; IS: 1387 |

### AMERICAN STANDARDS:

| | |
|---|---|
| Rolled steel sections and plates: | A 36 (yield stress, $F_y = 36$ ksi) |
| Bolts, nuts, and washers: | A325 F, A 490 N; A563 Grade A; F 436 Type 1 |
| Welding consumable: | AWS D 1.1 |

Materials are to be selected by the designer considering the overall economy and cost savings.

The use of high-strength steel for the axially loaded member is preferable to using it for girders where shape factor (section modulus) and depth govern bending stress. Similarly, the choice of suitable connection materials and minimizing the number of joints can save some costs. For example, a pipe bridge structure designed with angle member trusses may give less material weight compared with a structure of rolled steel joists, but overall cost saving may be negative due to the higher cost in fabrication and time of construction.

Efficient procurement of materials in matching lengths and sizes and an optimized cutting plan at the workshop can minimize loss due to the wastage of steel.

Above all, adopting a simple design detail saves time in construction. Timely completion of the job has a great contribution to overall project savings.

The selection of suitable member sections by the designer has a significant role in saving costs. The designers should use regularly available sizes of rolled steel sections and plates available in the market or produced by reputed manufacturers. This will ease the procurement procedure. Non-preferred sections in the design drawing call for a replacement with alternative sections during fabrication and thus affect the schedule and need additional weight. Similarly, if a designer selects a common dimension in sizing plated members, keeping in mind the standard rolling widths, the fabricator can optimize a cutting plan and thus minimize wastage.

## 3.15 BOLTS AND WELD

Standard codes for bolts, nuts, washers, and welding consumables are furnished as above. High strength bolts – bearing type or friction grip/non-slip type connections – should be used for permanent connections. Tightening torque should be specified in the drawings.

Joints of purlin, side girts, or temporary members can be done using mild steel bolts. HT bolts of 8.8 grade and mild steel bolts of 4.6 grade are commonly used. Hot-dip galvanized high tensile bolts, where used, should be galvanized by the bolt manufacturer at their shop or any other facility under their responsibility.

# 4 Design Basis Report

## 4.1 INTRODUCTION

The design basis report contains defined design criteria that will be followed by the design engineer in his or her design of a particular building or structure. When national building codes and technical specifications indicate different choices in the selection of materials and utility requirements in general, the engineer has the opportunity to describe the proposal in this document for approval by the authorities. Missing items or ambiguities within the contract specifications, if any, are specified and cleared by the engineer or contractor.

For example, if the owner's specification says space is needed for 20 people in an office room of a power plant, the designer can put the exact sizes of the office room floor area in this document giving length, breadth, and clear height of the office room to accommodate those 20 people. Similarly, the actual number of fans or air conditioning machines, lights and fixtures, tables, chairs, and cabinets can be mentioned in this document, even if the contract document only says well ventilated and illuminated office room. This will avoid conflicts at a later stage.

The design basis report is the first document prepared prior to the design of building structures. Sometimes, this document is prepared as a single document covering the design criteria of all the plant buildings, roads, drains, and other works in a project. It can also be prepared separately for an individual building or plant structure.

The codes and standards indicate general guidelines, but the design basis report will be specific to what the designer is going to follow in the design and construction of a building or plant. The contract specification may also give some design criteria that will be included in the design basis report.

Special structures such as chimneys, jetties and riverfront structures, water retaining structures, and multi-storied buildings have special design criteria.

## 4.2 ITEMS TO BE COVERED

In general, such documents include the following descriptions:

a. Description of the building and structure
b. Units and language
c. Site information and development
d. Soil characteristics and type of foundation
e. Codes and standards to be followed for loading and design
f. Loads and load combinations
g. Materials to be used

DOI: 10.1201/9781003211754-4

h. Design concept of the building structure
i. Protection against corrosion (subsoil and in the air)
j. Architectural works

### 4.2.1 Description of the Building and Structure

In this section, building and structure dimensions, the number of stories and bays, roof and floor construction, framing and cladding materials, and foundation type – pile or non-pile – are described.

### 4.2.2 Units and Language

SI units in the metric system or US customary units, FPS system. The text language in either English or another language is mentioned here.

### 4.2.3 Site Information and Development

Topography and location of the site, the approach from the nearest road or riverway, latitude and longitude, meteorological data, wind and seismic data, rainfall, environmental temperatures, relative humidity, high flood levels, elevation above mean sea level, formation of the site by filling or cutting or developed by a ground improvement system, and so on are indicated here.

### 4.2.4 Soil Characteristics and Type of Foundation

Description of subsoil stratum, groundwater levels, chloride and sulfate content in soil and water, and type of foundation system – pile or non-pile – as per the recommendation of the soil investigation report are stated here. Safe bearing capacity at the selected founding level and pile bearing capacities with suggested cut-off level are given for ready reference.

### 4.2.5 Codes and Standards to be Followed for Loads and Design

Whichever code is followed – Indian standards, American standards, British standards, other international codes, or local codes – is indicated here.

### 4.2.6 Load and Load Combinations

All types of loads that may stress the building or structure are indicated here. The definition of loads and their combinations are illustrated in the following sections.

#### 4.2.6.1 Load

*4.2.6.1.1 Dead Load (DL)*

| | |
|---|---|
| Reinforced concrete | 24 kN/m$^3$ |
| Plain concrete | 23 kN/m$^3$ |

# Design Basis Report

| Floor finish, plaster | 20 kN/m³ |
|---|---|
| Roof treatment with tile | 1.55 kN/m² |
| Brick masonry | 22 kN/m³ |
| Soil | 19 kN/m³ |

### 4.2.6.1.2  Live Load (LL)

This means a superimposed loading on floors and a movable roof. Minor equipment loads can also be included within this category. The equipment load is the total weight divided by its base area. The intensity of LL is provided by equipment supplied by the vendor in equipment layout drawings. This load includes erection and maintenance of the load on the floor. Reputed consultants have their own standards for live loads on floors for the safe design of industrial buildings. Unless stated otherwise, designers should follow the intensity of load as per national codes (IS:875; ASCE 7–10, UBC 1997, IBC ).

### 4.2.6.1.3  Equipment Load (DLE)

The weight of major equipment is shown here. The load data is provided by the equipment supplying vendor or manufacturer in their civil input drawing with the footprint, magnitude, and direction of load at its supports. This load is given for operating conditions, empty weight for erection, wind/seismic conditions, and for the worst case, if any, during a short circuit or other causes of failures.

### 4.2.6.1.4  Wind (WL)

The site-specific intensity of WL should be indicated here as per the general specification in the contract document. Otherwise, it should be taken from local codes of practice.

Design wind velocity and the method of computation are available in applicable national codes and practices. From there, design wind pressure at different heights should be calculated and presented in the design basis report. These values will be used in the design and analysis.

### 4.2.6.1.5  Seismic Load (SL)

Seismic coefficient and method of analysis (response spectra/force coefficient method) should be indicated here according to applicable national codes and practices unless specified otherwise in the general specification of contracts. The designer should indicate the procedure of computation that will be followed, for example, if the response spectra method is to be followed, then the definition of spectra should be indicated.

### 4.2.6.1.6  Temperature Load (TL)

Temperature load results from the thermal stress of framing members due to environmental temperature fluctuations. Since framing members are self-restrained, free thermal expansion or contraction is not possible, resulting in these thermal stresses. The design range is the average of the maximum and minimum operating temperatures as per contract specification meteorological data. For example, if the maximum temperature is 55°C and the minimum temperature is −5°C, the thermal range for strength design purposes should not be less than ±30°C.

#### 4.2.6.2 Combination of Loads

The designer should decide the number of combinations considering the worst possible cases. A few typical cases of combinations are shown as follows:

1. DL + LL
2. DL + WL
3. DL + SL
4. DL + LL + WL (no LL on roof)
5. DL + % LL +SL (no LL on roof)
6. DL + LL + TL
7. DL + WL + TL
8. DL + % LL +SL + TL
9. DL + LL + 50% WL

The reversal of transient loads (−WL, −SL, −TL) should be considered in addition to these load combinations. Major equipment load (DLE), if any, is included in DL as DL + DLE.

If there is crane load (CL) in the building, the number of combination cases should be increased with CL cases. For crane combinations, DL + CL + 50% WL is critical for some buildings.

### 4.2.7 Materials to Be Used

Type of cement, grade of concrete for the foundation and superstructure, and grade of reinforcement bars are stated here. Details of embedded items (galvanized or painted), if any, may be added here.

For steel structures, the grade of steel, type of connection materials (friction grip or load-bearing type), and grade of anchor bolts are shown in this document.

### 4.2.8 Design Concept of the Building Structure

Framing system, method of analysis (two-dimensional or three-dimensional space frame), and the type of computer software used for analysis or member design are explained here. Methods of design, the load factor of working stress, and members in the main frame and secondary members and type of connection joints (shear/moment) are shown.

### 4.2.9 Protection against Corrosion (Subsoil and in the Air)

The protection systems of the concrete foundations, grade beams, and other substructures in contact with the soil, as recommended in the soil report, are discussed

# Design Basis Report

here. Additionally, concrete cover and permissible crack width for structures in the air or in contact with the soil are covered.

### 4.2.10 Architectural Works

This section describes the details of roof treatment, wall masonry and cladding, doors and windows, floor finishes, plasters, painting, sanitary and plumbing, and so on.

Two typical design basis reports are used as examples for education about the engineering practices being followed in India and other countries.

## 4.3 DESIGN CRITERIA OF AN AUXILIARY PLANT BUILDING: RCC STRUCTURE WITHOUT A CRANE (INTERNATIONAL STANDARDS-BASED)

Let us discuss typical design basis criteria of a chemical house building to be constructed on a soft soil site in corrosive subsoil water. The groundwater is near to the finish grade level. The project site is flat land. Average rainfall intensity is 90 mm per hour during five to six soaking months a year. Average annual humidity is 84%.

American codes of practices will be followed in designing this building as per the contract specification.

### 4.3.1 Description of the Building

The chemical house building will be a two-story reinforced cement concrete (RCC) framed building resting on a pile foundation. The roof slab, the intermediate floor, and ground floors will be made of RCC slab. The level of the ground floor slab is 300 mm above the finish grade level. Side cladding will be cement concrete unit blocks masonry work. All pits for lime slaking and preparation, the alum preparation neutralization pit, sumps, and so on will be a watertight RCC construction and designed as an uncracked section.

### 4.3.2 Units and Language

All drawings and calculations will be in SI units – metric (English).

### 4.3.3 Site Information

The site is level ground. Finish grade level (FGL) is at RL (+) xxx M above mean sea level.

#### 4.3.3.1 Meteorological Data
i) Wind speed = 40 m/sec (peak gust in 3 sec; exposure category C, 50 years return period)
ii) Seismic criteria: Zone I as per UBC 1997

iii) Rainfall intensity = 90 mm per hour
iv) Air temperature: 36°C maximum and 12°C minimum

### 4.3.4 Soil Characteristics and Foundation System

Topsoil is 600 mm to 1 m layer of medium dense sand calcareous, silty fine-grained (SPT value 8–10). Below this is a 14-meter layer of sand; medium dense fine to coarse-grained with traces of silt, fine gravel, and shells (SPT 11–25) overlying a very dense sand layer (SPT > 50).

Pile foundation has been recommended in the soil report.

#### 4.3.4.1 Design Pile Capacity

RCC bore pile: diameter 400 mm, length 20 m below cut-off level (2.0 m below FGL).
Capacity: axial 600 kN; lateral 30 kN: uplift 120 kN.

### 4.3.5 Codes and Standards

ASCE 7–10: Minimum design loads for buildings and other structures.
Uniform Building Code (UBC 1997).
ACI 318-14 – American Concrete Institute, Building Code Requirements for Structural Concrete and Commentary.
ACI 350 – Code of Practice for Environmental Engineering Concrete Structures.

### 4.3.6 Load and Load Combinations

The building structure and foundation will be designed for all critical combinations of dead load, live load, equipment load, temperature load, wind load, and seismic load as shown in the following sections.

#### 4.3.6.1 Dead Load

| | |
|---|---|
| Reinforced concrete | 24.00 kN/m$^3$ |
| Plain concrete | 23.00 kN/m$^3$ |
| Floor finish, plaster, and screed concrete | 20.00 kN/m$^3$ |
| Steel grating floor (40 mm thick) | 0.50 kN/m$^2$ |
| Checkered plate cover (6 mm O/P) | 0.50 kN/m$^2$ |
| False floor | 1.00 kN/m$^2$ |
| False ceiling in control and elec. rooms | 0.4 kN/m$^2$ |
| Roof insulation and water proofing with Cement tile topping | 1.55 kN/m$^2$ |
| Brick masonry | 22.00 kN/m$^3$ |
| Soil | 19.00 kN/m$^3$ |

#### 4.3.6.2 Equipment Load

Included in dead load.

# Design Basis Report

### 4.3.6.3 Live Load

| | |
|---|---|
| Ground floor | 15.00 kN/m² |
| Intermediate floors | 15.00 kN/m² |
| Roof | 1.50 kN/m² |
| HVAC on roof | 3.50 kN/m² |

LL on floors includes pipes and cable tray loading and minor equipment.

### 4.3.6.4 Wind Load

Wind speed = 40 m/sec (Peak gust exposure category, 3 sec; 50 years).
Basic wind speed V = 54 m/sec, calculated as per ASCE 7–10.

Importance factor 1.15; topographic factor, $k_{zt}$ = 1.0; directionality factor, $k_d$ = 0.85; exposure coeff = $k_z$ for different heights.
Design wind pressure up to 10 m in height, $q = (0.613\, k_z.k_{zt}.k_d.V^2)/1000$ = 1.51 kN/m² (for $k_z$ = 1).

### 4.3.6.5 Seismic Load

For concrete structure design:
   $f_a$ = 0.107 g for strength design.
   $f_a$ = 0.107 g/1.4 = 0.076 g for working stress design.
   $f_a$ = Max horizontal seismic acceleration.
   Calculated as per UBC 1630.2.
   Seismic analysis will be done by the static force coefficient method with the above coefficients.

### 4.3.6.6 Temperature Load

Average maximum temperature = 36°C
Average minimum temperature = 12°C
Mean temperature = 24°C

Structures will be designed for a temperature variation of 12°C.
The coefficient of thermal expansion for concrete structure = $10 \times 10^{-6}$ per °C.

### 4.3.6.7 Load Combinations

The following load combinations will be considered in the design of structures and Foundations as per ASCE 7–10.

   DLS = Dead load that includes building weight only
   DLE = Equipment dead load that includes weight of major equipment
   DL = DLS + DLE
   LL = Live load on floors
   LLr = Live load on Roof
   WL = Wind load (normal wind; factored load as per ASCE-10)
   SL = Seismic load (strength design load)
   TL = Temperature load

### 4.3.6.7.1 For Structural Concrete Design Load Calculation (Load Factor Design)

Where the ultimate strength design method is used, structures and all portions thereof will be designed to resist the most critical effects from the following combinations of factored loads:

| | |
|---|---|
| Comb-01 | 1.4DL |
| Comb-02 | 1.2(DL+TL) + 1.6LL + 0.5LLr |
| Comb-03 | 0.9DL + (1.0WL or 1.0SL) |
| Comb-04 | 1.2DL + 1.0LL + 1.0SL |
| Comb-05 | 1.2(DL + TL) + 1.6LLr + 0.5WL |

Proper signs with (+) and/or (−) will be considered for the basic loads as applicable for the most critical load. Special load (SPL), if any, will be combined with the above combinations with the appropriate load factor depending on the type of load.

### 4.3.6.7.2 For Pile/Soil Bearing Design Load Calculation (Allowable Stress Design)

Where the allowable stress design method is used, structures and all portions thereof will be designed to resist the most critical effects from the following combinations of loads:

| | |
|---|---|
| Comb-01 | DL |
| Comb-02 | DL + LL + TL + LLr |
| Comb-03 | 0.60DL + (0.6WL or 0.7SL) |
| Comb-04 | 0.6DLs + 0.6WL |
| Comb-05 | DL + (0.6WL or 0.7SL) |
| Comb-06 | DL + 0.75(LL + TL + 0.6WL + LLr) |
| Comb-07 | DL + 0.75(LL + 0.7SL + LLr) |

Proper signs with (+) and/or (−) will be considered for the basic loads as applicable for the most critical load. Special load (SPL), if any, will be combined with the above combinations with appropriate load factor depending on the type of load.

### 4.3.7 Materials

Cement used for the foundation and substructure below ground floor level and in contact with soil shall be of OPC Type II conforming to ASTM C150. Superstructure concrete should be Type I OPC conforming to ASTM C150.

#### 4.3.7.1 Grade of Concrete

Foundations and substructures − 28 MPa (nominal 28 days compressive strength; cylinder test specimen).
Superstructures − 21 MPa.
Leveling and non-structure concrete − 10 MPa.

# Design Basis Report

### 4.3.7.2 Reinforcement Steel

Grade 60 bars (fy = 415 MPa) conforming to ASTM A615.

### 4.3.8 DESIGN CONCEPT

All RCC members shall be designed by load factor methods as per ACI 318.

The concrete frame structure should be a moment-resisting frame along both directions. Secondary beams that are not in column lines will be considered pin joints at the ends. Internal beams that are continuous over the support should be designed and reinforced for continuity at the supports.

### 4.3.9 CORROSION PROTECTION

#### 4.3.9.1 Concrete Cover

Foundations and surface below plinth and in contact with soil – 75 mm
Bored piles – 75 mm

All superstructures above plinth:

Beams and columns – 75 mm (external) and 40 mm (internal)
Slabs – 20

All miscellaneous steel embedded plates, corner angles, gratings, checkered plates, handrails, platform supporting structures, pipe and cable supports, and so on should be ASTM A36 (fy = 36 ksi). Gratings, platforms, and ladders shall be galvanized (458 g per sqm).

In addition, maximum crack width for the reinforced concrete structure is limited to the following:

Exposed to air and soil – 0.3 mm
Water retaining structure – 0.2 mm

### 4.3.10 ARCHITECTURE FINISHES

Not covered in this book.

[Note: Conceptual arch/mech/equipment layout drawing should be attached with the design basis report.]

## 4.4 DESIGN CRITERIA OF A STEAM TURBINE GENERATOR BUILDING: STEEL STRUCTURE WITH AN EOT CRANE (INDIAN STANDARDS-BASED)

In this section, we will detail the design basis of a steam turbine building with an electrical bay. The project site is in India.

Indian codes of practices will be followed in designing this building.

### 4.4.1 Description of the Building

This building will be made of a structural steel framework and RCC floor slabs supporting equipment and masonry walls. All steel columns will rest on RCC pedestals cast on a spread foundation resting directly on the soil. The building is 37 m wide and 120 m (2 × 60 m) long, consisting of two units of turbine generators.

The main bay is called the turbine bay (25 m wide) and it houses the steam turbine generator units. The ground floor of the turbine bay is for boiler feed pumps and various other pumps and equipment. It has a wide lay down area at the end bay (12 m bay) of the turbine hall. This is accessible by a railroad designed for unloading turbines and generators for erection. There is a mezzanine floor for heaters and critical piping for the turbine and then an operating floor at a higher level. The operating floor is a large hall for access to the steam turbine machine and an electrical control room. The roof over the turbine hall is at a higher elevation, giving clearance for an EOT crane running along the length of the turbine hall up to the end bay overlying laydown space at the ground floor. The mezzanine and operating floors below are open to the ground at the end bay for lifting turbines, generators, and other equipment by the overhead crane. The capacity of the EOT crane is 100 MT and is installed for the erection and maintenance of the steam turbine and generator.

The next bay running parallel to the turbine bay is called the electrical bay. This bay is 12 m wide. The ground floor is a cable spreader room, and above it is the MCC and switchgear room. There is another low-height cable spreader floor above this MCC and SWGR room for control and instrument cables. The main control room is above the cable spreader room. The control room floor is at the same level as the operating floor in the turbine hall. The roof over the control room floor runs from end to end of the building, covering the main control room area, chemical rooms, and offices with a toilet block, an elevator, and the main stairs. The major equipment deaerator is installed over this roof. Hence this roof is also called the deaerator floor. The deaerator is heavyweight equipment that rests on a steel stool directly connected to main frame girders.

The total length of this building is 120 meters. The building will be separated by an expansion joint of a 50 mm gap isolating two units (each 60 m long) at the center. There will be twin steel columns at this expansion joint having a centerline distance of 1250 mm in order to keep a 50 mm clear gap between the faces of twin foundation pedestals. These pedestals will rest on a single spread/raft footing, which will be designed for eccentric loads. The superstructure floor slabs will be extended from both units, keeping a clear gap of 50 mm across the building. Jointing of masonry walls and crane girders will be detailed to maintain this thermal expansion gap.

There are access stairs at all selected locations.

The cross-sectional view of this building is shown in Figure 4.1.

The steam turbine generator foundation is an RCC framed machine foundation structure, specially designed for this heavy rotary machine in accordance with

# Design Basis Report

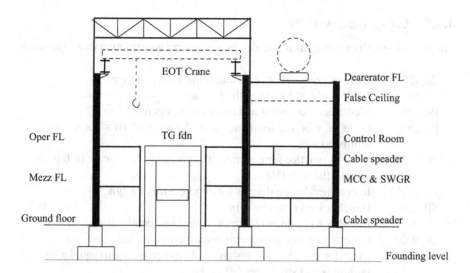

**FIGURE 4.1** A typical section of a steam turbine building.

relevant codes for machine foundations and recommendations of the manufacturer's permissible limits of frequencies and amplitudes. This foundation rests on soil separated from building floors by isolation joints. This section of the design basis report does not include the steam turbine foundation design.

### 4.4.2 Units and Language

All drawings and calculations will be in SI units – metric (English).

### 4.4.3 Site Information

The project site is 215 m above mean sea level and rests on a disintegrated medium to hard rock stratum. The site was prepared by cutting and filling existing land. The plot for the steam turbine building falls over the cutting area. The average safe bearing pressure is 250 kN/m². The groundwater is about 2 m below the finish grade level. Average rainfall intensity is 65 mm per hour. The humidity varies from 29 to 85% over the year. The seismic zone is III as per IS: 1893. The basic wind speed is 44 m/sec as per IS: 875 (Part 3). The maximum ambient temperature is 45°C and the minimum temperature is 5°C.

### 4.4.4 Soil Characteristics and Foundation System

The topsoil is medium dense silty sand with silty sand and rock fragments about 600 cm to 1.5 m of varying depth. Below this is a layer of 4–6 m disintegrated weathered rock overlying moderately weathered medium to coarse-grained augen gneiss.

Spread footing resting on the soil will be used for the foundation design.

## 4.4.5 CODES AND STANDARDS

The following Indian standards and codes for design and construction should be used:

| | |
|---|---|
| IS: 456 | Code of practice for plain and reinforced concrete |
| IS: 516 | Method of test for strength of concrete |
| IS: 1199 | Methods of sampling and analysis of concrete |
| IS: 2502 | Code of practice for bending and fixing of bars for concrete reinforcement |
| IS: 3370 | Code of practice for concrete structures for the storage of liquids (parts I through IV) |
| IS: 10262 | Recommended guidelines for concrete mix design |
| SP: 23 | Handbook on concrete mixes |
| SP: 34 | Handbook of concrete reinforcement and detailing |
| IS: 800 | Code of practice for general construction in steel |
| IS: 875 | Code for practice for design loads (other than earthquakes) for buildings and structures (all parts) |
| IS: 1893 | Criteria for earthquake resistant design of structures |

## 4.4.6 LOAD AND LOAD COMBINATIONS

### 4.4.6.1 Dead Load

The following unit weights will be considered in calculating the self-weight of the structure and foundation.

| | |
|---|---|
| Structural steel | 78.50 kN/M$^3$ |
| Reinforced concrete | 25.00 kN/M$^3$ |
| Plain concrete | 24.00 kN/M$^3$ |
| Floor finish, plaster, screed concrete | 20.00 kN/M$^3$ |
| Steel grating floor | 0.50 kN/M$^2$ |
| Checkered plate cover | 0.50 kN/M$^2$ |
| False floor | 1.00 kN/M$^2$ |
| False ceiling in control and elec. rooms | 0.4 kN/M$^2$ |
| Metal decking | 0.15 kN/M$^2$ |

Roof insulation and water proofing with

| | |
|---|---|
| cement tile on top | 1.55 kN/M$^2$ |
| Insulated metal siding | 0.25 kN/M$^2$ |
| Non-insulated metal siding | 0.15 kN/M$^2$ |
| Insulated metal roof | 0.25 kN/M$^2$ |

Masonry units with plaster on both sides

| | |
|---|---|
| 230 mm nominal thickness | 4.6 kN/M$^2$ |
| 115 mm nominal thickness | 2.3 kN/M$^2$ |
| Soil | 19.0 kN/M$^3$ |

# Design Basis Report

### 4.4.6.2 Live Load

| | |
|---|---|
| Ground floor | 15.00 kN/m² |
| Ground floor lay down area | 30.00 kN/m² |
| Mezzanine floor | 15.00 kN/m² |
| Operating floor | 20.00 kN/m² |
| SWGR and MCC rooms | 15.00 kN/m² |
| Cable spreader floor | 10.00 kN/m² |
| Deaerator floor cum roof (including piping) | 15.00 kN/m² |
| Intermediate floors | 10.00 kN/m² |
| Roof over turbine hall (accessible) | 2.50 kN/m² |
| HVAC area on roof | 3.50 kN/m² |
| Piping loads | 3.00 kN/m² |

### 4.4.6.3 Wind Load

To be calculated as per IS: 875 Part 3
Basic wind speed, $V_b = 44$ m/sec

$$\text{Design wind speed} = V_z = k_1 \cdot k_2 \cdot k_3 \cdot V_b$$

$k_1$ (Probability factor) = 1.07
$k_2$: Category 2; Class B (Table 4.2, IS: 875-1987)
$k_3$ (Topography factor) = 1.0

$$P_z = 0.6 V_z^2$$

Design Wind Pressure, Pz:

| | |
|---|---|
| Ground to 10 m height | 1.28 kN/m² |
| Ground to 10 m height | 1.38 kN/m² |
| Ground to 10 m height | 1.47 kN/m² |
| Ground to 10 m height | 1.61 kN/m² |

### 4.4.6.4 Seismic Load

Seismic loads should be calculated based on the specification and as per IS: 1893 (latest). The response spectrum method of analysis should be used.

### 4.4.6.5 Temperature Load

Average maximum temperature = 45°C
Average minimum temperature = 5°C
Mean temperature = 25°C

Structures will be designed for a temperature variation of 20°C.
The coefficient of thermal expansion for concrete structure = $10 \times 10^{-6}$ per °C
[There is another method of temperature load assessment that is recommended by some reputed consultants. In this method, the total temperature variation will be considered as two-thirds of the average maximum annual variation in temperature.

The average maximum annual variation in temperature for this purpose will be taken as the difference between the mean of the daily minimum temperature during the coldest month of the year and the mean of the daily maximum temperature during the hottest month of the year. The structure will be designed to withstand stresses of 50% of the total temperature variation.

Accordingly, total temperature variation = 2/3 × (45 − 5) = 26.6°C.

Design temperature = 50% of 26.6 = 13.3°C.]

### 4.4.6.6 Load Combinations

The steel framework will be designed in the following combinations.

*4.4.6.6.1 Basic Load Cases*

1. DLS [Only self-weight]
2. DLE [Self-weight of equipment]
3. LLR [Roof live load]
4. LLF [Floor live load]
5. CL-L+CLHx(+) [Maximum crane wheel load on left col. plus horizontal surge in (+) X-direction]
6. CL-L+CLHx(−) [Maximum crane wheel load on left col. plus horizontal surge in (−) X-direction]
7. CL-R+CLHx(+) [Maximum crane wheel load on right col. plus horizontal surge in (+) X-direction]
8. CL-R+CLHx(−) [Maximum crane wheel load on right col. plus horizontal surge in (−) X-direction]
9. WLx(+) [Wind load in (+) X-direction]
10. WLx(−) [Wind load in (−) X-direction]
11. WLy [Wind load in (−) Y-direction)
12. SLx [Seismic load in (+) X-direction]
13. SLy [Seismic load in (+) Y-direction]
14. CL [Crane self-weight]
15. TL [Temperature load]
16. CL-L+CLHx(+) [Maximum crane wheel load on left col with 50% lifting wt. Plus horizontal surge in (+) X-direction]
17. CL-L+CLHx(−) [Maximum crane wheel load on left col. with 50% lifting wt. Plus horizontal surge in (−) X-direction]
18. CL-R+CLHx(+) [Maximum crane wheel load on right col. with 50% lifting wt. Plus horizontal surge in (+) X-direction]
19. CL-R+CHHx(−) [Maximum crane wheel load on right col. with 50% lifting wt. Plus horizontal surge in (−) X-direction]

[Note: X-direction is toward the cross direction of the building from left to right.]

*4.4.6.6.2 Load Combination for Steel Framework and RCC Foundations*

For anchor bolts and uplift of foundation:

Comb 1 = 1

# Design Basis Report

$2 = 0.9(1) + 9$
$3 = 0.9(1) + 10$
$4 = 0.9(1) + 11$
$5 = 0.9(1) + 12$
$6 = 0.9(1) - 12$
$7 = 0.9(1) + 13$

For steel superstructure and foundation:

$8 = [(1+2)+(3+4)+16+9]$
$9 = [(1+2)+(3+4)+17+10]$
$10 = [(1+2)+(3+4)+18+9]$
$11 = [(1+2)+(3+4)+19+10]$
$12 = [(1+2)+(3+4)+5]$
$13 = [(1+2)+(3+4)+6]$
$14 = [(1+2)+(3+4)+7]$
$15 = [(1+2)+(3+4)+8]$
$16 = [(1+2)+0.5(4)+14+12]$
$17 = [(1+2)+0.5(4)+14-12]$
$18 = [(1+2)+0.5(4)+14+13]$
$19 = [(1+2)+(3+4)]$
$20 = [(1+2)+0.5(4)+14]$
$21 = [(1+2)+(3+4)+5+0.5(9)]$
$22 = [(1+2)+(3+4)+6+0.5(10)]$
$23 = [(1+2)+(3+4)+7+0.5(9)]$
$24 = [(1+2)+(3+4)+8+0.5(10)]$

[Note: Temperature loads (15) will be added with all the above load combinations.]

## 4.4.7 Materials

The following materials will be used for design and construction.

### 4.4.7.1 Structural Steel

| | |
|---|---|
| Rolled steel sections and plates | IS: 2062 (Yield stress, $F_y$ = 250 MPa) |
| Plates above 20 mm thickness | IS: 2062 (Grade B; $F_y$ = 250 MPa) |
| Bolts, nuts, and washers | IS: 1367; IS: 4000 |
| Welding consumables | IS: 814; IS: 1278; IS:1387 |
| Structural pipes | IS: 1161 (Yield stress, $F_y$ = 240 MPa) |
| Gratings and Checkered plates | IS: 2062; IS: 3502 |
| Anchor bolts | IS: 2062 ($F_y$=250MPa) |

### 4.4.7.2 Concrete Works

| | |
|---|---|
| Foundations, pits, and substructures below ground floor | M25 |
| Equipment foundations | M25 |
| Ground floor slab | M25 |

| | |
|---|---|
| RCC floors, roof, and superstructure works | M20 |
| Leveling concrete | M15 |
| Mass concrete | M10 |

### 4.4.7.3 Reinforcement Bars

High yield strength deformed bars   IS: 1786 (Fy = 415 MPa)

## 4.4.8 Design Concept

The building will be analyzed as a 3D structure using STAAD Pro computer software. The frames along the transverse direction will be moment-resisting joints and those along longitudinal directions will have vertical bracings at suitable locations.

The strength design of the steel members in the superstructure will be done in working stress methods following IS: 800 (latest). The sway and deflection criteria shall be considered as furnished in IS: 800.

Column foundations will be spread footings resting on the soil. The settlement of footings (total and differentials) will be checked, and the sizing of foundations revised, if necessary, to keep the deflection (DL + 50% LL) within the permissible limit as per the relevant IS code for shallow foundations.

All shop connections should be welded joints and field connections will be bolted. Except for column splices, all bolted connections will be a bearing type using high-strength bolts of 8.8 grade. Column splices will be friction grip connections. The minimum diameter of bolts will be 16 mm with the tightening torque mentioned in the drawing.

## 4.4.9 Corrosion Protection

### 4.4.9.1 Steel Structures

Painting with prime coats of red oxide zinc chromate (or phosphate) – one at the fabrication shop and another at the site before the application of the final coats. The final coats are two coats of synthetic enamel. The gratings will be galvanized.

### 4.4.9.2 Concrete Cover

Foundations and surface below plinth and in contact with soil – 75 mm
Bored Piles – 75 mm
All superstructures above the plinth:
Lintels and internal stub columns – 40 mm
Slabs and walls – 20
In addition, the maximum crack width for the reinforced concrete structure should be limited to 0.3 mm as per the recommendations in IS: 456.

## 4.4.10 Architecture Finishes

Not covered in this book.

[Note: The conceptual arch/mech/equipment layout drawing should be attached with the design basis report.]

## 4.5 CONCLUSIONS

The design criteria shown here are for guidance only and to highlight the minimum requirements of a building design. The engineer should read the contract documents, the scope of work, and technical specifications to comply with the owner's requirements before writing a design basis report for the design. There may be different options in the specifications and codes, therefore, this document is the place where the engineer or contractor should show the specifics of what he or she is going to design and supply.

# 5 Conceptual Design

## 5.1 INTRODUCTION

Conceptual design is the first task of a designer after he or she has read the contract document, the scope of work, meteorological data, and geotechnical information. The objective is to form a balanced structure considering the functional requirements and facilities available. Dimensional shape and size, selection of building materials, quality, safety, time of construction, and cost are the primary factors that should be considered when stepping into the basics of conceptual design work. A site visit helps the designer to get a feel of engineering and the aesthetic aspects specific to the site.

The structural framework of an industrial building or unit is developed according to a general arrangement plan and equipment layout drawing made by system engineers. The structural engineer must prepare a framework, which we call a basic drawing. This basic drawing will indicate the location of columns on grids, as well as framing beams and braces for withstanding gravity loads of people, machinery, and their own weight. The framework should also be flexible enough to resist natural forces generated during winds, storms, or earthquakes.

All the forces from the superstructure are transferred to the ground by a suitable foundation system consistent with the properties of the subsoil stratum. The whole structure from roof to foundation and its supporting soil stratum should be considered as **One** unit, which has harmonic integrity in geometrical property definition and ductility.

## 5.2 SELECTION OF THE FOUNDATION TYPE

The selection of the foundation type is one of the major steps in conceptual design. The engineer should be aware of the type of soil stratum over which the plant is going to be constructed.

The reader should study standard textbooks on soil mechanics and relevant codes of practice for in-depth knowledge of practical foundation engineering.

The concept of selecting an appropriate foundation depends on the intensity of loading that will be applied on the soil stratum and the safe bearing capacity of soil within a permissible settlement limit. The founding level of a plant building with column loads of medium to high magnitude is a soil stratum of medium-hard soil overlaying a layer of soft soil followed by dense sand. The foundation engineer has designed two types of foundation systems—one large based flexible mat foundation rests on soil and the other with isolated pile foundations resting on a dense sand layer. The first option is cost-efficient, but the estimated value of consolidation settlement exceeds the permissible limit over the design period. Hence, the designer needs to follow the second option with a pile foundation that meets the settlement criteria.

Similarly, a foundation engineer has to provide pile foundations for a pipe rack structure carrying steam pipes with a stringent settlement limit over a stretch of land between two buildings that are on pile foundations. This is to meet the allowable limit of differential settlement specified for the rack structure.

Industrial plants and structures are located at sites depending on the socio-economic conditions of a country. Hence, the foundation type must suit the site-specific requirement.

In Chapter 9, we will discuss the different types of foundation systems used to support buildings and structures on the ground.

## 5.3 TYPE OF STRUCTURE AND MATERIAL OF CONSTRUCTION

Industrial building structures are built to provide a safe operating life to the plant and machinery throughout the design life. These structures should be functionally efficient, maintenance-free, and stable against all probable combinations of service loads and forces of nature – flood, wind, or earthquake.

The member sizing should be slim to provide maximum clear spaces for equipment, piping, and operating personnel.

Time of construction and cost-effectiveness are the next major considerations.

The following are options for selecting materials for construction:

a) Reinforced concrete structure
b) Structural steelwork
c) Hybrid structure, i.e., a combination of both steel and concrete

*Reinforced cement concrete* (RCC) works are generally used to build buildings, roads and bridges, crossing works, flood protection barriers, sea wall and shore protection works, dams and river training works, canal lining, electrical poles, rail sleepers, high rise chimneys, and industrial structures. Since the reinforced concrete structure is less costly than the steel structure and needs minimal maintenance, it has become a popular choice.

The strength of reinforced concrete is much lower than steel, so the dimensions of building blocks like column and beam sections need to be robust compared with steel structures, as well as occupying a greater floor area. To eliminate this problem, advanced design methods use high-strength concrete and precast/prestressed concrete. Heavy depth bridge girders are replaced with lightweight and slim prestressed members.

The cement concrete structure does not require periodical maintenance like the painting of a steel structure. The use of sulfate resistance cement, pozzolanic cement, and mineral additions like ground granulated blast furnace slag and epoxy coated bars/corrosion-resistant bars enhance the durability and longevity of the concrete structure in extreme environments, including marine conditions.

# Conceptual Design

The quality of the concrete structure greatly depends on the fieldwork. While steel structures are manufactured in a factory under a set procedure of quality control, the production of concrete mix and its placement is greatly dependent on semi-skilled labor, site conditions, and weather.

For large projects, the quality of both materials and workmanship are monitored by qualified engineers. Modern equipment like a batching plant, transportation trucks fitted with a rolling mixture, concrete pumps, multi-facility cranes, and so on are used for construction. The design mix computation and production in the batching plant are done by engineers and skilled professionals using computer-guided equipment.

But in small works, concrete production is dependent on a hand-mix procedure or using a concrete mixture machine. The concrete mix is conveyed and placed by manual operation. The water:cement ratio, the proportion of concrete mix, and time of placement (within initial setting time) are not properly maintained in many construction sites; the concrete is poured in a slump that is suitable for workers, while ignoring the water:cement ratio.

The concrete structure is not a homogeneous product like a steel section. This is a cast-in product that works in combination with reinforcement steel bars and a cement concrete mix hardened to its characteristic strength. The hardened concrete bears compression and shearing forces, and flexural stress is taken by reinforcement steel bars. Shearing force in excess of concrete capacity is shared by reinforcement bars in the form of stirrups and diagonal bars. So, the cost of concrete is dependent on the ratio of unit prices of steel bars and the grade of concrete mix. A higher grade means higher strength, which needs more cement consumption in general. Hence, economy in overall cost can be achieved by balancing the number of reinforcement bars and the volume of concrete mix. High-strength concrete allows for a reduction in member sizes as well as less volume and weight in a structure. It gives a direct saving in foundation cost for dead load predominant structures. But one should remember that using high-strength concrete in construction needs perfect quality control.

Quality, cost economy, and time of construction are basic requirements that we should take into consideration when building any structure. Hence, the use of factory-made ready-mix concrete and precast and prestressed materials have become the obvious choice for many owners and constructors, including small residential house builders.

*Structural steelwork* for general building work and construction is valued more than other materials for its shorter time of construction and slender design. The material strength is higher than concrete so it can bear more load than an equivalent-sized concrete section despite using high-strength concrete or prestressed elements.

The cost of structural steelwork is higher than reinforced concrete. But saving time in construction, the reduction in foundation cost due to low weight superstructures, and the efficacy of spanning large distances while satisfying deflection limits and slender design have made it a competitive choice to owners.

Recent developments in using high-strength structural steel have made it more economic. Uses of the hybrid construction of steel and concrete sections in the same building have also become a cost-effective solution on occasion.

A field joint by using high-strength friction grip/non-slip connection bolts is the most popular solution to ensure quality and reliability. It is a standard practice in the United States and other countries.

Materials are to be selected by the designer considering the overall economy and cost savings. The use of high-strength steel for the axially loaded member is better than using it for girders where shape factor (section modulus) and depth govern bending stress. Similarly, the choice of connections and materials and minimizing the number of joints add to the savings. For example, a pipe bridge structure designed with lightweight angle member trusses may give a reduced total member weight compared with a simple beam-column structure of rolled steel joists; however, the overall saving may be negative due to the higher cost in fabrication and time of construction.

Efficient procurement of materials in matching lengths and sizes and an optimized cutting plan at the workshop can minimize losses due to the wastage of steel.

Above all, adopting a simple design detail to save time in construction, as well as timely completion of the job, has a great impact in saving costs in a project.

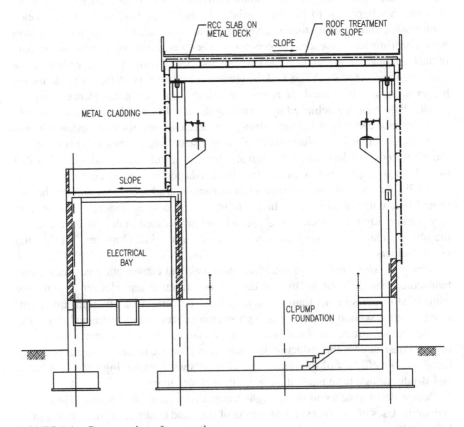

**FIGURE 5.1** Cross-section of a pump house.

# Conceptual Design

The selection of the member section by the designer has an important role in saving costs. The designers should use regularly available sizes of rolled steel sections and plates available in the market or produced by reputed manufacturers. This will save time in procurement and avoid the replacement procedure of alternative sections during fabrication. Similarly, if a designer selects a common width of plated members, keeping in mind the standard rolling widths produced by reputed manufacturers, the fabricator can optimize a cutting plan and thus minimize wastage.

## 5.4 HYBRID STRUCTURES

As stated earlier, this is a combination of RCC members and steel framework.

The common use of a steel frame roof structure is resting on top of RCC columns, where scaffolding work for casting concrete becomes extremely difficult and time-consuming as a result of the high altitude and large span. A pump house roof is an example of this type of construction.

Another use for a hybrid structure is a powerhouse or similar plant building. These buildings have RCC beam-column-slabs up to the operating floor and a steel

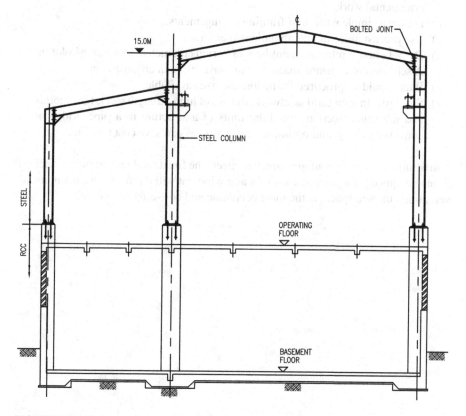

**FIGURE 5.2** Cross-section of a plant building.

framework above for the turbine hall. Riverbed pump houses have a steel superstructure above the operating floor cast on top of a concrete caisson.

In Indonesia, a mix of precast/prestressed concrete elements, a cast-in-situ foundation, and a steel roof is common practice. Pre-engineered buildings with a steel framework, lightweight concrete floor panels, and insulated metal cladding roof and side cladding are widely used in Middle Eastern countries.

Hybrid structures are often found to be a cost-effective and time-saving solution. Figures 5.1 and 5.2 are typical examples of plant buildings in hybrid constructions.

## 5.5 COST OPTIMIZATION

Cost economy is a time-dependent factor. Hence, a designer should look into how to save time in construction while meeting safety and functional requirements.

A few common options for cost optimizations are as follows:

a) Dimensioning beams and columns in groups of the same size as far as possible to allow reuse of formwork.
b) Avoid complicated detailing work, e.g., curved beams and architectural ornamental work.
c) Follow a simple pattern of framing arrangements.
d) Use precast units and combinations thereof.
e) Do not choose reinforcement bars of various diameters in a beam/column. A selection of common diameter bars saves time in construction.
f) Bars should be procured in the longest sizes available.
g) Similarly, in structural steelwork, the use of high-strength steel, fewer joints and fabricated sections, modular units of a structure in a pipe bridge or conveyor gallery, and optimized cutting plans can save cost and time.

A structure of a simple configuration that meets the functional requirements, such as giving adequate protection against rain and wind but without fancy ornamentals, as well as any unused space, is the most economic and cost-effective choice.

# 6 Procurement of Materials

## 6.1 GENERAL

The procurement of materials is one of the first activities after the award of a contract.

The design consultant prepares a document that contains a list of materials and their specifications.

Raw materials like structural steel, cement, and reinforcement bars are procured in advance to maintain the schedule of supply from the fabrication shop to field construction. These materials are procured in phases matching with the project schedule.

Since the finalization of the shop fabrication drawing or released for construction drawing is a time-dependent process, design consultants use their past experience to make a preliminary design for the assessment of quantities of raw materials like structural steel, cement, and reinforcement bars. The list of procurement is given in phases, extending over a time period that helps the designer to adjust the balance procurement order according to actual consumption corresponding to the approved drawing. These are parallel activities, and they are executed with proper coordination. It should be noted that all sections or types of material indicated in the list of orders may not be readily available in local markets within the scheduled time and they may have long lead times that will not be cost-effective. In such cases, the procurement list is revised with another alternative that supports the designer's requirements.

During the progress of the job, if there are shortages of certain sizes of material in the store or yard, then the designer suggests a replacement from the stored stock or another material that can be procured without delay. Similarly, at the end of the job, excess materials are also reviewed and used as alternative sections, thus changing the design. For example, there is a shortage of 25 mm diameter reinforcement bars in stock but bars of other diameters, say 20 and 16 mm, are available for use, the designer could suggest replacing 25 mm bars with 20 mm bars with closer spacing, leaving the total area of reinforcement unchanged. This is a continuous process of adjustment that goes on until the completion of the project.

Hence, the procurement list should be reviewed carefully by experienced engineers and monitored regularly.

The following sections show sets of standard charts used for initial procurement.

## 6.2 STRUCTURAL STEELWORK

The format used for this work will be as per the standard sheet of the engineering company entrusted with the job. The following charts are for general guidance only.

*Owner: M/s ABC Company Limited*
*Plant: 1×100 MW Power Plant*
*Document Title: Procurement list for Structural Steel Section* (Table 6.1)

DOI: 10.1201/9781003211754-6

## TABLE 6.1
### Summary Sheet

| | | Requirements in M Tons | | |
|---|---|---|---|---|
| | Materials | Phase I | Phases II | Phase III |
| A.1 | Rolled steel joists (ISMB) | | | |
| A.2 | Rolled steel channels (ISMC) | | | |
| A.3 | Mild steel plates | | | |
| A.4 | Rolled steel angles (ISA) | | | |
| A.5 | Mild steel flats | | | |
| A.6 | Mild steel checkered plate | | | |
| A.7 | Mild steel gratings | | | |
| A.8 | Mild steel black pipes | | | |
| B | Mild steel rounds | | | |
| | TOTAL | | | |

Phase I: For the period up to August 2020
Phase II: September to November 2020
Phase III: December 2020 to February 2021 (Tables 6.1–6.10)

**Notes:**

1. All steel material shall conform to the following specifications.
   a) Rolled steel sections and plates up to 20 mm thick: IS 2062 Grade E250 (Fe 410 W) A
   Plates above 20 mm: IS 2062 Grade E250 (Fe 410 W) B
   Plates for Crane Girder above 20 mm: IS 2062 Grade E250 (Fe 410 W) C
   b) IS: 806 – Code of practice for use of steel tubes in general building construction.
   c) IS: 1161 – Steel tubes for structural purposes.
   d) IS: 3502 – Steel checkered plate.
   e) All mild steel bars for anchor bolts and sag rods, etc. shall conform to IS: 432 or IS:2062.
2. All raw steel shall be of tested quality and test certificates should be obtained from the producers.
3. The requirement indicated is inclusive of the normal margin for indenting, wastage, etc., considering procurement of sections in standard sizes as available from producers.
4. The steel requirement is preliminary, and a more accurate estimate will be developed after the preparation of the fabrication drawing and bill of materials.
5. Any substitution in design sections will eventually mean a revision in the requirements of particular sizes or sections.
6. The above requirement does not include the quantities for the building/structures designed or supplied by other vendors.

## TABLE 6.2
### A.1 Rolled Steel Joists (ISMB)

| Section | Approx. Total Requirement (M Tons) | Requirement in M Tons | | |
|---|---|---|---|---|
| | | Phase I | Phase II | Phase III |
| ISMB 600 | | | | |
| ISMB 500 | | | | |
| ISMB 450 | | | | |
| ISMB 400 | | | | |
| ISMB 300 | | | | |
| ISMB 250 | | | | |
| ISMB 200 | | | | |
| ISMB 150 | | | | |
| TOTAL | | | | |

## TABLE 6.3
### A.2 Rolled Steel Channels (ISMC)

| Section | Approx. Total Requirement (M Tons) | Requirement in M Tons | | |
|---|---|---|---|---|
| | | Phase I | Phase II | Phase III |
| ISMC 250 | | | | |
| ISMC 200 | | | | |
| ISMC 150 | | | | |
| ISMC 125 | | | | |
| ISMC 100 | | | | |
| TOTAL | | | | |

## TABLE 6.4
### A.3 Mild Steel Plates

| Size (Thickness) | Approx. Total Requirement (M Tons) | Requirement in M Tons | | |
|---|---|---|---|---|
| | | Phase I | Phase II | Phase III |
| 50 mm | | | | |
| 45 mm | | | | |
| 40 mm | | | | |
| 36 mm | | | | |
| 32 mm | | | | |
| 28 mm | | | | |
| 25 mm | | | | |
| 20 mm | | | | |
| 16 mm | | | | |
| 12 mm | | | | |
| 10 mm | | | | |
| 8 mm | | | | |
| 6 mm | | | | |
| TOTAL | | | | |

## TABLE 6.5
### A.4 Rolled Steel Angle (ISA)

| Section | Approx. Total Requirement (M Tons) | Requirement in M Tons | | |
|---|---|---|---|---|
| | | Phase I | Phase II | Phase III |
| ISA 200×200×20 | | | | |
| ISA 150×150×16 | | | | |
| ISA 130×130×12 | | | | |
| ISA 130×130×10 | | | | |
| ISA 110×110×12 | | | | |
| ISA 100×100×10 | | | | |
| ISA 100×100×8 | | | | |
| ISA 90×90×8 | | | | |
| ISA 75×75×8 | | | | |
| ISA 75×75×6 | | | | |
| ISA 65×65×8 | | | | |
| ISA 65×65×6 | | | | |
| ISA 50×50×6 | | | | |
| TOTAL | | | | |

## TABLE 6.6
## A.5 Mild Steel Flat

| Size (mm×mm) | Approx. Total Requirement (M Tons) | Requirement in M Tons | | |
|---|---|---|---|---|
| | | Phase I | Phase II | Phase III |
| 100×8 | | | | |
| 50×6 | | | | |
| 25×6 | | | | |
| TOTAL | | | | |

## TABLE 6.7
## A.6 Mild Steel Checkered plate

| Size (Thickness) | Approx. Total Requirement (M Tons) | Requirement in M Tons | | |
|---|---|---|---|---|
| | | Phase I | Phase II | Phase III |
| 8 mm O/P | | | | |
| 6 mm O/P | | | | |
| TOTAL | | | | |

## TABLE 6.8
## A.7 Mild Steel Grating

| Size (Thickness) | Approx. Total Requirement (M Tons) | Requirement in M Tons | | |
|---|---|---|---|---|
| | | Phase I | Phase II | Phase III |
| 40 mm | | | | |
| 25 mm | | | | |
| TOTAL | | | | |

## TABLE 6.9
### A.8 Mild Steel Black Pipe – Medium Duty

| Size (Nominal Diameter) | Approx. Total Requirement (M Tons) | Requirement in M Tons | | |
|---|---|---|---|---|
| | | Phase I | Phase II | Phase III |
| 200 mm | | | | |
| 150 mm | | | | |
| 100 mm | | | | |
| 80 mm | | | | |
| 65 mm | | | | |
| 50 mm | | | | |
| 32 mm | | | | |
| TOTAL | | | | |

## TABLE 6.10
### B. Mild Steel Round for Anchor Bolts and Sag Rods, and Quantity Required for Cement Concrete Work

| Diameter | Approx. Total Requirement (M Tons) | Requirement in M Tons | | |
|---|---|---|---|---|
| | | Phase I | Phase II | Phase III |
| 63 mm | | | | |
| 50 mm | | | | |
| 40 mm | | | | |
| 36 mm | | | | |
| 32 mm | | | | |
| 28 mm | | | | |
| 24 mm | | | | |
| 20 mm | | | | |
| 16 mm | | | | |
| 12 mm | | | | |
| 8 mm | | | | |
| TOTAL | | | | |

## 6.3 CONCRETE STRUCTURE

### TABLE 6.11
### Phase Wise Requirement of Cement

| Approx. Total Requirement | Phase I | Phase II | Phase III | Phase IV |
|---|---|---|---|---|
| M Ton | M Ton | M Ton | M Ton | M Ton |

1. Phase I: For the period up to August 2020
2. Phase II: September to November 2020
3. Phase III: December 2020 to February 2021
4. Phase IV: March to May 2021 (Table 6.11)

**Notes:**

1. The requirement of cement indicated here is preliminary and a more accurate estimation will be given by the contractor after issuing design drawings.
2. Generally, cement shall be Ordinary Portland cement conforming to IS: 269. Portland slag cement conforming to IS: 455 and Portland pozzolana cement (fly ash based) conforming to IS: 1489 (Part I) may also be used.
3. The requirement is based on the normal allowance for wastage.
4. The above requirement does not include the quantities for the building/structures designed or supplied by other vendors (Table 6.12).

### TABLE 6.12
### Phase Wise Requirement of High Yield Strength Deformed Bars

| Diameter of Reinforcement Bar mm | Approx. Total Requirement M Ton | Phase I M Ton | Phase II M Ton | Phase III M Ton | Phase IV M Ton |
|---|---|---|---|---|---|
| 32 | | | | | |
| 28 | | | | | |
| 25 | | | | | |
| 20 | | | | | |
| 16 | | | | | |
| 12 | | | | | |
| 10 | | | | | |
| 8 | | | | | |
| TOTAL | | | | | |

1. Phase I: For the period up to August 2020
2. Phase II: September to November 2020
3. Phase III: December 2020 to February 2021
4. Phase IV: March to May 2021

**Notes:**

1. All high-strength deformed bars (HYSD) bars shall conform to IS: 1786.
2. All reinforcement bars shall be of tested quality and test certificates shall be obtained from the manufacturers.

3. The requirement is based on the normal allowance for laps and wastage considering the procurement in the maximum available sizes.
4. The requirement of reinforcement bars indicated here is preliminary and a more accurate estimation will be given by the contractor after preparation of the bar bending schedule.
5. Any substitutions in design sections/bar diameters will eventually mean revision in the requirement of the particular sizes or bar diameters.
6. The above requirement does not include the quantities for the buildings/structures designed or supplied by other vendors.

# 7 Design Office Procedures for Engineering and Drawing

## 7.1 TIMELINE PLANNING

Timeline planning is the chronological order of activities that helps the project progress systematically. The project manager controls the job according to this planning process. There are specific time allotments for engineering, drawing, procurement, fabrication, construction, and all such major activities in the form of bar charts in the project timeline. The designer follows the timeline for submissions of the deliverables. An efficient designer makes his or her own schedule of activities according to the timeline planning of the project.

The civil and structural designer's activities can be broadly classified in the following order:

a) Civil input data
b) Design basis report or design criteria
c) Conceptual design and framework preparation
d) Load assessment
e) Analysis and strength design
f) Drawing
g) Interdisciplinary checking of drawings
h) Incorporation of change requirements in drawings
i) Submit drawings with supporting calculations for approval (where needed)
j) Release for construction
k) Coordinate with field construction
l) As-built drawings

This list of activities covers the majority of the designer's work. In addition, they take part in the preparation of the material procurement list, design coordination meetings, the estimation of the bill of quantities, the review of fabrication drawings, and so on.

## 7.2 CIVIL INPUT DATA

Civil input data refers to basic system engineering drawings and the data being followed in the preparation of the civil and structural drawings for construction.

DOI: 10.1201/9781003211754-7

For industrial buildings, civil and structural input information includes the following:

a) Mechanical/electrical general arrangement drawings.
b) Equipment layout plan.
c) Floor loading intensity including the erection load into lay down areas, major equipment loads, and the outline of the footprint for the foundation pedestals, anchor bolts or pockets, floor cutouts, and insert plates.
d) Pipe and hanger loading, cable post support, or hangers.
e) Space or opening to be kept for the pipe corridor, cable trays and ventilation ducts, monorails, temporary openings, and bays to be reserved for erection hatches.
f) Provisions for a roof exhauster or fans, wall openings for piping, cables and ducts, etc.
g) For buildings with crane facilities – crane wheel load data, the number of wheels per side, the distance between wheels, the weight of the crab, and the lifting load and clearance diagram at the sides and overhead.
h) For ground floor civil input – layout of trenches and equipment, equipment drain pits, walkways for erection and maintenance, and floor load data.

There are specific details for equipment, systems, and load data for water treatment plants and chemical plants. Input for pipe racks indicates general distributed load and pipe support reactions including friction load. Conveyor gallery loads define point loads from conveyors inclusive of belt tension.

Civil input for large fans and machine foundations includes foundation outline drawings showing pocket and anchor bolt details, static and dynamic loads, short circuits or accidental loads, operating frequencies, criteria for vibration design limitations – permissible amplitudes and limiting frequencies – and so on. For large transformer foundations, the input data should include general arrangement drawings, the weight of the transformer with and without oil fill, the number of wheels, and its base dimensions or dimensions of skids, if skids are mounted. A large tank design input comprises general arrangement drawings, self-weight, liquid density with maximum and minimum filling heights, and holding down bolts, where necessary.

In general, all the above load details are from dead weight, live, and operating load. Wind and seismic load are to be determined by the building and foundation designer unless specified otherwise.

The above details are furnished as general guidelines only. Project-specific information for process plants and others will be available in respective civil input data.

All these input data may not be available at the beginning. The system engineer provides these data in stages and the final information comes at a later stage upon finalization of their design and ordering major equipment. For example, Level 1 type information gives column grid location, a basic outline for framing beams, and uniformly distributed floor load. Level 2 information may retain the same loading or add point loads. In this stage, floor cutouts and equipment footprints

# Design Office Procedures

with fixing details are indicated. Level 3 is the final confirmation of the above information.

Since civil work for foundations and the fabrication of steelwork begin in the early stages of work, the design engineer has to progress with his or her building design considering the general loading as per standard codes of practice and design criteria shown in the design basis report.

## 7.3 STRUCTURAL FRAMING, LOADING, AND ANALYSIS

We have explained the conceptual design, selection of the type of structure – RCC or steel – and a suitable foundation system in previous chapters. Definitions of load and load combinations have also been explained in detail in Chapter 4 for design basis criteria.

### 7.3.1 Structural Framing System

The framework of a building or structure consists of vertical members or columns in rows of a grid system. The columns are connected by horizontal members, i.e., beams at different floor levels in a horizontal plane along orthogonal directions. Beam members are also provided in between floor levels in the form of lintels over doors and windows around the periphery of buildings or as ties connecting columns. The floor slabs are plate members joining the beams at the top level and columns at the corners. The floor slab panels span over a grid of beams. There is a limitation of slope and deflections for the stability of floor slab panels against the gravity load, so intermediate beams are added to support the slab panels and are thus subdivided into small panels. Gravity loads from the slabs and the wall are carried by beam stiffness and then transferred to columns at joint nodes. Columns are sized to bear this load respecting the permissible limit of stresses as per the slenderness limit to avoid buckling.

Vertical bracings and moment-resisting connections at joints are provided to control horizontal deflection within allowable limits against wind or seismic generated forces.

Such a beam-column-brace system makes an integrated three-dimensional framework that is flexible enough to transfer the load onto the supporting foundation.

A masonry wall between floors acts like filled-up panels and its stiffness is not considered in frame stiffness except those that are cast-in-concrete around lift shafts or shear walls.

RCC frameworks are designed with fixed or moment joints in both directions. Vertical bracings are not conventional in RCC frame buildings.

Hence, columns in RCC frames should be designed for biaxial moments and axial force.

Steel frameworks are usually braced in a longitudinal direction and moment connection joints along a transverse direction. Tower-like structures and small buildings are designed with vertical bracings at all sides. Steel columns with moment connections along a transverse direction and braced in a longitudinal direction are designed for axial load and uniaxial moment. Bracings in steel structures are designed for axial forces – tension and compression.

Slab panels cast monolithically with beams give lateral stability to a framework in a horizontal plane. A floor resting on steel beams is not considered as laterally restrained unless a horizontal bracing system is provided at the top or the compression flange of beams is studded with shear lugs or partially embedded in slabs. The use of a metal deck form is popular in many constructions. The studding on the top flange of the steel beam is done by using welded (electro-forged) studs, for example, Nelson studs.

It is difficult to provide a regular pattern of framing arrangements in industrial floors maintaining the locations of equipment and openings for piping and cable hatches. Floor beam spacing and arrangement will depend on the layout of equipment and floor cut-out locations. Heavy equipment should have supporting beams directly below the load application points. Similarly, large cutouts should have beams along the edges.

Beams and girders are subjected to flexure stresses rather than axial stresses. These members should be given a shape that gives a higher moment of inertia. By optimizing the depth and width ratio, we can get a better economic position for flexural members.

Columns are primarily axial load-carrying members. The section should have a nearly equal radius of gyration about both the axes to get a favorable slenderness ratio (effective length/appropriate radius of gyration) and a higher value of permissible stress. The pattern of framing beams and girders shall be done in a way that columns are subject to more axial load than moments. An axially loaded member is found to be most effective because the total cross-section area is equally stressed.

Let us observe the types of framing patterns given in Figure 7.1.

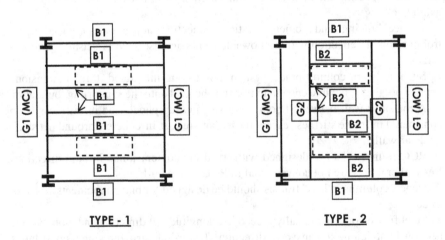

**FIGURE 7.1** Part plans of a floor beam arrangement (MC denotes moment connection joints at both ends).

# Design Office Procedures

In arrangement Type 1, the girder G1 is loaded by all the intermediate beams and transfers a large amount of fixed end moment and axial load to the columns, whereas, in Type 2 framing, the girder G1 carries less floor load than the G1 of Type 1. So, the columns in Type 2 get less moment but the same axial load.

Hence, the Type 2 arrangement is an economical choice for column design.

The bay lengths should be kept uniform in order to optimize the structural design. Standard bay lengths of 6 m, 7.5 m, 9 m, and 12 m are suitable for the best cutting plans with standard rolled steel sections and a minimized number of splice joints.

The unsupported length of a member plays a major role in the computation of the permissible stress of a compression member and the critical bending stress of a flexural member. Hence, columns and girders are to be tied along the weaker axis to reduce unsupported length. Similarly, for beams and other secondary members, care should be taken to provide lateral support and thus reduce unsupported length.

Top flanges of steel beams/girders are considered laterally restrained when embedded (10–15 mm) in concrete floor slabs or have welded shear lugs at the top flange for embedding into the slab.

Chequered plates or grating floors are not able to give effective lateral support to beams. Hence, plan bracings are provided at the top flange or just below the floor deck (within one-third depth of the beam) to give lateral support to beams.

The building structure is also made with load-bearing masonry work and the roof is made of slabs and intermediate beams.

### 7.3.2 Method of Analysis

The analysis of the frame can be done considering either a two-dimensional or three-dimensional frame model. Use of design software or static methods of analysis as per standard books of structural analysis can be done to obtain member forces and the strength design of members.

Commonly used software for analysis and designs are as follows:

a) STAAD. Pro
b) SAP 2000
c) ETABS
d) SAFE

Software for engineering modeling and drawing are as follows:

a) AutoCAD
b) Tekla
c) PDS/PDMS
d) SP 3D
e) AVEA E 3D

Workout examples showing the design of RCC and steel structures are provided in Chapter 12.

## 7.4 DRAWINGS FOR REINFORCED CONCRETE WORKS AND BAR BENDING SCHEDULES

Drawings for reinforced concrete works may be divided into the following categories:

1. Design drawings or engineering drawings
2. Detail working drawings
    i) Excavation drawings
    ii) Bar bending drawings and schedule
    iii) Formwork including scaffolding and supports
    iv) Fabrication drawings for anchor bolts and steel inserts

**Design drawings** include a general arrangement plan, sections and details, dimensioning of members, reinforcement bars details, insert plates, and all else that is necessary for construction. This drawing also includes standard notes indicating the grade of concrete, grade of reinforcement bars, concrete cover, lean concrete below foundations, and so on.

Design drawings are prepared generally at scales 1:100, 1:50, and 1:30. Details are made in 1:1, 1:5, 1:10, and 1:20 as required.

The information furnished in the design drawing is adequate for general construction work, but additional detailed drawings are prepared for large projects where the nature of the work needs more construction details in order to assist field construction and ensure safety and quality requirements.

**Excavation drawings** are prepared for a large area of construction where a number of foundations are close to each other and need a common pit. This drawing furnishes dimensions of the common pit, side slopes, and berm and ram/approaches of excavators that ensure safety and makes the construction fast and easy. This drawing is used for the payment of quantities.

For deep underground works, where shoring is done by steel sheet piles and braced walls, field construction drawings supported by design calculations are submitted to the owner for approval. The contractor prepares such drawings based on safety rules and standards in accordance with national or international codes to ensure safety.

**Bar bending drawings and schedules** show the shapes of each type of bar and the number, length, and weight. Each type of bar has a unique identification number in the drawing and schedule. Bars are fabricated at bar bending shops or yards and bundled in separate groups with identification tags. Bars are transported to the job site and laid on the formwork as per the bar bending drawing. Since the number and shapes are designed and worked out at the designer's desk, fieldwork can proceed smoothly, saving time in the placement. Bars shapes and standards for hooks, bends, and so on should conform with standard codes of practice (IS: 2502/ACI: 315).

**Formwork including a scaffolding and supports drawing** is prepared for casting slabs at higher elevations and tabletop foundations. The support system including planks or board holding the green concrete, erection load due to the movements of the equipment, and working personnel are designed for the required strength to meet the deflection limit. A weak scaffolding system and formwork may lead to a deformed structure or even serious accidents due to structural failure. Formwork and props are to be checked by the engineer prior to casting and should be kept under constant observation by a team of expert workers during concreting and for a period thereafter.

Removal of the formwork is very important. The striping time should be as stated in the codes of practices. The supports are to be dismantled in the proper sequence so that the load-bearing characteristics of the cast-in structure are not reversed. For example, a cantilever structure should have its props or supports removed starting from the end (free edge) toward the support so that the cast-in structure is stressed with tension at the top from its own weight.

**Fabrication drawings for anchor bolts and steel inserts** are required at the metal fabrication workshop for bolt assembly and inserts with anchor lugs.

## 7.5 DRAWINGS FOR STRUCTURAL STEELWORK

Drawings prepared for structural steelwork are of the following three types:

1. Design drawings
2. Fabrication or workshop drawings
3. Erection drawings including marking plans

Design drawings are the basic engineering drawings. They contain the general arrangement plan and sections and elevations of the building structural framework indicating basic parameters, for example, grid dimensions, number of bays and spans, member sizes, elevations, major connection details, column base details, key plans, and locating coordinates.

The fabrication drawing is prepared to show detailed information for workshops to prepare cutting plans, fabricate various components, and join them to an actual location. The shop drawing shows section sizes, cutting length, elevation of connection point/working point, shapes of gussets, cleat angle, bolt holes with spacing, weld size, length and type of weld (fillet or but), notch dimensions, splice location and details of bolt holes, and the material list in a table, which describes the number and weight of individual pieces and the summary of quantities.

Erection and marking plans show the item mark number and the location plan that is used for assembly and erection at the site.

Preparations of all the drawings are done in AutoCAD and other 3D drawing software that help to make the drawings clash free to a great extent. The popular software for preparing steel drawings is X-steel. This is 3D software that prepares automatic shop drawings and the bill of materials. The output is compatible with computer software and machines used in workshops for cutting to shape, drilling holes, and so on. Hence, the entire process from the design office to the workshop is automated.

## 7.6 INTERDISCIPLINARY COORDINATION

This is the part of design drawing preparation work that ensures correctness in the engineering drawing. The civil/structural engineer prepares his or her own drawing and checks the compatibility with the design calculations. The drawing is completed incorporating all the corrections needed. Then the drawing is sent to engineers of all the related disciplines –architecture, mechanical, electrical, and other system designers as applicable – for interdisciplinary review and confirmation. Upon receipt of the review comments, civil and structural engineers edit and modify the changes necessary. The drawing is then ready for submission for approval and release for construction.

All engineering companies have their own system of this interdisciplinary review procedure. Any errors or misses in their review may lead to incorrect construction that could cause a delay and potential rework.

## 7.7 QUANTITY ESTIMATION

Quantity estimation is particularly important for engineering projects. It is done at various stages as per the requirements. The estimated quantity at the budgetary stage is prepared by using data from similarly executed projects/buildings or done by a preliminary engineering design. After finalization of the engineering drawings, the required quantity is estimated in detail, which is then used for procurement and construction management. Finally, after construction, the built quantity is estimated for final payment in the item rate contract.

A typical format for quantity estimation is shown in Table 7.1 for guidance.

Structural steelwork is a costly item. Hence, procurement of materials in actual dimensions and minimizing wastage in fabrication and drawing materials should be the target of the steel constructor to control project cost.

Quantity estimations and the bill of quantity are efficient tools for this. The work is done in phases as work progresses and the final quantity is obtained after completion of the job.

### TABLE 7.1
### Estimation Sheet

| Item No. | Description | No. | Length (m) | Breadth (m) | Depth (m) | Area (SqM) | Volume (CuM) | Weight (M Ton) | Remarks |
|---|---|---|---|---|---|---|---|---|---|

# Design Office Procedures

The steps can be broadly defined as follows:

- Preliminary estimation for budgetary provision
- Section-wise estimation for procurement
- Estimation as per design drawing
- Bill of materials as per the fabrication drawing
- The final bill of materials incorporating changes and modifications during the final execution

## 7.8  DESIGN COORDINATION WITH THE CONSTRUCTION SITE

Coordination with the design office is a continual process for quality control and project monitoring. Designers visit the site to sort out technical issues or clarify any alternations that may arise due to unforeseen reasons.

## 7.9  AS-BUILT DRAWINGS

After completion of the work, the contractor at the site marks up changes or modifications done during construction in the drawings. This information is approved by the engineer-in-charge and sent to the design office for incorporation into the original drawing. The drawing is then updated with as-built changes and issued as an as-built drawing.

As-built drawings are preserved for archives for the record and future reference.

# 8 Steel Structures

## 8.1 GENERAL

In Chapter 5, we discussed the advantage of steel structures. Although the material cost is high, time saving and efficiency make steel structures popular in industrial construction.

Major activities for structural steelwork in design offices, fabrication shops/yards, and construction fields can be listed as follows:

a) Conceptual planning
b) Analysis and design
c) Design drawing
d) Material list for procurement
e) Fabrication drawing including the bill of quantity
f) Procurement of materials
g) Fabrication at the shop or yard
h) Primer coat of painting
i) Transportation to the building site
j) Erection and alignment
k) Final bolt tightening at filed joints after alignment
l) Grouting of column bases
m) Painting final coats/fire-resistant coat
n) Roof and side cladding
o) Handing over the structure for flooring, as applicable.

In this section, we will provide some worked-out examples of element design that will be useful for the reader to understand the basics of strength design.

## 8.2 PURLINS AND SIDE GIRTS

Purlins or roof beams are flexural members used to support roof slabs or a metal roof. These are generally considered secondary members, i.e., not a part of the main-frame structure. Purlins can be used for providing lateral restraint to the top chord of a roof truss and girders, and then designed for flexure and axial compression (maximum 2.5% of the chord force). Side girts or runners are flexural members supporting side sheet metal cladding to withstand lateral wind.

Secondary members like roof purlins and side girts/runners are tied with sag rods with double nuts at the ends (one nut at each side of the contact surface) to provide lateral support and control sagging in the plane perpendicular to the minor axis.

Rolled steel channels, joists, and cold-formed Z sections are used as the purlin section. Steel sheet metal roofs are fixed with self-tapping screws or metal hooks on the purlin section. Depending on span dimensions, it can be designed as a continuous member or a simply supported member. The ends should be fixed with a web cleat angle section, cut joists, or a bent plate effectively held in position and restrained against torsion.

The design of purlin members and side girts are done as flexure members spanning over roof trusses and columns. Design examples are furnished at the end of this chapter (Figure 8.1).

## 8.3 ROOF TRUSSES

Figure 8.2 shows a typical section of a roof truss commonly used in an industrial building roof. The configuration of the truss is prepared by the design engineer based on the roof span, and then panel width and depth are selected. The vertical depth of the truss is related to allowable deflections and the clear height above the floor. The sloping angle of diagonals is kept within 40 to 70 degrees. The roof truss should be fabricated as a single unit welded truss and field connected to columns by high tensile bolts. If the unit is fabricated in the shop, it can be made in parts of lengths to suit transportation facilities. The joints are welded/bolted at the site before erection.

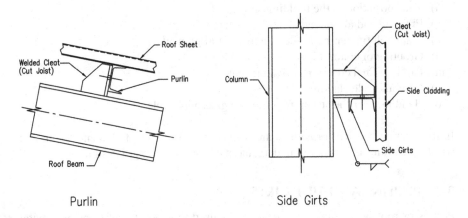

**FIGURE 8.1** Details of purlin and side girt.

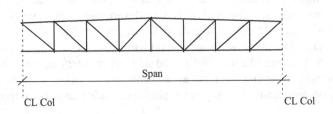

**FIGURE 8.2** Roof truss.

# Steel Structures

The chord members are usually formed with two angle sections back-to-back or spaced apart, say 250 mm back-to-back leaving adequate space for welding. The chord members may be of rolled steel channels or joists in case of a larger span and bearing medium to a heavy load, for example, cast-in-slab. The roof truss is connected to columns at the top and bottom chords. It makes a moment connection joint with the main frame columns.

The top chord and bottom chord of a truss are also called the rafter and tie member. Purlins should be placed at nodes as far apart as possible to avoid local bending of the top chord/rafter member.

Truss members are tied with plan bracings at the top chord/rafter level and the bottom chord or tie levels to provide lateral restraint.

Joints should be designed for the full strength of the member section unless the member forces are specified in the design drawing. The end connections are to be detailed with a common gusset plate welded to the end plate that is bolted to a column.

## 8.4 LATTICE GIRDERS

Lattice girders are trusses spanning over columns along a longitudinal direction to hold intermediate support posts in between the main column grid. Figure 8.3 is a lattice girder connected to the main frame columns at roof level to support an intermediate truss.

## 8.5 ROOF BRACINGS

The plan view shows a typical pattern of roof bracing used in rafter and tie levels (Figure 8.4). The purpose of this bracing framework is to provide lateral stability to the building framework, restrain truss members against lateral torsion bending, and transfer the gable end wind force to columns. Bracings are designed as axially loaded members for compression and tension. The member sections are to be selected considering allowable vertical deflection against their own weight. These members are erected aloft in single units; hence, they are provided with bolted connections.

## 8.6 BEAM AND GIRDERS

Beams and girders are generally used as single-span members of the framework. Girders span between columns along a transverse direction with moment connection

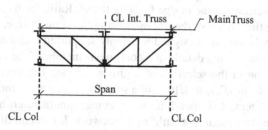

**FIGURE 8.3** Lattice girder.

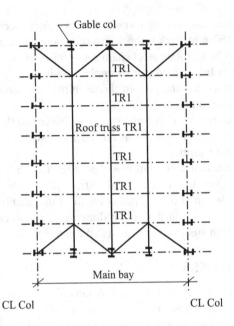

**FIGURE 8.4** Roof bracing.

joints at the ends. Beams are cross member spans between girders and connected to columns as ties (Figure 8.5).

In the framing plan, girders have moment connections that are fixed at the joints. These connections may be bolted or welded on the field. The joints should be flexible to resist moment and be distributed among the connecting members. There is an opening at the end representing an erection hatch over an equipment unloading bay. The floor slab is cast-in-situ reinforced concrete. The top of the girders and beams is embedded in slabs (10–15 mm) to gain compression flange restraint.

Worked-out design examples are shown at the end of this chapter.

## 8.7 CRANE GIRDERS

The design of the crane girder is discussed in the worked-out examples.

The crane girder or runway beams run in parallel along the length of a building to support the overhead crane bridge for long travel. Rails are fixed on the runway beams matching the wheels of the trolley bearing the crane bridge.

Crane runway beams are designed as simply supported members to avoid complications in the connection details of a continuous member. The vertical deflection is a major criterion in the selection of a girder section. The maximum allowable deflection varies from 1/750 to 1/1000 of a span for cranes up to and over 50 T crane lifting load. The lateral deflection limit set by crane manufacturers also gives a stringent condition in the design of a building framework. It is found that when the crab moves with a full load combined with a wind load (50% of maximum wind) surge in

# Steel Structures

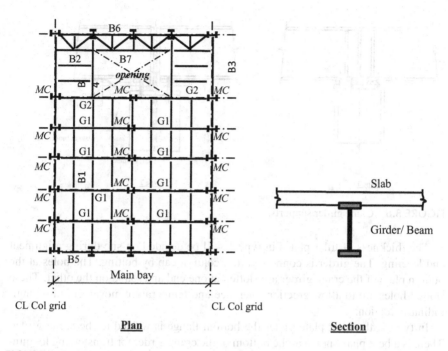

**FIGURE 8.5** Floor framing plan and section. [MC denotes moment connection at ends.]

a transverse direction, it becomes critical in the design of columns. Here again, the limiting deflection governs the sizing of the main columns in the crane bay.

The design of the crane girder (runway beam) is simple like the design of an ordinary beam. The girder section is designed for vertical wheel load, self-weight, and walkways if any. The wheel loads are placed in a position that gives maximum span moment. The web is checked for maximum end shear. Web stiffeners are necessary for noncompact webs and at closer spacing at the ends to avoid local buckling. The bending moment due to lateral surge from the wheels is calculated in proportion to the vertical wheel load moment. The top flange of the girder section is strengthened about the minor axis (in plane perpendicular to the major axis of the girder) by increasing the width of the flange plate or adding a channel section to resist this lateral moment. Maintenance walkways at the top of the crane girder level are provided in large plant buildings. This walkway structure takes part in resisting lateral surge, therefore, it is called a surge girder.

The ends of the crane girder rest on top of the crane leg column in buildings with twin-legged columns. Otherwise, it is placed on brackets projecting from a built-up fabricated column. The vertical reaction of the crane girder is transferred through end-bearing stiffeners. The horizontal reaction from the top chord of the crane runway beams goes to the columns by surge angles connected to the crane girder and column.

The longitudinal surge is transferred to the column by bolts at the bottom flange or bent plate. See Figure 8.6 for two types of end bearing. The gap (3 mm) between the girders at the joints is for allowing temperature expansion and contraction.

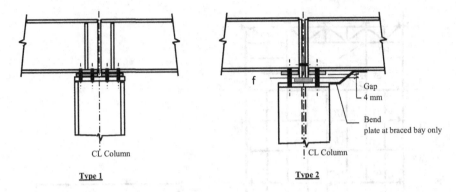

**FIGURE 8.6** Crane girder supports.

The thickness of filler plates in type 1 will be decided on-site during alignment and leveling. The girder is connected to a column cap by bolting. The holes at the bottom plate of the crane girder are slotted at one end and round on the other. These slotted holes are to allow erection tolerance and temperature movement in a longitudinal direction.

In type 2, the filler plate under the bottom flange is welded to the crane girder. There is a bent plate shown at the bottom of the crane girder for transferring longitudinal surge from girders to the column. This bent plate is only provided for girders at the braced bay. This bent plate comes as wired on with the crane girder and is welded to the underside of the crane girder and column cap on-site. The advantage of this end-bearing type connection is that the load is transferred to the column cap with minimum eccentricity. The underside of the end-bearing stiffener and the top of the sole plate are machined for perfect bearing. The gap between end-bearing stiffeners is filled with a separator plate at the bottom so that the longitudinal surge is carried over to the braced bay columns through the welded bent plate.

## 8.8 COLUMN BRACING AND TIE MEMBERS

Figure 8.7 shows a typical example of column line elevation with vertical bracing and tie members. The bracing members are part of a 3D framework and resist axial compression and tension resulting from lateral wind/seismic force. The tie members are provided to give lateral restraint to columns and the distribution of horizontal forces.

The common sections for bracings are single angle or double angles back-to-back or star sections. Box sections or wide flange beams are also used for bracings with heavy loads. The end connections are bolted or field welded. An engineer can often meet difficulties while placing bracing members in an industrial building due to interference from pipes, ducts, cable ways, or equipment. Therefore, the structural engineer needs to shift some sets asymmetrically as shown in Figure 8.7.

End connections should be designed for the full strength of members unless the forces are given in the drawings.

Design examples of bracing members are shown at the end of this chapter.

# Steel Structures

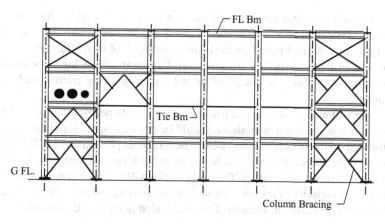

**FIGURE 8.7** Column line elevation.

## 8.9 COLUMN BASE PLATES AND ANCHOR BOLTS

The design of a steel column base for building columns or structural bridge supports on concrete pedestals should be prepared considering the transfer of all forces – compression, uplift, moment, and shear. End conditions (fixity or hinged) are to be maintained in the detailing of this joint.

This design has two parts – the steel column base and the concrete pedestal with anchor bolts.

The steel column base may be gusseted or machine faced for the bearing to transmit the load from the column shaft to the base plate. In the case of a gusseted base, where the contact faces are not finished for complete bearing, the weld to the base plate should have sufficient strength to transfer all the forces from the column shaft to the base plate. The gusset plates should have adequate height and be shaped for the uniform distribution of the column load.

The top of the concrete pedestals cannot be cast smooth or on true level for direct bearing. Hence, the top of the pedestal level is fixed at 50 to 75 mm below the design level of the bottom of the base plate. Steel columns are erected on steel screed bars placed on the concrete pedestal or alignment nuts fitted with anchor bolts. After final alignment and leveling, the gap between the base plate and the top of the pedestal is filled in with non-shrink cement grout. The compressive strength of this grout mix is higher than the grade of pedestal concrete and should be about two times the characteristic strength of the pedestal concrete. The nuts are finally tightened with predetermined torque after the grout mix reaches its final strength.

At this plane, the force transfer from the steel column base and pedestal will be similar to a reinforced concrete beam with reinforcement (here anchor bolts) carrying tensile force, and at the opposite end, compression load transferred by surface contact between base plate and grout.

Design calculations of base plates of different types are shown in worked-out calculations.

The transfer of shear stress at the column base is resisted by friction at the interface (vertical load × coefficient of friction), anchor bolts, and shear lugs. Shear lugs are designed based on the bearing strength of concrete, shear strength of concrete in front of the lug, and structural strength of the lug. It is advisable to design the shear lugs for complete shear transfer ignoring friction and anchor bolt capacity.

Now we will explain the second part, i.e., the concrete pedestal and anchor bolts.

The design of the concrete pedestal should be done using the standard procedure of reinforced concrete columns and slenderness ratio as per code (ACI 318/IS: 456).

The number and diameter of anchor bolts are determined using the reinforced concrete analogy as stated earlier. The anchor bolts will resist tension from uplift and moments. The capacity of anchor bolts should be lower than their structural strength and the concrete break-out strength of the embedded portion in the pedestal.

## 8.10 MISCELLANEOUS STRUCTURES: GABLE END STRUCTURES, STAIRS, AND PLATFORMS

Gable end structures comprise side girts, gable columns, and wind girders. The gable end structures are designed for stability against wind. The function of a wind girder is to provide lateral support to gable columns at intervals. Column bracings are provided where required for framing arrangement.

Stairs and platforms are secondary structures designed separately. The load from stairs and platforms is added to the main framework. Platform floors and treads of stairs are made of gratings or chequered plates. The use of a galvanized structure is popular for platforms and grating.

## 8.11 PIPE AND CABLE RACKS

A pipe and cable rack is an overhead structure used in petrochemical plants, process plants, fertilizer plants, power and steam generation plants, chemical plants, steel plants, and other industrial plants. Pipelines and cables run in separate tiers accessible by maintenance walkways and ladders from the ground. The number of tiers and the dimensions of the structure are fixed in compliance with the piping/system drawing. Cable racks are generally provided at the upper level or at the sides supported on brackets from columns.

The pipe rack structures may be a single-tier or a multiple-tier framework. The widths of structures are given in the piping/system drawing based on the number of pipes and cable racks, while also keeping space for walkways. The bays along longitudinal directions should be in multiples of 6 m, i.e., 6 m, 12 m, and so on for the economic use of rolled steel sections.

The material used for the superstructure is steel. The foundation is concrete pedestals resting on spread footing or pile cap. Prestressed concrete column frames and structural steel framing beams are also used for pipe racks. Walkways are made with steel grating.

The analysis of pipe rack structures can be done in either of two methods:

a) Simple frame with vertical bracings between all columns in transverse directions and a horizontal set of bracings in one or more bays along a longitudinal direction, or
b) Moment connection joints along a transverse direction and bracings along a longitudinal direction.

Plan bracings are provided for both the above methods to ensure lateral stability and transmit lateral forces to columns.

It is advisable to design frameworks in a modular concept. This will save the cost of fabrication and erection by adopting repeating typical designs wherever possible. Each model should be within permissible length, say 60 m long, with expansion joints at both ends and one set of column bracings at mid-length. This means the center bay of the module is fixed at the base and other columns on both sides are free to move and release stresses from temperature expansion.

For a 6 m bay structure, a straightforward design with rolled steel members for beams and columns is used to save time in fabrication and construction. For bays 12 m and above, horizontal members in the top and bottom tiers along column lines are connected by verticals and diagonals to form a truss spanning over the columns. In multiple-tier structures, this truss may be at the upper level or at the bottom which bears the vertical load of all the tiers. However, this should be the designer's choice.

The clear height of the overhead structure should be 5–6 m above the ground surface to leave enough headroom for traffic.

In case pipelines are supported on a single column, the distance between columns should be governed by the allowable deflection of the pipeline so that the free flow of water or condensate at a given longitudinal slope is not interfered with. For steam pipes, expansion loops or U-bends are installed in gaps between strain blocks. This bay supporting the loop should be spanned by simple supported beams that are free to move horizontally and allow thermal deformation.

The secondary steel or pipe supports of steam/hot pipelines may be of sliding, fixed, or hanger types. The structural beam bearing this support should be separately checked against local bending before the reaction is transferred to the mainframe. However, these support loads are a part of the design inputs that should be considered in the analysis of the framework.

The loading on pipe and cable rack structures can be classified as follows:

i. Dead load: the self-weight of structures including the weight of secondary steel for pipe supports.
ii. Pipe load: the weight at empty and water fill conditions.
iii. Frictional load: the load at supports for the thermal deformation of pipes. (This occurs for hot pipes only. There will be no thermal movement for pipes carrying cold fluid.)
iv. Cable load: the weight of cables including trays and supports.
v. Wind load: the wind load on pipes and cables considering shielding effects.
vi. Seismic load: the seismic load generated by the mass and support reaction given by the piping designer in the design report.
vii. Temperature load: the load due to the variation of seasonal temperature.

The connections of pipe rack structures should be simple for erection and alignment. Field connections should be bolted, and shop joints welded. Bracing joints should be for full strength member sections unless specified by the designer.

Painting should be suitable for the local environment. Aluminum paint or synthetic enamel paint is used in normal inland atmospheric conditions.

Worked-out examples of pipe and cable racks are shown in Chapter 12.

## 8.12 CONNECTIONS

The performance of a building or structure depends on the efficiency of its connection details. Hence, the engineer should take proper care in the design and detailing of steel member connections. All major connections should be shown in design drawings as a guideline for the preparation of workshop/fabrication drawings.

The details of connections should be simple and easy to approach. The designer should prepare the appropriate type of joint details matching the type of member end connection assumed in the analysis of the building framework, i.e., simply supported joint or hinged or fixed.

All connections should be designed for full strength member sections unless the member force is given in design drawing.

The following types of connections are used for structural steelwork:

A. Bolted connections:
    a) Bearing type
    b) Friction grip/non-slip/slip-resistant types

B. Welded connection:
    c) Fillet weld
    d) Butt weld
    e) Stud welding

In general, shop connections are welded joints and field connections are bolted joints. However, for some projects, field connections are also done by welding.

The minimum size of bolts should not be less than M16 and the minimum size of the fillet weld should not be less than 6 mm on all permanent connections.

Combinations of two types, i.e., bolt and weld should not be done at the same joint.

When using different fastening materials, for example, flange plates are joined by welding and webs by cover plates and friction grip bolts, compatibility of the deformation must be checked. However, these types of mixed jointing procedures should be avoided.

The basic rules of connections are as follows:

1. The center of gravity of the weld group (weld seam group) or bolt group should coincide with the gravity axis of the member parts to be connected.

If there is an offset, the connection weld/bolt group should be designed for the moment due to eccentricity.
2. The individual cross-sectional parts, for example, flanges or webs in a splice joint of the beam sections, should be joined individually in accordance with the proportionate forces being carried by the individual parts. If the connection is made in part, for example, only webs are joined at the splice location, the deflection of the member at that splice location should be checked as per the permissible limit.
3. The length of cover plates used at the splice location should be adequate to receive stress-satisfying moment requirements.
4. The diameter of bolts should not vary on the same joint.

### 8.12.1 BOLTED CONNECTIONS

#### 8.12.1.1 Bearing Type Connections

In this type of connection, bolts are stressed perpendicular to their axis. High tensile bolts of property class 8.8 grade or 10.9 are used without prestressing or with partial prestressing, i.e., the specified tightening torque.

Bolt capacity is to be computed as follows

- Shear capacity of the bolt, $Fs = As.m.\tau_s (KN)$
- Bearing capacity of the hole bearing area, $Fb = \sigma_1 . d . t (kN)$

Bolt value (Q) will be the minimum of the above of these two capacities.
So, number of bolts required, n = F/Q
when, F = Force to be transmitted, kN
As = Cross-sectional area of bolt shank, mm²
Q = Bolt value, permissible load per bolt. kN
d = Shank diameter of bolt in mm
n = Number of bolts
m = Number of shear planes (one for single shear, two for double shear)
t = smallest sum of the plate thickness with hole bearing pressure acting in the same direction in mm
$\sigma_1$ = Allowable hole bearing pressure between bolt and hole wall of the member to be connected, MPa
$\tau_s$ = Allowable Shear stress of bolt, MPa (Figure 8.8).

#### 8.12.1.2 Friction Grip or Non-Slip or Slip-Resistant Type

This type of connection is currently popular. Bolts are high tensile friction grip bolts (IS: 3757/ASTM A 325, A490). In these types of slip-resistance connections, all bolts are prestressed as per the defined tightening torque. In this way, a clamping force perpendicular to the bolt shank is transmitted in the contact surfaces of the elements being joined. Contact surfaces are prepared by sand/shot blasting or applying a slip-resistant coating.

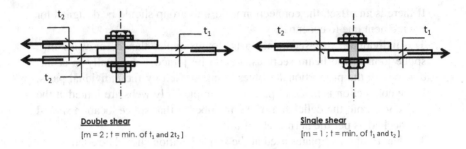

**FIGURE 8.8** Bolt shearing planes.

In slip-resistance connections, the bolt value with a clearance hole less than or equal to 2 mm is as follows:

$$Q = \mu \cdot Fv / v_g$$

Where, $\mu$ = Coefficient of friction of the contacted surface (= 0.5)
$Fv$ = Prestressing force in the bolt
$v_g$ = Coefficient of resistance to slip (1.10 at service load, 1.25 at ultimate load)

The above details are shown to give a general idea of bolted connections. The designer should follow the relevant codes of practice while designing connections.

### 8.12.2 Weld Connections

Welded joints are considered fixed and permanent joints that cannot be dismantled like bolted joints. These connections are considered rigid connections.

The welded joint is a process by which the metal parts are raised to a high temperature and metallurgically bonded by an electric arc created by a self-sacrificing metal electrode supplied with electrical power.

#### 8.12.2.1 Methods of Welding

There are three methods of welding used for structural steel construction

a) Manual arc welding
b) Submerged arc welding (SAW)
c) Metal inert gas welding (MIG)

**Manual arc welding** is done by using an electrode rod in a holder fitted with an electrical supply. The electrode rod is a basic metal wire coated with minerals that creates a gas shield around the weld spot. This gas shield maintains the stability of the arc and prevents the absorption of atmospheric gasses that damages the property of the weld metal. The job to be welded is connected to the earth clamp for grounding. When the welder brings the electrode to the jointing location, an

# Steel Structures

arc is struck between the electrode and the metal work piece. The flow of electricity creates enough heat to melt both the electrode and a superficial area of the work piece. The electromagnetic force helps to drop the molten electrode onto the melted area of the work piece where the two metals are fused together into a weld metal pool. These molten metals result in a permanent binding of the metals upon cooling. The minerals in the coating of the electrode, which are called flux, contribute to the metallurgical reaction and thus clean the molten weld metal bringing out the impurities in the form of slag that is deposited on the outer surface of the weld bead when cooled. This is removed by chipping hammers to expose the weld bead.

**Submerged arc welding** is a mechanical process of welding. Unlike the coated electrode used in manual arc welding, the bare electrode is supplied from a coil and the flux (granular bead powdery material) is fed continuously over the arc and weld pool. In this process, either the job is fixed on the base and the machine runs over the job to weld in a continuous motion, or the machine is fixed in position and the job moves under it. During the progress of the weld run, the arc is completely shielded by the granular flux, which is fed from the top, covering the tip of the bare electrode. The flux protects the molten metal by forming a protective layer of slag, which stabilizes the arc, cleans the molten metal, and controls the composition of the weld metal.

Since it is a mechanical process, the quality and productivity are high.

**Metal inert gas welding** is also a mechanical process and is very popular for welding in fabrication yards. The size of the machine is small, and it runs over the rail fitted on the job. In this process, a bare electrode is used, and the arc and weld pool are shielded by inert gas. The inert gas commonly used for structural steel welding is carbon dioxide ($CO_2$) or a mixture of argon and carbon dioxide.

There are three types of weld joints used in structural engineering: fillet weld (or lap weld), butt weld, and stud weld.

In the **fillet weld or lap weld**, metal pieces are overlapped to each other, and welding is done along the sides of the contact surface (Figure 8.9).

Figure 8.9 illustrates a typical fillet welded joint of a horizontal tie/bracing member connected to the column face with a gusset. All connecting weld runs are fillet welds. The sign convention of the fillet weld shown here is a triangle where the left side will be vertical. The details and rules of sign convention of all types of weld are available in codes of practice for structural steelwork (IS: 800; General Construction in Steel/AISC Manual of Steel Construction).

The strength of the fillet weld at the shop can be computed as follows:

$F_w$ = 0.7 × size of fillet × weld stress; for example, strength of a 6 mm size fillet weld is = 0.7 × 0.6 cm × 1100 kg/cm$^2$ = 462 kg per cm run = 0.462 T per cm. This is the shearing capacity of weld when the force is applied along the length of weld.

So, if a member carries 10 M-ton of axial load (tension or compression), the total length of filled weld required = 10 / 0.462 = 21.6 cm, say 220 mm.

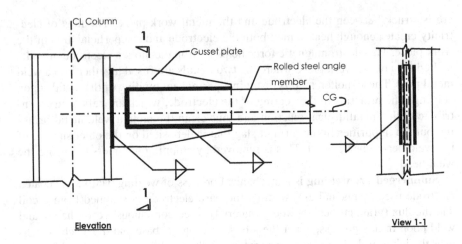

**FIGURE 8.9** Fillet weld joint.

This length shall be divided into two parts. The root side of the angle has a more effective area so it will take more force (70%): hence, the minimum length at the root side (heel side) will be equal to $0.7 \times 21.6 = 15.12$ cm = 150 mm. The toe side will have the rest, i.e., $220 - 150 = 70$ mm. But the practice is to weld full contact surface as before. The overlapping of the gusset should not be less than $150 + 20$ mm (extra for insufficient weld size at ends) = 170 mm.

At the column face, the gusset is fillet welded to the column. The applied force will be perpendicular to the weld run. The fillet weld on the column face will be in tension. If the capacity of the fillet weld in tension is assumed as 50% of its shear capacity, the total length of weld will be = $10 / (0.462 \times 0.5) = 43.29$ cm, say 434 mm. The gusset is welded on both sides, so length of the weld per side = $434 / 2 = 217 + 20$ mm extra = 237 mm minimum, say 250 mm.

The height of the gusset face at the column face is 250 mm and the length will be $220$ mm + $50$ mm gap = $270$ mm. A rectangular plate of $270 \times 250 \times 12$ thickness will be adequate for the fabrication of the gusset. The sectional area of the gusset near the column face should be verified with the member size and its cross-section.

The stress of the weld for joints on the field, i.e., on erected positions, is considered to be 80% of the shop weld run considering reliability factors and the quality of weld achievable.

This is how the joint details are worked out.

**Butt weld** joints are considered as full-strength joints because the parts are jointed face-to-face at contact points or forming a tee with full-faced contact of the one meeting at a right angle.

Details of some typical butt weld joints are shown in Figure 8.10.

The root gap indicated is the clear gap kept between the thin edges of the plate at the joints. This clear dimension varies from 0 to 6 mm in practice. The members are beveled in the form of a single V or double V, which is termed edge preparation. Back strips are provided on the opposite sides of the joints, which will not be welded.

# Steel Structures

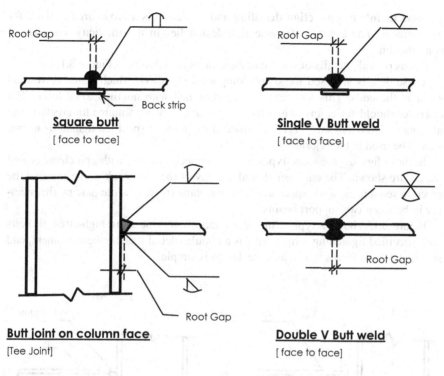

**FIGURE 8.10** Butt weld joint.

The weld pool is deposited in the notch layer by layer. In the first run, the thin root thickness is fused and is bonded with the molten weld metal deposited by the sacrificing electrode. After this root run, the slug is removed, and the welding process continues layer by layer. Since the deposit of metal is more in case of thick plated joints, excessive heat generates in the zone. The protection of the weld pool from cracks during cooling is very important. Low hydrogen electrodes are used to control the generation of hydrogen gas that invites the development of hydrogen cracks even 72 hours after cooling. The entire process of butt welding is critical and qualified welders are engaged for this type of weld.

But welding for long runs is generally done by a submerged arc welding process using machines as stated earlier. Column joints, which are done by a manual arc welding procedure, should be checked by 100% radiography. Spot radiography tests may be done for machine welded joints.

**Stud welding** is a mechanical process (Nelson stud), where the metal stud is welded to the flanges of the steel beams or girder. These studs are installed by an arc welding process using special equipment (stud welding gun) powered by electrical energy. The end of the stud is fused and gets bonded with steel flanges. These studs are used as shear keys in bridge girder construction or for steel purlin supporting roof slab cast on a metal deck. The stud welding gun fitted with a fastener weld can complete the work very fast.

**Steel member connection detailing** and its design is a broad area of work for an engineer. The essence of connection design lies in its simplicity and ease of construction.

Let us provide details of some common joints for reference (Figure 8.11).

These details are used for simply supported members, which transfers vertical shear at the ends. This joint can also be designed with nominal axial force. The engineer should select any one of these types and they are suitable for erection and alignment. High tension bolts of 8.8 grade or higher with specified tightening torque should be used at these joints (Figure 8.12).

In these figures, the same types of end connections, i.e., with end cleats or end plates, are shown. The engineer should use any of these types, keeping in mind the erection sequences of the space available for plane rotation while placing the member in between two support beams.

Figure 8.13 shows a typical moment connection joint using high-strength bolts with specified tightening torque. This is a popular detail that can bear moment, end shear, and axial forces if any, and the design is simple.

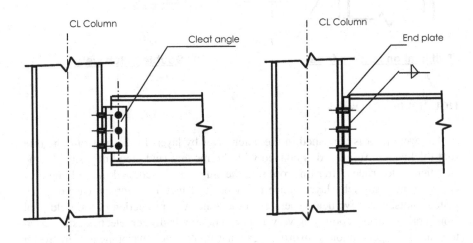

**FIGURE 8.11** Beam end shear connection with a column.

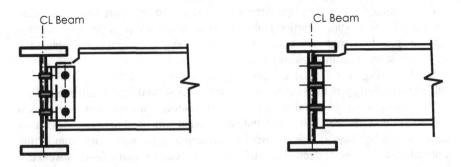

**FIGURE 8.12** Beam-to-beam shear connections.

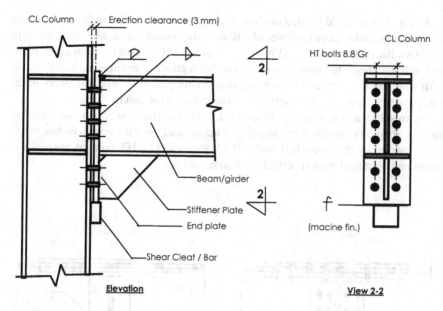

**FIGURE 8.13** Moment connection joint – bolted.

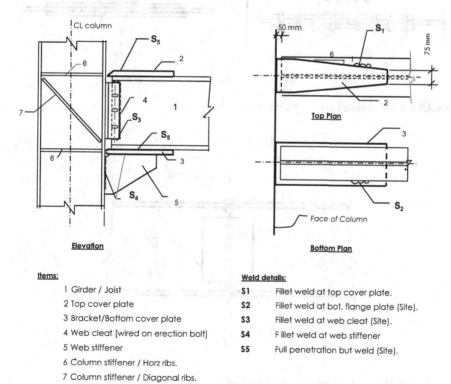

**Items:**
1. Girder / Joist
2. Top cover plate
3. Bracket/Bottom cover plate
4. Web cleat (wired on erection bolt)
5. Web stiffener
6. Column stiffener / Horz ribs.
7. Column stiffener / Diagonal ribs.

**Weld details:**
S1    Fillet weld at top cover plate.
S2    Fillet weld at bot. flange plate (Site).
S3    Fillet weld at web cleat (Site).
S4    Fillet weld at web stiffener
S5    Full penetration but weld (Site).

**FIGURE 8.14** Moment connection joint – welded.

Figure 8.14 is a field welded moment connection joint carried on by engineers for heavy beams and girders in power plants and other industrial buildings. In this type of detail, the column is erected with the shop welded bracket (bottom cover plate) and web stiffener. The moment connection (MC) girder or beam is delivered on-site with wired on the cleat and a top cover plate. After placement and alignment of the MC girder, the top plate and web cleats are welded to the column face.

Figures 8.15 and 8.16 are for beam/girder splices. The connection bolts/welding fastener should be designed individually in accordance with the proportionate forces being carried by the individual parts. The bolts should be HT bolts of grade 8.8 or above and tightened with specified tightening torque.

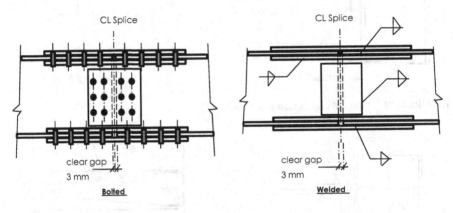

**FIGURE 8.15** Splice joints in beams or girders.

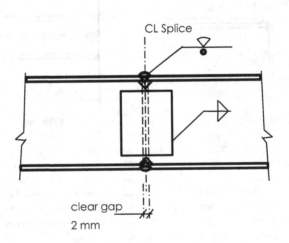

**FIGURE 8.16** Splice joint in a rolled steel joist.

# Steel Structures

Splices for column sections should be similar to beam/girder splices. Column splices should be designed for the following loading conditions, whichever gives the worst case:

a) 1.5 times of design moment, shear and axial load at splice locations, or sectional capacity, whichever is minimum.
b) 60% of the shear capacity of the section and tensile capacity.

Direct bearing type connections are also adopted in column splice joints and at base plate joints. The surfaces are milled to the required percentage of bearing. Additional cover plates are added to compensate for the balance load.

Connections for bracing and ties should be designed for full capacity unless the design load is indicated in drawings. The end of all bracing connections (bolted type) should have a minimum of two M16 bolts.

All fillet welds should be a minimum of 6 mm in size.

Shim plates are used to make up gaps in misalignment and so on in bolted connection joints. For shims less than 6 mm, no allowance is necessary, but for shims 6 to 20 mm (max) the connection design needs to be reassessed.

### 8.12.2.2 Connection to RCC Members

Figure 8.17 shows a mild steel insert plate embedded in a concrete structure. The connecting steel beam is welded to the plate at the site through a web cleat. The web cleat is joined to the beam by bolts. Sometimes, slotted holes are provided to allow movement for expansion. The lugs and their weld to the insert plate should be designed to carry moment, shear, and axial forces.

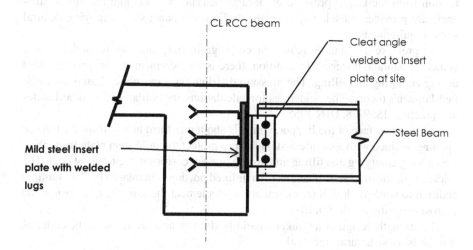

**FIGURE 8.17** Insert plate for connection to RCC members.

## 8.13 COAL BUNKER

Raw coal bunkers are used in power and steam generation plants to temporarily store and feed coal to mills for pulverization and thus supply fuel to the boiler. Bunker structures are also used in coal mines, steel plants, fertilizer complexes, cement plants, and process plants.

The bunker structure is a large structure where bulk material is filled from the top and discharged from the bottom. It is made in two parts: the top portion is vertical and the bottom part is a hopper. The shape and size are determined by the storage volume of the bulk material. It is supported on a separate column structure or by the building structure. Power plants and mill buildings house coal bunker units as integrated parts of the main building structure.

The coal bunkers in a power plant have multiple hoppers in a unit. The shape of hoppers and the vertical portion may be a combination of cone and cylinder or an upside-down truncated pyramid and a container of rectangular or square cross-sections. The rectangular bunkers may have a long length that serves as a common trough over multiple hoppers or be separated by partition walls to control fire hazards.

Structurally, the hopper part is connected to the vertical portion like a hanging unit. A girder is formed at this joint that bears the load of the suspended structure as well as supports the vertical wall. This girder is called a crutch girder. This girder holds the bunker structure around the perimeter or at four sides. In rectangular or square-shaped structures, the vertical wall may also be designed at the deep beam, which bears the total vertical load. The web panel of this vertical wall or deep girder is strengthened with intermediate stiffeners to resist the lateral pressure of storage material. These girders are connected to building columns and share the forces. Hopper plates are subjected to meridian tension for holding vertical loads and hoop tension from the lateral pressure of storage material. Vertical and horizontal stiffeners are provided outside hopper plates to limit the panel size, satisfying flexural stress requirements.

The pressure distribution is influenced by grain size, unit weight, and moisture content of storage material. In addition, there is an increment of this pressure load during emptying and filling. The pressure distribution is calculated based on modified Janssen's theory. The guidelines for calculations are available in standard codes and practice (IS: 9178; DIN 1055).

The inner surface of the hopper, which is about one-third height from the bottom opening, is lined with a stainless-steel liner to protect the mild steel plate from abrasion during emptying and filling and also to prevent corrosion due to moist coal. The thickness of the mild steel plate over the unlined portion is increased by 2 to 4 mm in addition to strength design requirements to supplement the loss due to abrasion and corrosion during the design life.

The strength design of a bunker should be done in accordance with the codes of standards for structural steelwork.

The structural arrangement of a typical set of bunkers in a power plant is presented below for reference and guidelines (Figures 8.18–8.25).

# Steel Structures

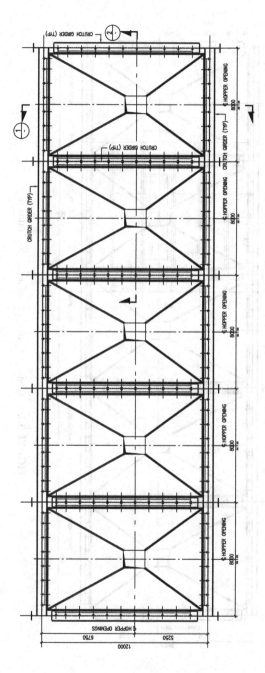

**FIGURE 8.18** Crutch girder plan of a raw coal bunker.

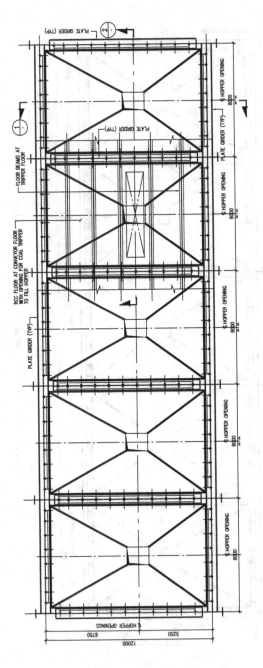

**FIGURE 8.19** Conveyor floor plan of a raw coal bunker.

# Steel Structures

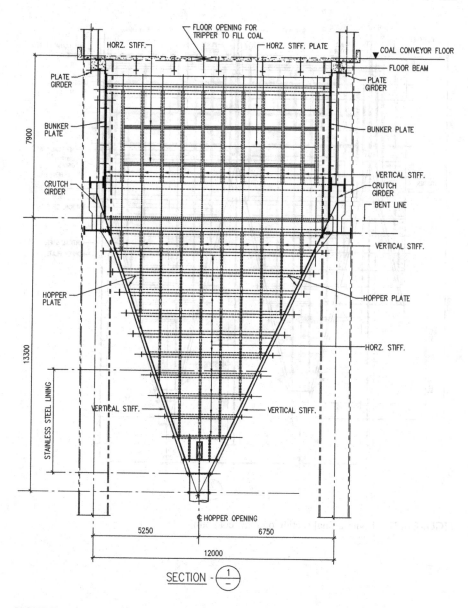

**FIGURE 8.20** Cross-section of a raw coal bunker.

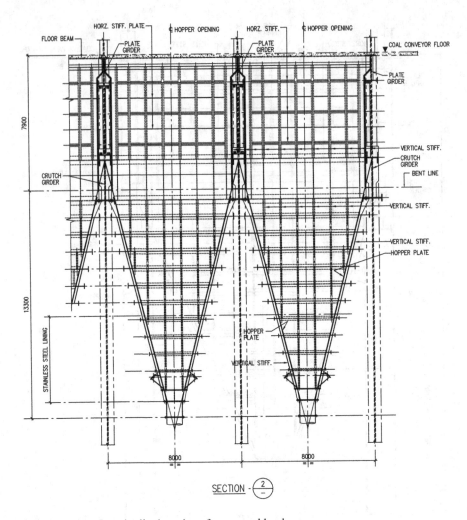

**FIGURE 8.21** Longitudinal section of a raw coal bunker.

# Steel Structures

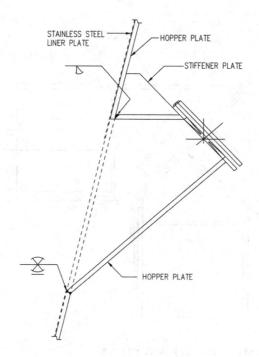

**FIGURE 8.22** Detail of a poking hole in a raw coal bunker.

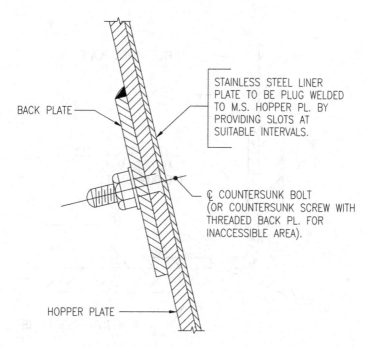

**FIGURE 8.23** Detail of SS plate liner inside a raw coal bunker.

**124**                                                    Design of Industrial Structures

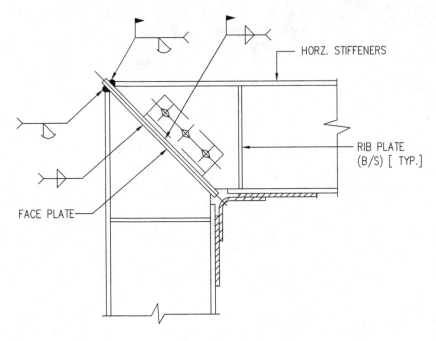

**FIGURE 8.24**   Detail of a stiffener joint at a corner.

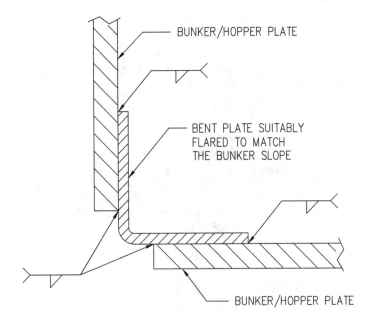

**FIGURE 8.25**   Detail of hopper plate joint at a corner.

Steel Structures                                                                    125

## 8.14  DESIGN EXAMPLES

In the section, we will furnish some worked-out examples as guidelines for member strength design in accordance with AISC (14 Ed) and IS: 800 2007.

***Example 8.14.1 : Design of a Beam with plated section***
[ Beams with equal flanges and bent about the axis of max. strength x-x axis ]

**References**
        1.0 AISC Steel construction manual - 14th Edition
        2.0 ANSI/AISC 360-16 Specification for Structural steel building
        3.0 Companion to the AISC Steel Construction manual - Vol 1 15.1 version

**Input Data.**

**Material**

| ASTM A 36 | | Tables 2-4 | AISC manual | |
|---|---|---|---|---|
| Yield Stress | $F_y$ | 36 | ksi | 250 | MPa |
| Tensile Stress | $F_u$ | 65 | ksi | 450 | MPa |
| Elasticity | E | | | 200000 | MPa |

**Member Loading**
**LRFD**

| Load case | Description | Axial KN P | $M_{vertical}$ KN m $M_x$ | $M_{horz}$ KN m $M_y$ | $V_{vertical}$ KN $V_x$ | $V_{horz}$ KN $V_y$ | | Nature of load |
|---|---|---|---|---|---|---|---|---|
| 1 | 1.4 D | | | | | | | |
| 2 | 1.2D + 1.6L +0.5(Lr or S or R) | | 594 | | 396 | | | |
| 3 | 1.2D +1.6(Lr os S or R)+(0.5Lor0.5W) | | | | | | | |
| 4 | 1.2D + 1.0W + 0.5L +0.5(Lr or S or R) | | | | | | | |
| 5 | 1.2D + 1.0E + 0.5L + 0.2S | | | | | | | |
| 6 | 0.9D + 1.0W | | | | | | | |
| 7 | 0.9D + 1.0E | | | | | | | |

**ASD**

| Load case | Description | Axial KN P | $M_{vertical}$ KN m $M_x$ | $M_{horz}$ KN m $M_y$ | $V_{vertical}$ KN $V_x$ | $V_{horz}$ KN $V_y$ | | Nature of load |
|---|---|---|---|---|---|---|---|---|
| 1 | D | | | | | | | |
| 2 | D + L | | 405 | | 270 | | | |
| 3 | D + (Lr or S or R) | | | | | | | |
| 4 | D + 0.75L + 0.75(Lr or S or R) | | | | | | | |
| 5 | D + (0.6W or 0.7E) | | | | | | | |
| 6 | D + 0.75L+0.75(0.6W)+0.75(Lr/S/R) | | | | | | | |
| 7 | D + 0.75L+ 0.75(0.7E) + 0.75 S | | | | | | | |
| 8 | 0.6D + 0.6W | | | | | | | |
| 9 | 0.6D + 0.7E | | | | | | | |

## Methods of Design

(AISC 360 -16 Chapter F)

1 LRFD :    Load and Resistance factor design    [$P_u < \phi P_n$,    $\phi$ = resistance factor]
           Nominal strength = Pn, Mn and Vn
           Design strength = $\phi$ Pn, $\phi$ Mn and $\phi$ Vn
           $\phi$ = 0.9   for limit states involving yielding

2 ASD :    Allowable strength design    [$P_u <= P_n / \Omega$,    $\Omega$ = factor of safety]
           Nominal strength = Pn, Mn and Vn
           Design strength = Pn/$\Omega$, Mn/$\Omega$ and Vn/$\Omega$
           $\Omega$ = 1.67   for limit states involving yielding

## Load, Load Factors, and Load Combinations    (AISC Chapter 2/ ASCE/SEI 7 Section 2.4)

D = Dead load
L = Live load
Lr = Roof live load
S = Snow load
R = initial rain/ice
W = Wind load
E = Earthquake load

### LRFD : Load combinations - factored.
1. 1.4 D
2. 1.2D + 1.6L +0.5(Lr or S or R)
3. 1.2D +1.6(Lr os S or R) + (0.5L or 0.5 W)
4. 1.2D + 1.0W + 0.5L +0.5(Lr or S or R)
5. 1.2D + 1.0E + 0.5L + 0.2S
6. 0.9D + 1.0W
7. 0.9D + 1.0E

### ASD : Load combinations - unfactored.
1. D
2. D + L
3. D + (Lr or S or R)
4. D + 0.75L + 0.75(Lr or S or R)
5. D + (0.6W or 0.7E)
6. D + 0.75L+ 0.75(0.6W) + 0.75(Lr or S or R)
7. D + 0.75L+ 0.75(0.7E) + 0.75 S
8. 0.6D + 0.6W
9. 0.6D + 0.7E

## Design Load

| | | | | |
|---|---|---|---|---|
| span L= | 6 M | | | |
| spcg = | 6 M | | | |

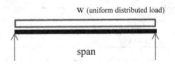
W (uniform distributed load)

span

Floor load intensity:
| | | | | |
|---|---|---|---|---|
| DL = | 5 kN/m2 | $w_{DL}$ = | 30 | kN/m |
| LL = | 10 kN/m2 | $w_{LL}$ = | 60 | kN/m |
| | | w = | 90 | kN/m |

# Steel Structures

LRFD : w =    1.2 x 30 + 1.6 x 60 =    132 kN/m    Mu =    594 kNm    Vu =    396 kN
ASD :    w =    30 + 60 =    90 kN/m    Mu =    405 kNm    Vu =    270 kN

Laterally Unsupported length    Lb =    2 M
Spacing of web stiffener    a =    1.5 m

## Section Properties

| PLATED SECTION | | |
|---|---|---|
| FLANGE PL | 200 mm. | bf |
|  | 20 mm. | tf |
| WEB PL | 710 mm. | h |
|  | 12 mm. | tw |
| WT/M | 129.7 | kg. |
| Area | 16520 | mm2. |
| Ix | 1423977667 | mm4. |
| Iy | 26768907 | mm4. |
| rx | 294 | mm. |
| ry | 40 | mm. |
| tw | 12 | mm |
| zx (Sx) | 3797274 | mm3 |
| J    Σ bt3/3 | 1475627 | mm4 |
| h₀    h+tf | 730 | mm |

## Deflection
*(AISC 360 Section L3)*

$\delta = 5wL^4 / 384 EI =$    5 x 90 x 6000 ^4 / 384 x 200000 x 1423977667 =    5.33 mm

Allowable deflection limit L /360 =    16.67 mm

OK.

## Proportioning Limit

A.    a =    1500 mm    spacing of stiffener
     h =    710 mm
     tw =    12 mm      a/h =    1500 / 710 =    2.1127    > 1.5.    *(AISC 360-16 Eqn F13.2)*

(h/tw) max =    0.4 E/Fy =    0.4 x 200000 / 250 =    320    *(AISC 360-16 Eqn F13.4)*
     h/ tw = 710 / 12 =    59    OK < 320

[ if a/h <= 1.5 , (h/tw) max = 12x(E/Fy)^0.5 =    339    ]    *(AISC 360-16 Eqnn F13.3)*

B.    Ratio of web area to the compression flange shall not exceed 10

     h.tw / bf.tf =    710 x 12 / 200 x 20 =    2.13    < 10 OK.    *(AISC 360-16 Eqn F13.4)*

Check for Web Slenderness  (AISC 360-16 Table B4.1b)

| | | | | |
|---|---|---|---|---|
| h / tw = | 710 / 12 = | 59.167 | < 106 | Compact. |

*Limiting depth to thickness ratio:*
Compact/ noncompact, $\lambda p$ =   3.76 $(E/Fy)^{0.5}$ =   106
Noncompact / slender, $\lambda r$ =   5.70 $(E/Fy)^{0.5}$ =   161

Check for flange  (AISC 360-16 Table B4.1b)

| | | | | |
|---|---|---|---|---|
| 0.5 b / f = | 100 / 20 = | 5 | < 11 | Compact. |

*Limiting width to thickness ratio:*
Compact/ noncompact, $\lambda p$ =   0.38 $(E/Fy)^{0.5}$ =   11
Noncompact / slender, $\lambda r$ =   1.0 $(E/Fy)^{0.5}$ =   28

*Compact I shaped member; Critical bending stress will be as per Eqn F2-1 AISC360-16.*

### Nominal flexural strength, Mn

Mn shall be the lower value obtained according to the limit state of yielding (plastic moment) and lateral torsional buckling.

### a) Yielding

$$Mn = Mp = Fy\, Zx \qquad \text{(AISC 360-16 Eqn F2-1)}$$

Fy = Yield stress =   250 Mpa

Zx = plastic sectional modulus about the x axis = $\Sigma A y$
  = 2 x 200 x 20 x 730/ 2 + 2 x (710/2) x 12 x (710/4)
  = 4432300   mm4

Mn =   250 x 4432300/1000000 =   **1109** kNm.   (= Mp)

### a) Lateral-torsional buckling

Limiting unbraced length for the limit state of yielding = Lp
Limiting unbraced length for the limit state of inelastic lateral torsional buckling = Lr

# Steel Structures

$L_p = 1.76\, r_y\, (E/F_y)^{0.5} =$    $1.76 \times 40 \times (200000 / 250)^{0.5}$      (AISC 360-16 Eqn F2-5)
$= 2003$ mm
$= \mathbf{2.003}$ **M.**

$L_r = 1.95\, r_{ts}\, (E / 0.7F_y)\, [\,(J/S_x.h_0) + \{\,(J/S_x.h_o)^{\wedge 2} + 6.76\,(0.7F_y/E)^{\wedge 2}\,\}^{\wedge}0.5\,]^{\wedge}0.5$

(AISC 360-16 Eqn F2-6)

| | | | | |
|---|---|---|---|---|
| $r_{ts} =$ | 50 mm | ( See calc below) | | |
| $E/0.7F_y =$ | 1142 | | $0.7F_y/E =$ | 0.00088 |
| $J / S_x.h_0 =$ | 0.0005 | | | |

$\mathbf{X} = \{\,(J/S_x.h_o)^{\wedge 2} + 6.76\,(0.7F_y/E)^{\wedge 2}\,\}^{\wedge}0.5 =$    0.0023
$[\,(J/S_x.h_0) + \mathbf{X}\,]^{\wedge}0.5 =$    0.0536

$\mathbf{L_r} =$    5920 mm    **5.92 M**


(i) when $L_b < L_p$,    the limit state of lateral buckling does not apply.

(ii) when $L_p < L_b <= L_r$      (AISC 360-16 Eqn F2-2)

     $M_n = C_b\,[\,M_p - (M_p - 0.7\, F_y\, S_x)\,((L_b-L_p)/(L_r - L_p))\,] <= M_p$
     $= 1[1109 - (1109 - 0.7 \times 250 \times 3797274/1000000) \times ((2-2.003)/(5.92 - 2.003)$
     $= 1109.6$ kNm    $<= (M_p =$    1109 kNm )
     $M_n =$    **1109** kNm

(iii) when $L_b > L_r$      (AISC 360-16 Eqn F2-3)

     $M_n = F_{cr}\, S_x <= M_p$

     $F_{cr} = [C_b\, \pi^2\, E / (L_b / r_{ts})^2]\,.\,[1 + 0.078\,(J_c /(S_x.h_o))\,.(L_b / r_{ts})^2]^{\wedge}0.5$      (AISC 360-16 Eqn F2-4)

| | | | | |
|---|---|---|---|---|
| $C_b =$ | 1 | | $J_c =$   1475627 | mm4 |
| $E =$ | 200000 Mpa | | $S_x =$   3797274 | mm3 |
| $L_b =$ | 2000 mm | | $h_o =$   7300 | mm |

$r_{ts}$ = radius of gyration of compression flange plus one sixth of the web

Iyc = 20x 200^3 / 12 + (710/6) x 12^3 / 12 =   13350373   mm4
Ayc = 200 x 20 + (710/6) x 12=   5420 mm2
$r_{ts}$ = ( Iyc/Ayc )^0.5 =   50 mm

Fcr =   [1 x 3.14^2 x 200000 / (2000 / 50)^2] x [1 + 0.078(1475627 / (3797274 + 7300) x (2000/50)^2]^0.5
        1218 Mpa

Now,   **Mn** = Fcr.Sx =   1218 x 3797274/1000000 =   4627 kNm.
                                    and Mp =   1109 kNm.
                        So, the effective, **Mn** =   **1109** kNm

| Lb = | 2.0 M | Lp = | 2.003 M | Lr = | 5.920 M | span = | 6 | M |
|---|---|---|---|---|---|---|---|---|

In this example of beam section, **Mn** =   **1109 Mpa.**   Lb < Lp

**Nominal Shear Strength, Vn**

$$Vn = 0.6Fy\ Aw\ Cv1 \qquad \text{(AISC 360-16 Eqn G2-1)}$$

*To find out web shear coefficient, Cv1*

(a) For webs of rolled I shapes
      h/tw <= 2.24 (E/Fy)^0.5          Cv1=   1      (AISC 360-16 Eqn G2-2)
            <=    63
(b) For all other I-shaped members and channels

(i) when,       h/tw <=1.10 (kv E/Fy)^0.5        Cv1 =   1      (AISC 360-16 Eqn G2-3)

(ii) when,      h/tw > 1.10 (kv E / Fy)^0.5                     (AISC 360-16 Eqn G2-4)

$$Cv1 = 1.10\ (kv\ E\ /\ Fy)^{0.5}\ /\ (h/tw)$$

where, Aw = area of web, h = clear dist between flanges and tw = web thickness.

      kv = shear buckling co-efficient                          (AISC 360-16 Eqn G2-5)

(i)     for web without stiffeners ,           kv =   5.34
(ii)    for web with stiffeners :              kv =   5 + 5/(a/h)^2
                                                  =   5.34   for a/h > 3

      where, a = clear distance between transverse stiffeners (mm)

*Transverse stiffener*

A.    Transverse stiffeners are not required where,    h/tw <= 2.46 (E/Fy) ^0.5
                                                            <=    70
      Or, if the web shear provided > Required Shear stress, Vu

B.    Moment of inertia of transverse stiffener, Ist          (AISC 360-2010 Eqn G2-7)
        Ist > = b tw ^ 3.  j
        j = [2.5 / (a/h)^2]  -  2   >= 0.5                    (AISC 360-2010 Eqn G2-8)
      [ stiffeners to be provided where required to meet the design web shear]

# Steel Structures

*For this design section*

| | | | | | |
|---|---|---|---|---|---|
| a = | 1500 mm | h/tw = | 710 / 12 = | | **59.167** |
| h = | 710 mm | a/h = | | | 2.11 |
| tw = | 12 mm | kv = | 5+5/(1500/710)^2= | | 6.1202 |
| Fy = | 250 Mpa | 1.10 (kv E/Fy)^0.5 = | | | **77** |
| E = | 200000 Mpa | | | Cv1 = | **1.0** |
| Aw=h.tw = | 8520 mm2 | | | | |

$\mathbf{Vn}$ = 0.6Fy Aw Cv1 = 0.6x250x8520x1/1000 = **1279** kN

## Strength Design

| | LRFD | | | ASD | | |
|---|---|---|---|---|---|---|
| | Load Case | 2 | | Load Case | 2 | |
| | $\phi$ = resistance factor = 0.9 | | | $\Omega$ = factor of safety = 1.67 | | |
| | Mu < $\phi$ Mn | | | Mu <= Mn / $\Omega$, | | |
| | Vu < $\phi$ Vn | | | Vu <= Vn / $\Omega$, | | |
| Load w = | 1.2D + 1.6L = 1.2 x 30 + 1.6 x 60 = **132** | | kN/m | D + L = 30 + 60 = **90** | | kN/m |
| Span End | L = 6 simply supported | | M | L = 6 simply supported | | M |
| Moment | Mu = 132 x 6^2 / 8 **594** | | kNm kNm | Mu = 90 x 6^2 / 8 **405** | | kNm kNm |
| | Mn = 1109 $\phi$ Mn = **998** > Mu; Safe. | | kNm kNm | Mn = 1109 Mn/$\Omega$ = **664** > Mu; Safe. | | kNm kNm |
| Shear | Vu = 132 x 6 / 2 **396** | | kN kN | Vu = 90 x 6 / 2 **270** | | kN kN |
| | Vn = 1279 $\phi$ Vn = **1151** > Vu; Safe. | | kN kN | Vn = 1279 Vn/$\Omega$ = **766** > Vu; Safe. | | kN kN |
| weld betn flange & web | Required Shear, F = Vu Q/Ix | | | Required Shear, F = Vu Q/Ix | | |
| | Q = A y = 200x20x365= Ix = 1423977667 mm4 F = 396000 x 1460000 / 1423977667 = 406 N/mm | | 1460000 mm3 | Q = A y = Ix = 1423977667 mm4 F = 270000 x 1460000 / 1423977667 = 277 N/mm | | 1460000 mm3 |
| | Provide Fillet weld on both sides of web Force per mm run = 406/2 = 203 N < w; OK. Capacity of 6 mm fillet weld, w : w =0.7x0.6x1100= 462 kg/cm = 462 N/mm weld stress = 1100 kg/cm2 | | | Provide Fillet weld on both sides of web Force per mm run = 277/2 = 139 N < w; OK. Capacity of 6 mm fillet weld, w : w =0.7x0.6x1100= 462 kg/cm = 462 N/mm weld stress = 1100 kg/cm2 | | |

End

### Example 8.14.2 : Design of a Column with axial load and bending about major axes.
[ Column with Equal flanges and Bent about the axis of max. strength x-x axis ]

**Reference:**
    1.0 AISC Steel construction manual - 14 Edition
    2.0 ANSI/AISC 360-16 Specification for Structural steel building
    3.0 Companion to the AISC Steel Construction manual - Vol 1 15.1 version

**Material**

| ASTM A 36 | | Table 2-4 | AISC manual | |
|---|---|---|---|---|
| Yield Stress | $F_y$ | 36 | ksi | 250 MPa |
| Tensile Stress | $F_u$ | 65 | ksi | 450 MPa |
| Elasticity | E | | | 200000 MPa |

**Member Loading**
**LRFD**

| Load case | Description | Axial KN $P$ | $M_{vertical}$ KN m $M_x$ | $M_{horz}$ KN m $M_y$ | $V_{vertical}$ KN $V_x$ | $V_{horz}$ KN $V_y$ | | Nature of load |
|---|---|---|---|---|---|---|---|---|
| 1 | 1.4 D | | | | | | | |
| 2 | 1.2D + 1.6L +0.5(Lr or S or R) | 1355 | 160 | 5 | | 62 | | |
| 3 | 1.2D +1.6(Lr os S or R)+(0.5Lor0.5W) | | | | | | | |
| 4 | 1.2D + 1.0W + 0.5L +0.5(Lr or S or R) | | | | | | | |
| 5 | 1.2D + 1.0E + 0.5L + 0.2S | | | | | | | |
| 6 | 0.9D + 1.0W | | | | | | | |
| 7 | 0.9D + 1.0E | | | | | | | |

**ASD**

| Load case | Description | Axial KN $P$ | $M_{vertical}$ KN m $M_x$ | $M_{horz}$ KN m $M_y$ | $V_{vertical}$ KN $V_x$ | $V_{horz}$ KN $V_y$ | | Nature of load |
|---|---|---|---|---|---|---|---|---|
| 1 | D | | | | | | | |
| 2 | D + L | 469 | 133 | 5 | | 40 | | |
| 3 | D + (Lr or S or R) | | | | | | | |
| 4 | D + 0.75L + 0.75(Lr or S or R) | | | | | | | |
| 5 | D + (0.6W or 0.7E) | | | | | | | |
| 6 | D + 0.75L +0.75(0.6W)+0.75(Lr/S/R) | | | | | | | |
| 7 | D + 0.75L+ 0.75(0.7E) + 0.75 S | | | | | | | |
| 8 | 0.6D + 0.6W | | | | | | | |
| 9 | 0.6D + 0.7E | | | | | | | |

**Methods of design**                                (AISC 360-16 Chapter E)

    1 LRFD :      Load and Resistance factor design    [Pu < φ Pn,    φ = resistance factor]
                           Nominal strength = Pn, Mn and Vn    Design strength = φ Pn, φ Mn and φ Vn
                           φ = 0.9 for limit states involving yielding

    2 ASD :       Allowable strength design            [Pu <= Pn / Ω,    Ω = factor of safety]
                           Nominal strength = Pn, Mn and Vn    Design strength = Pn/Ω, Mn/Ω and Vn/Ω
                         Ω = 1.67 for limit states involving yielding

**Dimensions**

**Effective Lengths, Lc :**

| Laterally Unsupported length | Lx = | 6 M | Kx = | 0.85 | KLx = | 5.1 M |
|---|---|---|---|---|---|---|
| | Ly = | 3 M | Ky = | 1 | Kly = | 3.00 M |

Laterally Unsupported length    Lbx =    6 M

# Steel Structures

**Section properties:**

| PLATED SECTION | | | |
|---|---|---|---|
| FLANGE PL | | 250 mm. | bf |
| | | 20 mm. | tf |
| WEB PL | | 450 mm. | h |
| | | 12 mm. | tw |
| WT/M | | 120.9 | kg. |
| Iy | | 52148133 | mm4. |
| rx | | 204 | mm. |
| ry | | 58 | mm. |
| tw | | 12.00 | mm |
| zx (Sx) | | ###### | mm3 |
| zy (Sy) | | 427467 | mm3 |
| J | S bt3/3 | ###### | mm4 |
| $h_0$ | h+tf | 470 | mm |

### Slenderness of member section

**Check for flange**      (AISC 360-16 Table B4.1b)

$0.5\ b/f =$    125 / 20 =    6.25      Nonslender.
*Limiting width to thickness ratio:*
nonslender/slender, $\lambda f =$    $0.56\ (E/Fy)^{0.5} =$    16

**Check for Web**      (AISC 360-16 Table B4.1b)

$h/tw =$    450 / 12 =    38      Nonslender.
*Limiting width to thickness ratio:*
nonslender/slender, $\lambda f =$    $1.49\ (E/Fy)^{0.5} =$    42

*Slenderness ratio of column:*

KLx / rx =    5100 /204 =    24.9
KLy / ry =    3000 /58 =    51.6    $Lc/r_{max} =$    **51.6** [ = Lc/r ]

### Sectional capacity/Nominal strength:

*Nominal Compressive strength, Pn (= Pc) :*

Pn = Fcr. Ag,    where, Fcr = Critical stress and Ag = gross c/s area.    (AISC 360-16 Eqn E3-1)

when, $Lc/r <= 4.71\ (E/Fy)^{0.5} =$    $4.71 \times (200000/250)^{0.5}$    (AISC 360-16 Eqn E3-2)
       =    133
$Fcr = [0.658^{\wedge}(Fy/Fe)]\cdot Fy =$    $0.658\ ^{\wedge} (250 / 742) \times 250$    =    217 Mpa

and    $Lc/r > 4.71\ (E/Fy)^{0.5} =$    133    (AISC 360-16 Eqn E3-3)

Fcr = 0.877 Fe    =    0.877 x 742    =    651 Mpa

Fe = elastic buckling stress = $\pi^{\wedge}2\ E / (Lc/r)^{\wedge}2$
      =    $3.14^{\wedge}2 \times 200000 / 52^{\wedge}2 =$    742 Mpa
Here    E =    200000 Mpa
       Fy =    250 Mpa
       Ag =    15400 mm2
       Lc / r =    51.6    < 133    Fcr =    217 Mpa

**Pn** =    Fcr . Ag =    217 x 15400/ 1000 =    **3347** kN    ( = Fca)

*Nominal flexural strength of doubly symmetric Compact I shaped and channels member bent about major axis, Mnx (= Mcx) :*

Mnx shall be the lower value obtained according to the limit state of yielding (plastic moment) and lateral torsional buckling

#### a) Yielding

     Mnx = Mpx = Fy Zx      (AISC 360-16 Eqn F2-1)

$F_y$ = Yield stress = 250 Mpa

$Z_x$ = plastic sectional modulas about the x axis = $\Sigma A y$
= 2 x 250 x 20 x 470/ 2 + 2 x (450/2) x 12 x (450/4)
= 3E+06    mm4

$M_{nx}$ = 250 x 2957500/1000000 = **740** kNm.    (= $M_p$)

## a) Lateral-torsional buckling

Limiting unbraced length for the limit state of yielding = $L_p$
Limiting unbraced length for the limit state of inealstic lateral torsional buckling = $L_r$

**$L_p$** = 1.76 $r_y$ (E/$F_y$) ^0.5 =    1.76 x 58 x (200000 / 250) ^0.5    (AISC 360-16 Eqn F2-5)
= 2895 mm
= **2.895 M.**

**$L_r$** = 1.95 $r_{ts}$ (E / 0.7$F_y$) [ (J/$S_x.h_0$) + { (J/$S_x.h_o$)$^2$ + 6.76 (0.7$F_y$/E)$^2$}^0.5]^0.5
(AISC 360-16 Eqn F2-6)

$r_{ts}$ =    **66** mm    ( See calc below)
E/0.7$F_y$ =    1142    0.7$F_y$/E =    0.00088
J / $S_x.h_0$ =    0.0013

**X** = { (J/$S_x.h_o$)$^2$ + 6.76 (0.7$F_y$/E)$^2$}^0.5 =    0.0026
[ (J/$S_x.h_0$) + **X** ]^0.5 =    0.0625

**$L_r$** =    9246 mm    **9.25 M**

(i) when $L_b$ < $L_p$,    the limit state of lateral buckling does not apply.

(ii) when $L_p$ < $L_b$ <= $L_r$    (AISC 360-16 Eqn F2-2)

$M_n$ = $C_b$ [ $M_p$ - ($M_p$ - 0.7 $F_y$ $S_x$) (($L_b$-$L_p$)/($L_r$ - $L_p$))] <= $M_p$
= 1[740 - (740 - 0.7 x 250 x 2627381/1000000) x ((6-2.895) /(9.246 - 2.895))
= 603 kNm    <= ($M_p$ =    740 kNm )
$M_n$ =    **603 kNm**

(iii) when $L_b$ > $L_r$    (AISC 360-16 Eqn F2-3)

$M_n$= $F_{cr}$ $S_x$ <= $M_p$

$F_{cr}$ = [$C_b$ $\pi^2$ E / ($L_b$ / $r_{ts}$)$^2$]. [1 + 0.078 (Jc /($S_x.h_o$)). ($L_b$ / $r_{ts}$)$^2$]^0.5    (AISC 360-16 Eqn F2-4)

$C_b$ =    1        $J_c$ =    1592533    mm4
E =    200000 Mpa        $S_x$ =    2627381    mm3
$L_b$ =    6000 mm        $h_o$ =    470    mm

# Steel Structures

$r_{ts}$ = radius of gyration of compression flange plus one sixth of the web

Iyc = $20 \times 250^3 / 12 + (450/6) \times 12^3 / 12$ = 26052467 mm4
Ayc = $250 \times 20 + (450/6) \times 12$ = 5900 mm2
$r_{ts}$ = ( Iyc/Ayc )^0.5 = 66 mm

Fcr = $[1 \times 3.14^2 \times 200000 / (6000/66)^2] \times [1 + 0.078(1592533 / (2627381 \times 470)) \times (6000/66)^2]^{0.5}$
= 326 Mpa
Now, **Mnx** = Fcr.Sx = $326 \times 2627381 / 1000000$ = 857 kNm.
and Mp = 740 kNm.
So, the effective, **Mnx** = 740 kNm

| Lbx = | 6.0 M | Lp = | 2.895 M | Lr = | 9.246 M | height = | 6 | M |
|---|---|---|---|---|---|---|---|---|

In this example of column section, **Mnx** = 603 Mpa. Lp< Lbx < Lr

*Nominal flexural strength of doubly symmetric Compact I shaped and channels member bent about minor axis, Mny ( = Mcy) :*

### a) Yielding

$$Mny = Mpy = Fy\, Zy \leq 1.6\, Fy\, Sy \qquad \text{(AISC 360-16 Eqn F6-1)}$$

Fy = 250 Mpa
Sy = Elastic section Mod about y axis = 427467 mm3
Zy = Plastic section Mod about Y axis (AY') =
 = $4 \times [(20 \times 125) \times (250/4)]$ = 625000 mm3

Mny = Fy . Zy = $250 \times 625000 / 1000000$ = 156 kNm
1.6 Fy Sy = $1.6 \times 250 \times 427467/1000000$ = 171 kNm
**Mny =** **156 kNm**

### b) Flange local buckling             (AISC 360-16 F6-2a )

For sections with compact flanges, the limit state of flange local bucking does not apply.

*Nominal Shear strength of member*

**Vn** = 0.6Fy Aw Cv1 = $0.6 \times 250 \times 450 \times 12 \times 1/1000$ = **811 kN**
Cv1 = 1

### Design check for Axial compression and Flexural moments    (AISC 360-16 Chapter H 1 )

a) when Pr / Pc >= 0.2

$$Pr/Pc + 8/9\,[\,Mrx / Mcx + Mry/Mcy\,] \leq 1.0 \qquad \text{(AISC 360-16 Eqn H1-1a )}$$

b) when Pr / Pc < 0.2

$$Pr/ 2Pc + [\,Mrx / Mcx + Mry/Mcy\,] \leq 1.0 \qquad \text{(AISC 360-16 Eqn H1-1b )}$$

where,
$P_r$ = required axial strength (N) using LRFD or ASD load combinations
$P_c$ = available axial strength = $\phi c\ P_n$ (LRFD) and $P_n/\Omega$ (ASD)
$M_{rx}$ = required flexural strength (kNm) about major axis (x) using LRFD or ASD load combinations
$M_{cx}$ = available flexural strength (kNm) about major axis (x) = $M_{nx}$ = $\phi c\ M_{nx}$ (LRFD) and $M_{nx}/\Omega$ (ASD)
$M_{ry}$ = required flexural strength (kNm) about minor axis (y) using LRFD or ASD load combinations
$M_{cy}$ = available flexural strength (kNm) about minor axis (y) = $M_{ny}$ = $\phi c\ M_{ny}$ (LRFD) and $M_{ny}/\Omega$ (ASD)

For LRFD
$\phi c$ = resistance factor for compression = 0.9        (AISC 360-16 Section B.3.1)
$\phi b$ = resistance factor for flexure = 0.9

For ASD
$\Omega c$ = safety factor for compression = 1.67         (AISC 360-16 Section B.3.2)
$\Omega b$ = Safety factor for flexure = 1.67

**Strength design**                                        (AISC 360-16 Chapter H)

| | LRFD | | | ASD | | |
|---|---|---|---|---|---|---|
| | $\phi c$ = resist. factor comp. = 0.9 <br> $\phi b$ = resist. factor flexure = 0.9 <br> $P_r < \phi c\ P_n\ (= P_c)$ <br> $M_{rx} < \phi b\ M_{nx}\ (= M_{cx})$ <br> $M_{ry} < \phi b\ M_{ny}\ (= M_{cy})$ <br> $V_u < \phi V_n$ | | | $\Omega$ = FOS for Comp. = 1.67 <br> $\Omega$ = FOS for flexure = 1.67 <br> $P_r \leq P_n/\Omega\ (= P_c)$ <br> $M_{rx} \leq M_{nx}/\Omega,\ (= M_{cx})$ <br> $M_{ry} \leq M_{ny}/\Omega\ (= M_{cy})$ <br> $V_u \leq V_n/\Omega,$ | | |
| Load | *Axial Comp. and Flexure* <br> 2 **1.2D + 1.6L + 0.5(Lr or S or R)** | | | *Axial Comp. and Flexure* <br> 2 **D + L** | | |
| | $P_r$ = | 1355 kN | | $P_r$ = | 469 kN | |
| | $M_{rx}$ = | 160 kNm | | $M_{rx}$ = | 133 kNm | |
| | $M_{ry}$ = | 5 kNm | | $M_{ry}$ = | 5 kNm | |
| Comp & flexure | Now from the calculation above | | | | | |
| | $P_n$ = 3347 | $P_c = \phi P_n$ = | 3012 kN | $P_n$ = 3347 | $P_c = P_n/\Omega$ = | 2004 kN |
| | $M_{nx}$ = 603 | $M_{cx} = \phi M_{nx}$ = | 543 kNm | $M_{nx}$ = 603 | $M_{cx} = M_{nx}/\Omega$ = | 361 kNm |
| | $M_{ny}$ = 156 | $M_{cy} = \phi M_{ny}$ = | 141 kNm | $M_{ny}$ = 156 | $M_{cy} = M_{ny}/\Omega$ = | 94 kNm |
| | *Interaction factor:* | | | *Interaction factor:* | | |
| | $P_r/P_c$ = | 1355/3012 = | 0.450 | $P_r/P_c$ = | 469/2004 = | 0.234 |
| | | | | | | 0.1 |
| | $P_r/P_c + 8/9\ [M_{rx}/M_{cx} + M_{ry}/M_{cy}]$ = <br> 1355/3012 + 8/9 x [160/543 + 5/141] | | 0.74 **Safe.** | $P_r/P_c + 8/9\ [M_{rx}/M_{cx} + M_{ry}/M_{cy}]$ = <br> 469/2004 + 8/9 x [133/361 + 5/94] | | 0.61 **Safe.** |
| shear | $V_r$ = | 62 kN | | $V_r$ = | 40 kN | |
| | $V_n$ = | 811 kN | | $V_n$ = | 811 kN | |
| | $\phi V_n$ = | 730 kN | **Safe.** | $V_n/\Omega$ = | 486 kN | **Safe.** |

End

# Steel Structures

## Example 8.14.3 : Design of Axial loaded bracing members        Compression /Tension

### 8.14.3.1  Single angle compression members with Equal legs

End connection: Welding or Bolting with a minimum of two bolts

**Reference.**
  1.0    ANSI/AISC 360-16 Specification for Structural steel building

**Material**

| ASTM A 36 | | Table 2-4 | AISC manual | |
|---|---|---|---|---|
| Yield Stress | $F_y$ | 36 ksi | 250 | MPa |
| Tensile Stress | $F_u$ | 65 ksi | 450 | MPa |
| Elasticity | E | | 200000 | MPa |

**Member Loading**
**LRFD**

| Load case | Description | Axial KN P | $M_{vertical}$ KN m $M_x$ | $M_{horz}$ KN m $M_y$ | $V_{vertical}$ KN $V_x$ | $V_{horz}$ KN $V_y$ | | Nature of load |
|---|---|---|---|---|---|---|---|---|
| 1 | 1.4 D | | | | | | | |
| 2 | 1.2D + 1.6L +0.5(Lr or S or R) | 75 | 0 | 0 | | 0 | | |
| 3 | 1.2D +1.6(Lr os S or R)+(0.5Lor0.5W) | | | | | | | |
| 4 | 1.2D + 1.0W + 0.5L +0.5(Lr or S or R) | | | | | | | |
| 5 | 1.2D + 1.0E + 0.5L + 0.2S | | | | | | | |
| 6 | 0.9D + 1.0W | | | | | | | |
| 7 | 0.9D + 1.0E | | | | | | | |

**ASD**

| Load case | Description | Axial KN P | $M_{vertical}$ KN m $M_x$ | $M_{horz}$ KN m $M_y$ | $V_{vertical}$ KN $V_x$ | $V_{horz}$ KN $V_y$ | | Nature of load |
|---|---|---|---|---|---|---|---|---|
| 1 | D | | | | | | | |
| 2 | D + L | 50 | 0 | 0 | | 0 | | |
| 3 | D + (Lr or S or R) | | | | | | | |
| 4 | D + 0.75L + 0.75(Lr or S or R) | | | | | | | |
| 5 | D + (0.6W or 0.7E) | | | | | | | |
| 6 | D + 0.75L+0.75(0.6W)+0.75(Lr/S/R) | | | | | | | |
| 7 | D + 0.75L+ 0.75(0.7E) + 0.75 S | | | | | | | |
| 8 | 0.6D + 0.6W | | | | | | | |
| 9 | 0.6D + 0.7E | | | | | | | |

**Methods of design**                                              (AISC 360-16 Chapter E-5)
  1 LRFD :        Load and Resistance factor design    [Pu < $\phi$ Pn,    $\phi$ = resistance factor]
                  $\phi$ =    0.9
  2 ASD :         Allowable strength design            [Pu <= Pn / $\Omega$,    $\Omega$ = factor of safety]
                  $\Omega$ =    1.67

**Section properties:**

| Section Provided: | wt/m (kg) | $A_g$ (mm$^2$) | $r_x$ (mm) | $r_y$ (mm) | $r_v$ (mm) | $I_x$ (mm$^4$) | $I_y$ (mm$^4$) | $C_{yy}$ (mm) |
|---|---|---|---|---|---|---|---|---|
| ISA 100 100 10 | 14.9 | 1903 | 30.50 | 30.5 | 19.4 | 1770000 | 1770000 | 28.4 |

| Leg sizes (mm) | | |
|---|---|---|
| $b_l$ | $b_s$ | t |
| 100 | 100 | 10 |

*Check for slenderness:* (AISC 360-16 Table B 4-1a)

b/t < 0.45* ( E / Fy)^0.5        b/t =        10        Non Slender.
                        0.45* ( E / Fy)^0.5 =    13

b/t < 0.71* ( E / Fy)^0.5        b/t =        10        *Flextural-torsional buckling check Not reqd.*
                        0.45* ( E / Fy)^0.5 =    20

**Effective Lengths (Lc) and Slenderness ratio ( Lc/r) :**

Laterally Unsupported length    Lx =    4 M    Kx =    0.85    KLx =    3.40 M
                                Ly =    2 M    Ky =    1       Kly =    2.00 M

a)  Maxm slenderness ratio (Lc/r) shall not exceed        **200**    (AISC 360-16 E5 )
    Here,   KLx/rx =    3400/30.5 =    111
            KLy/rv =    2000/19.4 =    103  Lc/r =  <200, OK.

b)  when L/ra = Lx / rx <= 80,  Lc/r = Kly/rv = 72+0.75 (Lx/rx) =        170    (AISC 360-16 E5-1)

    when L/ra = Lx / rx > 80,  Lc/r = Kly / rv = 32+1.25 (Lx/rx) =       196    (AISC 360-16 E5-2)

    L = Length of member between work point = Lx
    Lc = Effective length for buckling about minor axis =KLy
    ra = radius of gyration about x axis parallel to connected leg = rx
    rz =ry radius of gyration about minor axis

    Here,   Lx /rx =        4000 / 30.5 =    131

    So,    **Lc/r =    196**

**Nominal Compressive strength, Pn :**

Pn = Fcr. Ag,    where, Fcr = Critical stress and Ag = gross c/s area.    (AISC 360-16 Eqn E3-1)

# Steel Structures

when , $L_c/r \le 4.71 (E/F_y)^{0.5}$ =     $4.71 \times (200000/250)^{0.5}$     (AISC 360-16 Eqn E3-2)
= 99
$F_{cr} = [0.658^{(F_y/F_e)}] \cdot F_y$     = $0.658^{(250/51)} \times 250$     = 33 Mpa

and if, $L_c/r > 4.71 (E/F_y)^{0.5}$ =     99     (AISC 360-16 Eqn E3-3)

$F_{cr} = 0.877 F_e$     = $0.877 \times 51$     = 45 Mpa

$F_e$ = elastic buckling stress = $\pi^2 E / (L_c/r)^2$
= $3.14^2 \times 200000 / 196^2$ =     51 Mpa

Here    E =     200000 Mpa
        $F_y$ =     250 Mpa
        $A_g$ =     1903 mm2
        $L_c / r$ =     196     > 99         $F_{cr}$ =     45 Mpa

**Pn** =    $F_{cr} \cdot A_g$ =     $45 \times 1903 / 1000$ =     **86** kN     ( = $F_{ca}$ )

**Strength design**                                                (AISC 360-16 Chapter E)

| LRFD | | ASD | |
|---|---|---|---|
| $\phi_c$ = resist. factor comp. = | 0.9 | $\Omega$ = FOS for Comp. = | 1.67 |
| $P_r < \phi_c P_n$ | | $P_r < P_n / \Omega$ | |
| $P_r$ =    75 kN | | $P_r$ =    50 kN | |
| $P_n$ =    86 kN | | $P_n$ =    86 kN | |
| $\phi_c P_n = 0.9 \times 86$ = | 77 kN | $P_n / \Omega = 86 / 1.67$ = | 51 kN |
| | Safe. | | Safe. |

## 8.14.3.2 Single angle Tension members with Equal legs

(AISC 360-16 Chapter D)

**Material**

| ASTM A 36 | | Table 2-4 AISC manual | |
|---|---|---|---|
| Yield Stress | $F_y$ | 36 ksi | 250 MPa |
| Tensile Stress | $F_u$ | 65 ksi | 450 MPa |
| Elasticity | E | | 200000 MPa |

The design tensile strength = $\phi t\, Pn$  → LRFD  $\phi t =$ 0.9
The allowable tensile strength = $Pn / \Omega$  → ASD  $\Omega t =$ 1.67

### Member Loading

**LRFD**

| Load case | Description | Axial | $M_{vertical}$ | $M_{horz}$ | $V_{vertical}$ | $V_{horz}$ | | Nature of load |
|---|---|---|---|---|---|---|---|---|
| | | KN | KN m | KN m | KN | KN | | |
| | | T | $M_x$ | $M_y$ | $V_x$ | $V_y$ | | |
| 1 | 1.4 D | | | | | | | |
| 2 | 1.2D + 1.6L +0.5(Lr or S or R) | 75 | 0 | 0 | | 0 | | |
| 3 | 1.2D +1.6(Lr os S or R)+(0.5Lor0.5W) | | | | | | | |
| 4 | 1.2D + 1.0W + 0.5L +0.5(Lr or S or R) | | | | | | | |
| 5 | 1.2D + 1.0E + 0.5L + 0.2S | | | | | | | |
| 6 | 0.9D + 1.0W | | | | | | | |
| 7 | 0.9D + 1.0E | | | | | | | |

**ASD**

| Load case | Description | Axial | $M_{vertical}$ | $M_{horz}$ | $V_{vertical}$ | $V_{horz}$ | | Nature of load |
|---|---|---|---|---|---|---|---|---|
| | | KN | KN m | KN m | KN | KN | | |
| | | T | $M_x$ | $M_y$ | $V_x$ | $V_y$ | | |
| 1 | D | | | | | | | |
| 2 | D + L | 50 | 0 | 0 | | 0 | | |
| 3 | D + (Lr or S or R) | | | | | | | |
| 4 | D + 0.75L + 0.75(Lr or S or R) | | | | | | | |
| 5 | D + (0.6W or 0.7E) | | | | | | | |
| 6 | D + 0.75L+0.75(0.6W)+0.75(Lr/S/R) | | | | | | | |
| 7 | D + 0.75L+ 0.75(0.7E) + 0.75 S | | | | | | | |
| 8 | 0.6D + 0.6W | | | | | | | |
| 9 | 0.6D + 0.7E | | | | | | | |

# Steel Structures

**Section properties:** all dimensions are in mm

| Section Provided: | wt/m (kg) | $A_g$ (mm$^2$) | $r_x$ (mm) | $r_y$ (mm) | $r_v$ (mm) | $I_x$ (mm$^4$) | $I_y$ (mm$^4$) | $C_{yy}$ (mm) |
|---|---|---|---|---|---|---|---|---|
| ISA 100 100 10 | 14.9 | 1903 | 30.50 | 30.5 | 19.4 | 1770000 | 1770000 | 28.4 |
| | Leg sizes (mm) | | | Connection bolt | | | An | Ae |
| | $b_l$ | $b_s$ | t | $\phi$dia | $\phi$hole | rows | mm2 | mm2 |
| | 100 | 100 | 10 | 16 | 18 | 1 | 1723 | 1179 |

Ae = An.U , where U is shear lag factor = 0.68  (AISC 360-16 Table D3.1)
where Ae = effective area and An = net area after deduction of bolt holes.
     An =   Ag - (no of rows x $\phi$ hole x t)

*Shear lag factor U*

The value of U, need not be less than the ratio of the gross area of connected leg to the member gross area according to AISC spec 360-16 D3.

Case 1 : One leg i.e half of the member is connected , so    U =   0.5
Cases 2 :    U = 1 - x/L    wher x = y = CG dist Cyy =    28.4 mm
         L = length of connection , say for bolts   3    90 mm   U=   0.68
         [ refer Table D3.1]

                                Therefore, use the larger U= 0.68

**Effective Lengths (L) and Slenderness ratio ( L/r ) :**

Laterally Unsupported length    Lx   6 M    Kx   1    KLx   6.00 M
                                 Ly   3 M    Ky   1    Kly   3.00 M

     Maxm slenderness ratio (L/r) shall not exceed    300    (AISC 360-16 D1 )
     Here,   KLx/rx =   6000/30.5 =    197
            KLy/rv =   3000/19.4 =    155
                   L/r =   197    <300, OK.

**Tensile strength, Pn**                                                     (AISC 360-16 D2 )

a) For tension yielding of gross section , Pn = Fy Ag =    250 x 1903/1000 =    476 kN

b) For tensile rupture in net section, Pn = Fu Ae =    450 x 1179/1000 =    531 kN

                                                           Pn =   476 kN

**Strength design** (AISC 360-16 Chapter D)

| LRFD | ASD |
|---|---|
| $\phi t$ = resist. factor tension. = 0.9 | $\Omega t$ = FOS for tension. = 1.67 |
| $T < \phi t\, Pn$ | $T < Pn / \Omega$ |
| T = 75 kN | T = 50 kN |
| Pn = 476 kN | Pn = 476 kN |
| $\phi t\, Pn = 0.9 \times 476 =$ 429 kN | $Pn / \Omega = 476 / 1.67 =$ 285 kN |
| Safe. | Safe. |

*8.14.3.3 Double angle compression members with Equal legs Back/back*

End connection: Welding or Bolting with a minimum of two bolts

**Reference.**
    1.0 ANSI/AISC 360-16 Specification for Structural steel building

**Material**

| ASTM A 36 | | Table 2-4 AISC manual | | |
|---|---|---|---|---|
| Yield Stress | $F_y$ | 36 | ksi | 250 MPa |
| Tensile Stress | $F_u$ | 65 | ksi | 450 MPa |
| Elasticity | E | | | 200000 MPa |

**Member Loading**

**LRFD**

| Load case | Description | Axial KN P | $M_{vertical}$ KN m $M_x$ | $M_{horz}$ KN m $M_y$ | $V_{vertical}$ KN $V_x$ | $V_{horz}$ KN $V_y$ | | Nature of load |
|---|---|---|---|---|---|---|---|---|
| 1 | 1.4 D | | | | | | | |
| 2 | 1.2D + 1.6L +0.5(Lr or S or R) | | | | | | | |
| 3 | 1.2D +1.6(Lr os S or R)+(0.5Lor0.5W) | | | | | | | |
| 4 | 1.2D + 1.0W + 0.5L +0.5(Lr or S or R) | 150 | 0 | 0 | | 0 | | |
| 5 | 1.2D + 1.0E + 0.5L + 0.2S | | | | | | | |
| 6 | 0.9D + 1.0W | | | | | | | |
| 7 | 0.9D + 1.0E | | | | | | | |

**ASD**

| Load case | Description | Axial KN P | $M_{vertical}$ KN m $M_x$ | $M_{horz}$ KN m $M_y$ | $V_{vertical}$ KN $V_x$ | $V_{horz}$ KN $V_y$ | | Nature of load |
|---|---|---|---|---|---|---|---|---|
| 1 | D | | | | | | | |
| 2 | D + L | | | | | | | |
| 3 | D + (Lr or S or R) | | | | | | | |
| 4 | D + 0.75L + 0.75(Lr or S or R) | | | | | | | |
| 5 | D + (0.6W or 0.7E) | 100 | 0 | 0 | | 0 | | |
| 6 | D + 0.75L+0.75(0.6W)+0.75(Lr/S/R) | | | | | | | |
| 7 | D + 0.75L+ 0.75(0.7E) + 0.75 S | | | | | | | |
| 8 | 0.6D + 0.6W | | | | | | | |
| 9 | 0.6D + 0.7E | | | | | | | |

# Steel Structures

**Methods of design**

| | | | |
|---|---|---|---|
| 1 LRFD : | Load and Resistance factor design | $[P_u < \phi P_n,$ | $\phi$ = resistance factor] |
| | $\phi =$ 0.9 | | (AISC 360-16 Chapter E-6) |
| 2 ASD : | Allowable strength design | $[P_u <= P_n / \Omega,$ | $\Omega$ = factor of safety] |
| | $\Omega =$ 1.67 | | |

**Section properties:**

| Section Provided: | | s | | Built-up member dimensions | | | | | | |
|---|---|---|---|---|---|---|---|---|---|---|
| | | | | wt/m (kg) | $A_g$ (mm$^2$) | $r_x$ (mm) | $r_y$ (mm) | | $I_x$ (mm$^4$) | $I_y$ (mm$^4$) |
| 2 | ISA 100 100 10 | 12 | mm b/b | 29.8 | 3806 | 30.50 | 46.0 | | 3540000 | 8043868 |
| | | | | Single Leg dimensions | | | | | | |
| | | | | $b_l$ | $b_s$ | t | | rv (ri) | a | Ki | Cyy |
| | | | | mm | mm | mm | | mm | mm | | mm |
| | | | | 100 | 100 | 10 | | 19.4 | 450 | 0.5 | 28.4 |

Distance between connectors (welded spacer plate) =   a   (shall not exceed 0.75 (ri . L/r)

| Effective length factor, **Ki** | |
|---|---|
| Angle back to back | 0.5 |
| Channel back to back | 0.75 |
| For all other cases | 0.86 |

*Check for slenderness:*                                                                 (AISC 360-16 Table B 4-1a)

$b/t < 0.45 * ( E / F_y )^{0.5}$       b/t = 10       Non Slender.
                      $0.45 * ( E / F_y )^{0.5} =$ 13

$b/t < 0.71 * ( E / F_y )^{0.5}$       b/t = 10       *Flextural-torsional buckling check Not reqd.*

$0.45 \times (E/Fy)^{0.5} =$   20

### Effective Lengths (Lc) and Slenderness ratio ( Lc/r ) :

Laterally Unsupported length    Lx    6 M    Kx    0.85    KLx    5.10 M
                                Ly    6 M    Ky    1       Kly    6.00 M

a)  Maxm slenderness ratio (Lc/r) shall not exceed        **200**    (AISC 360-16 E5 )
    Here,  KLx/rx =    5100/30.5 =    167
           KLy/ry =    6000/46 =      131

    Maxm Slenderness ratio of built-up member =    Lc/r =    167  < 200, OK.
    $(= Lc/r)_o$

*Modified slenderness ratio of built-up member, $(Lc/r)m$*        (AISC 360-16 E5-2)

a)  When  **a / ri** <= 40    $(Lc/r)_m = (Lc/r)_o$    =    167
b)  When  **a / ri** > 40     $(Lc/r)_m = ( (Lc/r)_o^2 + (Ki.a/ri)^2 )^{0.5}$
                              =    $(167^2 + (0.5 \times 450/19.4)^2)^{0.5}$ =    167.6

    here    **a /ri** =    450 / 19.4 =    23.20      | Lc/r =    167 |

Note:   Value of "a" shall not exceed 0.75 (ri . L/r) =    0.75 x 19.4 x 167 =    2433    (AISC 360-16 E6-2)

### Nominal Compressive strength, Pn :

Pn = Fcr . Ag,   where, Fcr = Critical stress and Ag = gross c/s area.    (AISC 360-16 Eqn E3-1)

when , Lc/r <= 4.71 $(E/Fy)^{0.5}$ =    4.71 x $(200000/250)^{0.5}$    (AISC 360-16 Eqn E3-2)
                                  =    133
Fcr = $[0.658^{(Fy/Fe)}]$ . Fy    =    $0.658^{(250/71)}$ x 250    =    57 Mpa

and     Lc/r > 4.71 $(E/Fy)^{0.5}$ =    133    (AISC 360-16 Eqn E3-3)

Fcr = 0.877 Fe    =    0.877 x 71    =    62 Mpa

Fe = elastic buckling stress = $\pi^2 E / (Lc/r)^2$
                             =    $3.14^2$ x 200000 / $167^2$ =    71 Mpa

Here    E =    200000 Mpa
        Fy =   250 Mpa
        Ag =   3806 mm2

Lc / r =    167    >133    Fcr =    62 Mpa

**Pn =**    Fcr . Ag =    62 x 3806/ 1000 =    **235 kN**    ( = Fca)

# Steel Structures

**Strength design** (AISC 360-16 Chapter E)

| LRFD | | ASD | |
|---|---|---|---|
| $\phi c$ = resist. factor comp. = | 0.9 | $\Omega$ = FOS for Comp. = | 1.67 |
| $Pr < \phi c\, Pn$ | | $Pr < Pn / \Omega$ | |
| $Pr$ = 150 kN | | $Pr$ = 100 kN | |
| $Pn$ = 235 kN | | $Pn$ = 235 kN | |
| $\phi c\, Pn = 0.9 \times 235$ = | 212 kN | $Pn / \Omega = 235 / 1.67$ = | 141 kN |
| | Safe. | | Safe. |

### *8.14.3.4 Double angle Tension members with Equal legs* (AISC 360-16 Chapter D)

**Material**

| ASTM A 36 | | Table 2-4 | AISC manual | |
|---|---|---|---|---|
| Yield Stress | $F_y$ | 36 | ksi | 250 MPa |
| Tensile Stress | $F_u$ | 65 | ksi | 450 MPa |
| Elasticity | E | | | 200000 MPa |

The design tensile strength = $\phi t\, Pn$    | LRFD |    $\phi t$ =    0.9
The allowable tensile strength = $Pn / \Omega$    | ASD |    $\Omega t$ =    1.67

**Member Loading**
**LRFD**

| Load case | Description | Axial KN T | $M_{vertical}$ KN m $M_x$ | $M_{horz}$ KN m $M_y$ | $V_{vertical}$ KN $V_x$ | $V_{horz}$ KN $V_y$ | | Nature of load | |
|---|---|---|---|---|---|---|---|---|---|
| 1 | 1.4 D | | | | | | | | |
| 2 | 1.2D + 1.6L +0.5(Lr or S or R) | | | | | | | | |
| 3 | 1.2D +1.6(Lr os S or R)+(0.5Lor0.5W) | | | | | | | | |
| 4 | 1.2D + 1.0W + 0.5L +0.5(Lr or S or R) | 150 | 0 | 0 | | 0 | | | |
| 5 | 1.2D + 1.0E + 0.5L + 0.2S | | | | | | | | |
| 6 | 0.9D + 1.0W | | | | | | | | |
| 7 | 0.9D + 1.0E | | | | | | | | |

# Design of Industrial Structures

**ASD**

| Load case | Description | Axial KN T | $M_{vertical}$ KN m $M_x$ | $M_{horz}$ KN m $M_y$ | $V_{vertical}$ KN $V_x$ | $V_{horz}$ KN $V_y$ | | Nature of load |
|---|---|---|---|---|---|---|---|---|
| 1 | D | | | | | | | |
| 2 | D + L | | | | | | | |
| 3 | D + (Lr or S or R) | | | | | | | |
| 4 | D + 0.75L + 0.75(Lr or S or R) | | | | | | | |
| 5 | D + (0.6W or 0.7E) | 100 | 0 | 0 | | 0 | | |
| 6 | D + 0.75L+0.75(0.6W)+0.75(Lr/S/R) | | | | | | | |
| 7 | D + 0.75L+ 0.75(0.7E) + 0.75 S | | | | | | | |
| 8 | 0.6D + 0.6W | | | | | | | |
| 9 | 0.6D + 0.7E | | | | | | | |

**Section properties:**         all dimensions are in mm

| Section Provided: | | s | | wt/m (kg) | $A_g$ ($mm^2$) | $r_x$ (mm) | $r_y$ (mm) | $r_{v/leg}$ (mm) | $I_x$ ($mm^4$) | $I_y$ ($mm^4$) | $C_{yy}$ (mm) |
|---|---|---|---|---|---|---|---|---|---|---|---|
| 2 ISA 100 100 10 | | 12 | mm b/b | 29.8 | 3806 | 30.50 | 46.0 | 19.4 | 3540000 | 8043868 | 28.4 |
| | | | | Leg sizes (mm) | | | Connection bolt | | | An | Ae |
| | | | | $b_l$ | $b_s$ | t | φdia | φhole | rows | mm2 | mm2 |
| | | s | | 100 | 100 | 10 | 16 | 18 | 1 | 3626 | 2482 |

Ae = An.U    , where U is shear lug factor =    0.68         (AISC 360-16 Table D3.1)
where Ae = effective area and An = net area after deduction of bolt holes.
    An =    Ag - (no of rows x φ hole x  t)

*Shear lag factor U*

U, need not be less than the ratio of the gross area of connected leg to the member gross area according to AISC spec 360-16 D3.

Case 1 : One leg i.e half of the member is connected , so                    U =        0.5
Cases 2 :        U = 1 - x/L        wher x = y = CG dist Cyy =         28.4 mm
               L = length of connection , say for bolts       3      90 mm    U=    0.68
               [ refer Table D3.1]
                                     Therefore, use the larger   U=    0.68

**Effective Lengths (L) and Slenderness ratio ( L/r) :**

Laterally Unsupported length       Lx       6 M        Kx       1        KLx        6.00 M
                                         Ly       6 M        Ky       1        Kly        6.00 M

       Maxm slenderness ratio (L/r) shall not exceed                **300**        (AISC 360-16 D1 )
       Here,    KLx/rx =       6000/30.5 =       197
               KLy/ry =       6000/46 =         131
                           L/r =            197         <300, OK.

# Steel Structures

**Tensile strength, Pn** (AISC 360-16 D2)

a) For tension yielding of gross section, $P_n = F_y A_g =$   250 x 3806/1000 =   953 kN

b) For tensile rupture in net section, $P_n = F_u A_e =$   450 x 2482/1000 =   1117 kN

| | |
|---|---|
| Pn = | 953 kN |

**Strength design** (AISC 360-16 Chapter D)

| LRFD | ASD |
|---|---|
| $\phi_t$ = resist. factor tension. = 0.9 | $\Omega_t$ = FOS for tension. = 1.67 |
| $T < \phi_t P_n$ | $T < P_n / \Omega$ |
| T = 150 kN | T = 100 kN |
| Pn = 953 kN | Pn = 953 kN |
| $\phi_t P_n$ = 0.9 x 953 = 857 kN | $P_n / \Omega$ = 953 / 1.67 = 570 kN |
| Safe. | Safe. |

End

## Example 8.14.4 : Design of Roof Purlin with rolled section.   (AISC 360 -16 Chapter F2)
[Doubly symmetric Compact Beams and Channels bent about the axis of max. strength x-x axis ]

### Reference.
1.0 AISC Steel construction manual - 14 Edition
2.0 ANSI/AISC 360-16 Specification for Structural steel building
3.0 Companion to the AISC Steel Construction manual - Vol 1 15.1 version

### Material

| ASTM A 36      |       | Table 2-4 | AISC manual |        |     |
|----------------|-------|-----------|-------------|--------|-----|
| Yield Stress   | $F_y$ |        50 | ksi         |    345 | MPa |
| Tensile Stress | $F_u$ |        65 | ksi         |    450 | MPa |
| Elasticity     | E     |           |             | 200000 | MPa |

### Member Loading
#### LRFD

| Load case | Description | Axial KN P | $M_{vertical}$ KN m $M_x$ | $M_{horz}$ KN m $M_y$ | $V_{vertical}$ KN $V_x$ | $V_{horz}$ KN $V_y$ | | Nature of load |
|---|---|---|---|---|---|---|---|---|
| 1 | 1.4 D | | | | | | | |
| 2 | 1.2D + 1.6L +0.5(Lr or S or R) | | 54 | 36 | | | | |
| 3 | 1.2D +1.6(Lr os S or R)+(0.5Lor0.5W) | | | | | | | |
| 4 | 1.2D + 1.0W + 0.5L +0.5(Lr or S or R) | | | | | | | |
| 5 | 1.2D + 1.0E + 0.5L + 0.2S | | | | | | | |
| 6 | 0.9D + 1.0W | | | | | | | |
| 7 | 0.9D + 1.0E | | | | | | | |

#### ASD

| Load case | Description | Axial KN P | $M_{vertical}$ KN m $M_x$ | $M_{horz}$ KN m $M_y$ | $V_{vertical}$ KN $V_x$ | $V_{horz}$ KN $V_y$ | | Nature of load |
|---|---|---|---|---|---|---|---|---|
| 1 | D | | | | | | | |
| 2 | D + L | | 41 | | 27 | | | |
| 3 | D + (Lr or S or R) | | | | | | | |
| 4 | D + 0.75L + 0.75(Lr or S or R) | | | | | | | |
| 5 | D + (0.6W or 0.7E) | | | | | | | |
| 6 | D + 0.75L+0.75(0.6W)+0.75(Lr/S/R) | | | | | | | |
| 7 | D + 0.75L+ 0.75(0.7E) + 0.75 S | | | | | | | |
| 8 | 0.6D + 0.6W | | | | | | | |
| 9 | 0.6D + 0.7E | | | | | | | |

### Methods of design   (AISC 360 -16 Chapter F2)

1 LRFD :   Load and Resistance factor design     [Pu < $\phi$ Pn,    $\phi$ = resistance factor]
          Nominal strength = Pn, Mn and Vn
          Design strength = $\phi$ Pn, $\phi$ Mn and $\phi$ Vn
          $\phi$ = 0.9 for limit states involving yielding

2 ASD :    Allowable strength design     [Pu <= Pn / $\Omega$,   $\Omega$ = factor of safety]
          Nominal strength = Pn, Mn and Vn
          Design strength = Pn/$\Omega$, Mn/$\Omega$ and Vn/$\Omega$
          $\Omega$ = 1.67 for limit states involving yielding

# Steel Structures

**Load, Load factors and Load combinations**  (AISC Chapter 2/ ASCE/SEI 7 Section 2.4)
D = Dead load
L = Live load
Lr = Roof live load
S = Snow load
R = initial rain / ice
W = Wind load
E = Earthquake load

LRFD : Load combinations - factored.
    1  1.4 D
    2  1.2D + 1.6L +0.5(Lr or S or R)
    3  1.2D +1.6(Lr os S or R) + (0.5L or 0.5 W)
    4  1.2D + 1.0W + 0.5L +0.5(Lr or S or R)
    5  1.2D + 1.0E + 0.5L + 0.2S
    6  0.9D + 1.0W
    7  0.9D + 1.0E

ASD : Load combinations - unfactored.
    1  D
    2  D + L
    3  D + (Lr or S or R)
    4  D + 0.75L + 0.75(Lr or S or R)
    5  D + (0.6W or 0.7E)
    6  D + 0.75L+ 0.75(0.6W) + 0.75(Lr or S or R)
    7  D + 0.75L+ 0.75(0.7E) + 0.75 S
    8  0.6D + 0.6W
    9  0.6D + 0.7E

**Design Load :**

span L=     6 M
spcg =     2 M

w (uniform distributed load)

span

Load intensity:
DL =    3 kN/m2    $w_{DL}$ =    6 kN/m
LL =    1.5 kN/m2    $w_{LL}$ =    3 kN/m
                                   w =    9 kN/m

*LRFD :*  w =   1.2 x 6 + 1.6 x 3 =      12 kN/m   Mu =    54 kNm   Vu =    36 kN
*ASD :*  w =   6 + 3 =      9 kN/m   Mu =    41 kNm   Vu =    27 kN

| | | | | | |
|---|---|---|---|---|---|
| Laterally Unsupported length | Lb | | 2 M | with sag rod | |
| Spacing of web stiffener | a | | 0 m | Not required | |

**Section properties:**

| Designation | C12X30 | | | | |
|---|---|---|---|---|---|
| FLANGE PL | | 3.17 in | | 81 mm. | bf |
| | | 0.501 in | | 13 mm. | tf |
| WEB PL | | 12 in | | 305 mm. | h |
| | | 0.51 in | | 13 mm. | tw |
| WT/M | | 30 lb/ft | | 45 | kg/m. |
| Area | | 8.82 in2 | | 5689 | mm2. |
| Ix | | 162 in4 | | 67429491 | mm4. |
| Iy | | 5.14 in4 | | 2139430 | mm4. |
| rx | | 4.29 in | | 109 | mm. |
| ry | | 0.763 in | | 19 | mm. |
| tw | | 0.51 in | | 13 | mm |
| zx (Sx) | | 27 in3 | | 442451 | mm3 |
| J | $\Sigma$ bt3/3 | 0.796 in4 | | 342000 | mm4 |
| $h_0$ | h+tf | 12.501 in | | 318 | mm |

**Deflection:** $\hspace{6cm}$ (AISC 360 Section L3)

$\delta = 5 \text{ w } L^4 / 384 \text{ EI} =$ $\quad$ 5 x 9 x 6000 ^4 / 384 x 200000 x 67429491 = $\quad$ 11.26 mm

$\hspace{6cm}$ Allowable deflection limit L /360 = $\quad$ 16.67 mm

$\hspace{12cm}$ OK.

# Steel Structures

**Proportioning limit**

Check for Web Slenderness  (AISC 360-16 Table B4.1b)

$h/t_w =$  305 / 13 =  23.462   Compact.
*Limiting width to thickness ratio:*
Compact/ noncompact, $\lambda_p =$  3.76 $(E/F_y)^{0.5} =$  91
Noncompact / slender, $\lambda_r =$  5.70 $(E/F_y)^{0.5} =$  137

Check for flange  (AISC 360-16 Table B4.1b)

$b/f =$  81 / 13 =  6   Compact.
*Limiting width to thickness ratio:*
Compact/ noncompact, $\lambda_p =$  0.38 $(E/F_y)^{0.5} =$  9
Noncompact / slender, $\lambda_r =$  1.0 $(E/F_y)^{0.5} =$  24

*Compact I shaped member; Critical bending stress will be as per Eqn F2-1 AISC360-16.*

**Nominal flexural strength, Mn**

Mn shall be the lower value obtained according to the limit state of yielding (plastic moment) and lateral torsional buckling.

**a) Yielding**

$$M_n = M_p = F_y Z_x$$  (AISC 360-16 Eqn F2-1)

$F_y =$ Yield stress =  345 Mpa

$Z_x =$ plastic sectional modulus about the x axis $= \Sigma A y$
$=$ 2 x 81 x 13 x 318/ 2 + 2 x (305/2) x 13 x (305/4)
$=$ 637185  mm4

$M_n =$ 345 x 637185/1000000 =  **220** kNm.  (= $M_p$)

## a) Lateral-torsional buckling

Limiting unbraced length for the limit state of yielding = Lp
Limiting unbraced length for the limit state of inealstic lateral torsional buckling = Lr

$Lp = 1.76 \, r_y \, (E/Fy)^{0.5}$ =     $1.76 \times 19 \times (200000/345)^{0.5}$     (AISC 360-16 Eqn F2-5)
                       =     805 mm
                       =     **0.805 M.**

$Lr = 1.95 \, r_{ts} \, (E/0.7Fy) \, [ (J/S_x.h_0) + \{ (J/S_x.h_0)^2 + 6.76 \, (0.7Fy/E)^2 \}^{0.5}]^{0.5}$

                                                                              (AISC 360-16 Eqn F2-6)

$r_{ts}$ =      18 mm      ( See calc below)
$E/0.7Fy$ =      829      $0.7Fy/E$ =      0.00121
$J/S_x.h_0$ =      0.0024

$X = \{ (J/S_x.h_0)^2 + 6.76 \, (0.7Fy/E)^2 \}^{0.5}$ =      0.004
$[ (J/S_x.h_0) + X ]^{0.5}$ =      0.08

**Lr** =      2388 mm      **2.39 M**

(i) when Lb < Lp,     the limit state of lateral buckling does not apply.

(ii) when Lp < Lb <= Lr                                                         (AISC 360-16 Eqn F2-2)

        $Mn = Cb \, [ \, Mp - (Mp - 0.7 \, Fy \, Sx) \, ((Lb-Lp)/(Lr - Lp))] <= Mp$
              =      1[220 - (220 - 0.7 x 345 x 442451/1000000) x ((2-0.805) /(2.388 - 0.805)]
              =      134.49 kNm      <= (Mp =      220 kNm )
        Mn =      **134 kNm**

(iii) when Lb > Lr                                                                     (AISC 360-16 Eqn F2-3)

        $Mn = Fcr \, Sx <= Mp$

        $Fcr = [Cb \, \pi^2 \, E / (Lb/r_{ts})^2] \cdot [1 + 0.078 \, (Jc/(Sx.ho)) \cdot (Lb/r_{ts})^2]^{0.5}$      (AISC 360-16 Eqn F2-4)

           Cb =      1               Jc =      342000      mm4
           E =      200000 Mpa      Sx =      442451      mm3
           Lb =      2000 mm      ho =      3180      mm

# Steel Structures

$r_{ts}$ = radius of gyration of compression flange plus one sixth of the web
Iyc = 13x 81^3 / 12 + (305/6) x 13^3 / 12 = 585034 mm4
Ayc = 81 x 13 + (305/6) x 13= 1713.8 mm2
$r_{ts}$ = ( Iyc/Ayc )^0.5 = 18 mm     5.291

Fcr = [1 x 3.14^2 x 200000 / (2000 / 18)^2] x [1 + 0.078(342000 / (442451 + 3180) x (2000/18)^2]^0.5
186 Mpa
Now,  **Mn** = Fcr.Sx = 186 x 442451/1000000 = 82 kNm.
and Mp = 220 kNm.
So, the effective, **Mn** = 82 kNm

| Lb = | 2.0 M | Lp = | 0.805 M | Lr = | 2.388 M | span = | 6 M |
|---|---|---|---|---|---|---|---|

In this example of beam section, **Mn** = **134 Mpa.**   Lb > Lp

## Nominal Shear strength, Vn

Vn = 0.6Fy Aw Cv1     (AISC 360-16 Eqn G2-1)

*To find out web shear coefficient, Cv1*

(a) For webs of rolled I shapes
h/tw <= 2.24 (E/Fy)^0.5     Cv1 = 1     (AISC 360-16 Eqn G2-2)
<= 54

(b) For all other I-shaped members and Channels

(i) when,   h/tw <=1.10 (kv E/Fy)^0.5   Cv1 = 1   (AISC 360-16 Eqn G2-3)

(ii) when,  h/tw > 1.10 (kv E / Fy)^0.5     (AISC 360-16 Eqn G2-4)

Cv1 = 1.10 (kv E / Fy)^0.5 / (h/tw)

where, Aw = Area of web, h = clear dist between flanges and tw = web thickness.

kv = shear buckling co-efficient     (AISC 360-16 Eqn G2-5)

(i)   for web without stiffeners ,     kv = 5.34

*For this design section*
a = 0 mm        h/tw = 305 / 13 = **23.462**
h = 305 mm      a/h = 0.00

| | | | | | |
|---|---|---|---|---|---|
| tw = | 13 mm | kv = | | | 5.34 |
| Fy = | 345 Mpa | 1.10 (kv E/Fy)^0.5 = | | | |
| E = | 200000 Mpa | 1.1(5.34 x 200000 / 345)^0.5 = | | | 61 |
| Aw=h.tw = | 3965 mm2 | Cvl = | | | 1.0 |

**Vn** = 0.6Fy Aw Cv1 = 0.6 x 345 x 3965 x 1 / 1000 =     **820** kN

## Strength design

| | | LRFD | | | ASD | | |
|---|---|---|---|---|---|---|---|
| | | φ = resistance factor = | 0.9 | | Ω = factor of safety = | 1.67 | |
| | | Mu < φ Mn | | | Mu <= Mn / Ω, | | |
| | | Vu < φ Vn | | | Vu <= Vn / Ω, | | |
| Load w = | | 1.2D + 1.6L = 1.2 x 6 + 1.6 x 3 = 12 | | kN/m | D + L = 6 + 3 = 9 | | kN/m |
| Span End | L = | 6 simply supported | | M | L = 6 simply supported | | M |
| Moment | Mu = | 12 x 6^2 / 8 **54** | | kNm kNm | Mu = 9 x 6^2 / 8 **41** | | kNm kNm |
| | Mn = φ Mn = | 134.49 **121** > Mu; Safe. | | kNm kNm | Mn = 134.49 Mn/ Ω = **81** > Mu; Safe. | | kNm kNm |
| Shear | Vu = | 12 x 6 / 2 **36** | | kN kN | Vu = 9 x 6 / 2 **27** | | kN kN |
| | Vn = φ Vn = | 820 **738** > Vu; Safe. | | kN kN | Vn = 820 Vn/ Ω = **491** > Vu; Safe. | | kN kN |

End

# Steel Structures

## Example 8.14.5 : Design of Plate girder with built up plated section (IS 800 -2007)

[ Beams with Equal flanges and Bent about the axis of max. strength x-x axis ]

### Reference.
    1.0 IS 800 : 2007   General construction in steel - Code of practice

### Input Data.

### Material

| IS 2062 | | | | | |
|---|---|---|---|---|---|
| Grade | E 250 (Fe 410 W) A | | | | |
| Yield Stress | $f_y$ | 250 | 240 | 230 | MPa |
| d or t | | <20 | 20-40 | >40 | |
| Tensile Stress | $f_u$ | 410 | 410 | 410 | MPa |
| Elasticity | E | | | 200000 | MPa |
| Mat. safety fac | $\gamma_{m0}$ | | | 1.10 | |

(IS 800:2007 - Table 1)

(IS 800:2007 - 2.2.4.1)
(IS 800:2007 - Table 5)

### Member Loading    *Unfactored*

| Load case | Description | Axial kN P | $M_{vertical}$ kN m $M_x$ | $M_{horz}$ kN m $M_y$ | $V_{vertical}$ kN $V_x$ | $V_{horz}$ kN $V_y$ | | Nature of load |
|---|---|---|---|---|---|---|---|---|
| 1 | DL + LL | | 405 | | 270 | | | |
| 2 | DL + LL + WL | | | | | | | |
| 3 | DL + LL + EL | | | | | | | |
| 4 | DL + WL | | | | | | | |
| 5 | DL + EL | | | | | | | |
| | | | | | | | | |
| | | | | | | | | |

### Methods of design
    1 Limit State design
    2 Working stress design

### Load, Load factors and Load combinations

Load description                          Load factors
D = Dead load                             As per IS : 800 2007
L = Live load
LLr = Roof live load                    Load Combination
WL = Wind load                         As per desgin basis reports and referenced codes of practice
EL = Earthquake load                ( IS 875, IS 800)

# Design of Industrial Structures

## Dimensioning

| | | | |
|---|---|---|---|
| EffectiveSpan | L = | 6 m | (IS 800:2007 - 8.1.1) |
| Spacing | s = | 6 m | |
| Laterally unsupported length | $L_{LT}$ = | 2 m | *(for lateral torsional buckling)* (IS 800:2007 - 8.3) |

*Floor load intensity:*

| | | | | | |
|---|---|---|---|---|---|
| DL = | 5 kN/m2 | $w_{DL}$ = | 30 kN/m | | |
| LL = | 10 kN/m2 | $w_{LL}$ = | 60 kN/m | | |
| | | w = | 90 kN/m | | |

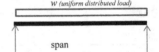

W (uniform distributed load)

span

### Design Load        Unfactored

| Load Case | P | Mx | My | Vx | Vy |
|---|---|---|---|---|---|
| | kN | kNm | kNm | kN | kN |
| 1 | | 405 | | 270 | |

Mx = 90 x 6^2/8 = 405 kNm.
Vx = 90 x 6/2 = 270 kN.

### Section properties:

| PLATED SECTION | | | |
|---|---|---|---|
| FLANGE PL | 200 | mm. | b |
| | 20 | mm. | tf |
| WEB PL | 710 | mm. | d |
| | 12 | mm. | tw |
| | | | |
| WT/M | 129.7 | | kg. |
| Area | 16520 | | mm2. |
| Ix | 1423977667 | | mm4. |
| Iy | 26768907 | | mm4. |
| rx | 294 | | mm. |
| ry | 40 | | mm. |
| tw | 12 | | mm |
| Zx (Ze) | 3797274 | | mm3 |
| J (It)  Σ bt3/3 | 1475627 | | mm4 |
| $h_f$   d+tf | 730 | | mm |
| for welded section | $\alpha_{LT}$ = | 0.49 | (IS 800:2007 - 8.2.2) |
| for Rolled section | | 0.21 | |

| Web stiffener on b/s of web | | | |
|---|---|---|---|
| Thick | tq | 8 | mm |
| wide | tb | 75 | mm |
| spcg | c | 900 | mm |

Note:   Shear buckling check not required.

### Deflection:  (IS 800:2007 - 5.6.1)

$\delta = 5 w L^4 / 384 EI$ =   5 x 90 x 6000 ^4 / 384 x 200000 x 1423977667 =   5.33 mm
Allowable deflection limit L /300 =   20.00 mm   OK.

### Proportioning limit  (IS 800:2007 - Table 2)

$\varepsilon = (250 / fy )^{0.5}$ =   $(250 / 240)^{0.5}$ =   1.02

# Steel Structures

## Check for flange

**0.5 b / tf =**     100 / 20 =     **5**     Compact     (IS 800:2007 - 3.7.2)
*Limiting width to thickness ratio:*
Plastic, $\lambda r =$            $8.4\varepsilon = 8.4 \times 1.02 =$     9
Compact, $\lambda p =$       $9.4\varepsilon = 9.4 \times 1.02 =$     10
Semi compact, $\lambda sp =$    $13.6\varepsilon = 13.6 \times 1.02 =$     14

## Check for web

**d / tw =**     710 / 12 =     **59**     Plastic.     (IS 800:2007 - 3.7.2)
*Limiting depth to thickness ratio:*
Plastic, $\lambda r =$            $84\varepsilon = 84 \times 1.02 =$     86
Compact, $\lambda p =$       $105\varepsilon = 105 \times 1.02 =$     107
Semi compact, $\lambda sp =$    $126\varepsilon = 126 \times 1.02 =$     129

## *Shear buckling check*     (IS 800:2007 - 8.4.2)

Resistance to shear buckling to be checked,
(i)     when    d / tw >     67 $\varepsilon w$     for web without stiffeners
                                    = 67 x 1.02 =     **68.34**

(ii)    when    d / tw >     67 $\varepsilon w(Kv/5.35)^{0.5}$     for web with stiffeners
                                    = 67 x $1.02(7.839/5.35)^{0.5}$
                                    = **79.99**
where Kv = $5.35 + 4/(c/d)^2$    for c/d >= 1
c = spcg. of stiffener and d is depth of web.     c/d =    900 / 710 =     **1.27**
so, let us assume, Kv =    $5.35 + 4 / (900/710)^2 =$     7.8394
OR,    Kv = $4 + 5.35/(c/d)^2$ for c/d < 1
          Kv =    $4 + 5.35 / (900/710)^2 =$     7.33
                                       **Kv =**    **7.8394**

In this case,     Shear buckling check not required.

## Design Shear strength of web , Vd     (IS 800:2007 - 8.4)

The nominal shear strength of web, Vn is goverered by plastic shear resistance.

Vn =    Vp = Av fyw / $(3)^{0.5}$     (IS 800:2007 - 8.4.1)

Shear Area, Av = d x tw =     710 x 12 =     8520 mm2
Yield strength of web, $f_{yw}$ = fy =                250 Mpa

Vn = Vp =     8520 x 250 $/3^{0.5}/1000 =$     1230 kN.

Vd = Vn / $\gamma_{m0}$     (IS 800:2007 - 8.4)

**Vd =**     1230 / 1.1 =     **1118** kN

## Design flexural strength, Md  (IS 800:2007 - 8.2)

Md shall be the lower value obtained according to the limit state of yielding (plastic moment) and lateral torsional buckling.

### a) Laterally supported beam ( *web not susceptible for shear buckling; design shear ≤ 0.6Vd* )

Nominal flexural strength, $M_n = M_p = f_y Z_p$ = Plastic moment

$f_y$ = Yield stress = 240 Mpa
$Z_p$ = plastic sectional modulus about the x axis = $\Sigma A y$
  = 2 x 200 x 20 x 730/ 2 + 2 x (710/2) x 12 x (710/4)
  = 4432300   mm$^3$

$M_n$ = 240 x 4432300/1000000 = 1064 kNm.  (= $M_p$)

Design flexural strength = $M_d = \beta \cdot M_n / \gamma_{mo}$ < 1.2 Ze $f_y /\gamma_{mo}$   (IS 800:2007 - 8.2.1.2)

$M_d$ = 1 x 1064 / 1.1 = 967 kNm.  where, $\beta = 1$ for plastic and compact section

1.2 Ze $f_y /\gamma_{mo}$ = 1.2 x 3797274 x 240 / 1.1/1000000 = 994.2 kNm  > Md; Okay.

Design flextural strength, **Md** = 967 kNm

### b) Lateraly unsupported beam ( *susceptible to Lateral-torsional buckling* )

Design flexural strength = $M_d = \beta \cdot Z_p \cdot f_{bd}$   (IS 800:2007 - 8.2.2)

The expressions are

$f_{bd} = \chi_{LT} f_y / \gamma_{mo}$

$\chi_{LT}$ = $1 / [\phi_{LT} + (\phi_{LT}^2 - \lambda_{LT}^2)^{0.5}]$  <= 1

$\phi_{LT} = 0.5 [1 + \alpha_{LT} (\lambda_{LT} - 0.2) + \lambda_{LT}^2]$

$\lambda_{LT} = (\beta Z_p f_y / M_{cr})^{0.5}$  <= $(1.2 Z_e f_y / M_{cr})^{0.5}$

$M_{cr} = (\pi^2 E I_y h_f / 2 L_{LT}^2) \cdot [1 + (1/20) \cdot ((L_{LT} / r_y) / (h_f/t_f))^2]^{0.5}$

# Steel Structures

Now, putting following values into above equation

| | | | | | | | |
|---|---|---|---|---|---|---|---|
| hf = | 730 mm | Iy = | 26768907 | mm4 | | $L_{LT}$ = | 2000 mm |
| tf = | 20 mm | E = | 200000 | Mpa | | | |
| ry = | 40 mm | | | | | | |
| β = | 1 | $α_{LT}$ | 0.49 | | | (IS 800:2007 - 8.2.2) | |
| fy = | 240 Mpa | Zp = | 4432300 | mm3 | | | |
| $γ_{m0}$ = | 1.10 | Ze = | 3797274 | mm3 | | | |

we get,

$Mcr = ( π^2 EIy\ hf / 2\ L_{LT}^2)\ .\ [\ 1 + (1/20)\ .\ ((L_{LT}\ /\ ry)\ /\ (hf/tf))^2\ ]^{0.5}$

$π^2\ EIy\ hf / 2\ L_{LT}^2 =$   3.14^2 x 200000 x 26768907 x 730 / 2 x 2000^2 =     4816735557

$((L_{LT}\ /\ ry) / (hf/tf))^2 =$     ((2000/40) / (730/20) )^2 =     1.88

Mcr =   4816735557 x [1 + (1/20) x 1.88]^0.5 =     5038038301 Nmm

$λ_{LT} = (β\ Zp\ fy / Mcr)^{0.5}$  <=  $(1.2\ Ze\ fy / Mcr)^{0.5}$

$(β\ Zp\ fy / Mcr)^{0.5}$ =   (1 x 4432300 x 240 / 5038038301)^0.5 =     0.4595

$(1.2\ Ze\ fy / Mcr)^{0.5}$ =   (1.2 x 3797274 x 240 / 5038038301)^0.5=     0.4659

$λ_{LT}$ =   0.4595  =     **0.46**

$\phi_{LT} = 0.5 [ 1 + \alpha_{LT} (\lambda_{LT} - 0.2) + \lambda_{LT}2] \quad = 0.5[1+0.49(0.46-0.2)+0.46^2] = \quad 0.6695$

$\chi_{LT} = 1 / [ \phi_{LT} + (\phi_{LT}2 - \lambda_{LT}2)^{0.5} ] <= 1$
$\phantom{\chi_{LT}} = 1 / [0.6695 + (0.6695^2 - 0.46^2)^{0.5}]$
$\phantom{\chi_{LT}} = 0.865$

$f_{bd} = \chi_{LT} \, f_y / \gamma_{mo} \quad = 0.865 \times 240 / 1.1 = \quad$ 189 Mpa

Design flexural strength = $M_d = \beta \cdot Z_p \cdot f_{bd} = \quad 1 \times 4432300 \times 189/10^6 = \quad$ 838 kNm

*If the value of $\lambda_{LT}$ is less than **0.4**, Lateral torsional buckling need not be checked and member shall be treated as Laterally supported beam (IS 800:2007 8.2.1.2)*

Hence, the Design flexural strength will be **$M_d$ = 838 kNm**

### Intermediate web stiffeners         Web stiffener on b/s of web         (IS 800:2007 - 8.7.2)

| size | wide | tb | 75 mm | $\varepsilon = (250/f_y)^{0.5} =$ | 1.02 |
|---|---|---|---|---|---|
|  | thick | tq | 8 mm | depth of web, d = | 710 mm |
|  | spcg | c | 900 mm | web thickness, tw = | 12 mm |

design     Max allowable width = 20.tq.ε         (IS 800:2007 - 8.7.1.2)
$\phantom{design \quad Max allowable width}= 20 \times 8 \times 1.02$
$\phantom{design \quad Max allowable width}= 163$ mm     **tb < allowable limit; OK.**

Moment of area required, $Is_{rqd} >= 1.5 \, d^3 \, tw^3 / c^2$         (IS 800:2007 - 8.7.2.4)

$Is_{rqd} = 1.5 \times 710^3 \times 12^3 / 900^2 = \quad$ 1145315     mm4

Is provided = $8 \times (2 \times 75)^3 / 12 = \quad$ 2250000     mm4

**Is prov > Is reqd.; OK.**

# Steel Structures

## Strength design

| | Limit state design<br>(IS 800:2007 - Section 8) | Working stress design<br>(IS 800:2007 - Section 11) |
|---|---|---|
| | Factored design moment = M<br>Design flexural strength of the section = Md<br>Factored design shear = V<br>Design shear strength of the section = Vd | Actual design moment = Ms<br>Permissible flexural strength = Ma<br>Actual design shear = Vs<br>Permissible shear strength = Va |
| | M <= Md<br>V <= Vd | Ms <= Ma<br>Vs <= Va |
| Load<br>w = | 1.5D + 1.5L<br>= 1.5 x 30 + 1.5 x 60 =     135 kN/m | D + L<br>= 30 + 60 =     90 kN/m |
| Span<br>End | L =     6 M<br>simply supported | L =     6 M<br>simply supported |
| Moment | M = 135 x 6^2 / 8 =     608     kNm<br>Md =           838     kNm<br>**Md > M; Safe.** | Ms =     90 x 6^2 / 8 =     **405**     kNm<br><br>_Allowable flexural strength, Ma :_     (IS 800-11.4.1)<br>a) For laterally supported beam:<br>fabc = fabt = Ma/Zx     (Zx top = Zx bot)<br>Ma = fabc. Zx = 0.66 fy Zx     (fabc = fabt)<br>     =     0.66x240x3797274/10^6<br>     =     601.49 kNm<br><br>b) _For laterally Unsupported beam:_<br>fabc = fabt = 0.6 Md/Zx     (Zx top = Zx bot)<br>Md =     838 knM<br>Ma = fabc. Zx = (0.6Md/Zx) Zx     (fabc = fabt)<br>     =     0.6 Md<br>     =     0.6 x 838 =     502.8 kNm<br><br>Allowable Flexural mom, **Ma** =     **503** kNm.<br>Actual Design Moment, Ms =     405 kNm.<br>**Ma > Ms; Safe.** |
| Shear | V =     135 x 6 / 2 =     405 kN<br>Vd =           1118 kN<br>**Vd > V; Safe.** | Vs =     90 x 6 / 2 =     **270** kN<br><br>_Shear strength, Va :_     (IS 800-11.4.2)<br>a) Pure shear<br>Va = $\tau_{ab}$ . Av ,     where Av shear area<br>     =     0.4 fy x Av<br>     =     0.4 x 250 x 710 x 12/1000<br>     =     852 kN<br>b) when subject to shear buckling<br>Va = $\tau_{ab}$ . Av ,     where Av shear area<br>     =     (0.7 Vd / Av) x Av<br>     =     0.7 Vn     [contd] |

|  | |  |
|---|---|---|
|  |  | = 0.7 x 1118 |
|  |  | = 783 kN |
|  |  | Allowable Shear, **Va** = 783 kNm. |
|  |  | Actual design shear, Vs = 270 kNm. |
|  |  | **Va > Vs; Safe.** |
| weld betn flange & web | Required Shear at joint, F = V Q / Ix | Required Shear, F = Vu Q/Ix |
|  | Q = A y =   200x20x365= 1460000 mm3 | Q = A y =   1460000 mm3 |
|  | Ix =   1423977667  mm4 | Ix =   1423977667  mm4 |
|  | F =   405000 x 1460000 / 1423977667 | F =   270000 x 1460000 / 1423977667 |
|  | =   415 N/mm | =   277 N/mm |
|  | Provide Fillet weld on both sides of web | Provide Fillet weld on both sides of web |
|  | Force per mm run =   415/2 =   207.5 N | Force per mm run =   277/2 =   139 N |
|  |                                 < w; OK. |                                 < w; OK. |
|  | Capacity of   6 mm fillet weld, w : | Capacity of   6 mm fillet weld, w : |
|  | w =0.7x0.6x1100=   462 kg/cm =   462 N/mm | w =0.7x0.6x1100=   462 kg/cm =   462 N/mm |
|  | weld stress = 110( kg/cm2 | weld stress = 110( kg/cm2 |

End

# Steel Structures

***Example 8.14.6:*** ***Design of Column with axial load and bending about major axes. (IS 800 - 2007)***
[ Column with Equal flanges and Bent about the axis of max. strength z-z axis ]

**Reference.**
    1.0 IS 800 : 2007   General construction in steel - Code of practice

**Input Data.**

**Material**

| IS 2062 | | | | | | |
|---|---|---|---|---|---|---|
| Grade | | E 250 (Fe 410 W) A | | | | |
| Yield Stress | $f_y$ | 250 | 240 | 230 | MPa | (IS 800:2007 - Table 1) |
| d or t | | <20 | 20-40 | >40 | | |
| Tensile Stress | $f_u$ | 410 | 410 | 410 | MPa | |
| Elasticity | E | | | 200000 | MPa | (IS 800:2007 - 2.2.4.1) |
| Mat. safety fac | $\gamma_{m0}$ | | | 1.10 | | (IS 800:2007 - Table 5) |

**Member Loading**       ***Unfactored load***

| Load case | Description | Axial kN P | $M_{vertical}$ kN m $M_z$ | $M_{horz}$ kN m $M_y$ | $V_{vertical}$ kN $V_z$ | $V_{horz}$ kN $V_y$ | | Nature of load |
|---|---|---|---|---|---|---|---|---|
| 1 | DL + LL | 469 | 133 | 5 | | 40 | | |
| 2 | DL + LL + WL | | | | | | | |
| 3 | DL + LL + EL | | | | | | | |
| 4 | DL + WL | | | | | | | |
| 5 | DL + EL | | | | | | | |
| | | | | | | | | |
| | | | | | | | | |

**Methods of design**
    1  Limit State design
    2  Working stress design

**Load, Load factors and Load combinations**

Load description
D = Dead load
L = Live load
LLr = Roof live load
WL = Wind load
EL = Earthquake load

Load factors
As per IS : 800 2007

Load Combination
As per design basis reports and referenced codes of practice
( IS 875, IS 800)

## Dimensions :

### Effective Lengths, KL :
(IS 800:2007 - Table 11)

| | | | | | | |
|---|---|---|---|---|---|---|
| Laterally Unsupported length | Lz = | 6 M | Kz = | 0.8 | KLz = | 4.8 M |
| | Ly = | 3 M | Ky = | 1 | Kly = | 3.00 M |

Effective length for lateral torsional buckling     $L_{LT}$ =   3.9 M    (assumed 0.65L)

### Section properties:

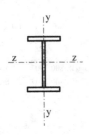

| PLATED SECTION | | | |
|---|---|---|---|
| FLANGE PL | 250 | mm. | bf |
| | 20 | mm. | tf |
| WEB PL | 450 | mm. | d |
| | 12 | mm. | tw |
| WT/M | | 120.9 | kg. |
| Ae (Ag) | | 15400 | mm2. |
| Iz (Ix) | | 643708333 | mm4. |
| Iy | | 52148133 | mm4. |
| rz (rx) | | 204 | mm. |
| ry | | 58 | mm. |
| tw | | 12.00 | mm |
| Zz (Ze) | | 2627381 | mm3 |
| Zy | | 427467 | mm3 |
| J | S bt3/3 | ###### | mm4 |
| $h_f$ | d+tf | 470 | mm |
| Buckling abt z-z | b | α = | 0.34 | (IS 800:2007 - Table 7) |
| Buckling abt y-y | c | α = | 0.49 | |
| for welded section | | $α_{LT}$ = | 0.49 | (IS 800:2007 - 8.2.2) |
| for Rolled section | | | 0.21 | |

### Buckling class
     tf <= 40     Buckling class =   b    for buckling about major axis    (IS 800:2007 - Table 10)
               Imperfaction factor, α =   0.34                                    (IS 800:2007 - Table 7)
                       Buckling class =   c    for buckling about minor axis
               Imperfaction factor, α =   0.49

### Proportioning limit
(IS 800:2007 - Table 2)

     ε = (250 / fy )^0.5 =    (250 / 240)^0.5=        1.02

# Steel Structures

### Check for flange
$0.5b / tf =$  $\quad 125 / 20 =$  $\boxed{6.25}$  $\quad$ Semi compact $\quad$ (IS 800:2007 - 3.7.2)

*Limiting width to thickness ratio:*
Plastic, $\lambda r =$ $\quad\quad 8.4\varepsilon = 8.4 \times 1.02 =$ $\quad\quad$ 9
Compact, $\lambda p =$ $\quad\quad 9.4e = 9.4 \times 1.02 =$ $\quad\quad$ 10
Semi compact, $\lambda sp =$ $\quad 13.6e = 13.6 \times 1.02 =$ $\quad\quad$ 14

### Check for web

$d / tw =$ $\quad 450 / 12 =$ $\quad \boxed{38}$ $\quad$ Plastic. $\quad$ (IS 800:2007 - 3.7.2)

*Limiting depth to thickness ratio:*
Plastic, $\lambda r =$ $\quad\quad\quad 84e = 84 \times 1.02 =$ $\quad\quad$ 86
Compact, $\lambda p =$ $\quad\quad 105\varepsilon = 105 \times 1.02 =$ $\quad\quad$ 107
Semi compact, $\lambda sp =$ $\quad 126\varepsilon = 126 \times 1.02 =$ $\quad$ 129

### *Slenderness ratio of column, KL/r :*

$KLz / rz =$ $\quad 4800 / 204 =$ $\quad$ 23
$KLy / ry =$ $\quad 3000 / 58 =$ $\quad$ 52
$K_{LT} / r_{min} =$ $\quad 3900 / 58 =$ $\quad$ 67 $\quad [KL/r]_{max} =$ $\quad$ **67** $\quad$ OK

Maximum permissible value of $KL/r =$ $\quad$ 180 for DL+LL $\quad$ (IS 800:2007 - Table 3)
$\quad\quad\quad\quad\quad\quad\quad\quad\quad\quad\quad\quad\quad\quad$ 250 for DL+LL+WL/EL

## **Sectional capacity / Design strength:**

### *(i) Compression due to yielding and bending about major axis* $\quad\quad$ (IS 800:2007 - 7.1)

Design compression strength, $Pd = Ae \cdot fcd$ $\quad\quad$ where, $Ae =$ effective sectional area
$\quad\quad\quad\quad\quad\quad\quad\quad\quad\quad\quad\quad\quad\quad\quad\quad\quad\quad\quad\quad fcd =$ design compressive stress

$$fcd = \chi\, fy / \gamma_{mo} \quad <= fy / \gamma_{mo} \quad\quad \text{(IS 800:2007 7.1.2.1)}$$

$$\chi = 1 / [\, \phi + (\phi^2 - \lambda^2)\wedge 0.5\,]$$

$$\phi = 0.5\, [\, 1 + \alpha(\lambda - 0.2) + \lambda^2\,]$$

$$\lambda = (fy / fcc)\wedge 0.5$$

$$fcc = \pi^2 E / (KL/r)\wedge 2 \quad\quad \text{Euler buckling stress}$$

Now, putting following values into above equation

$\alpha =$ $\quad$ 0.34 $\quad\quad E =$ $\quad$ 200000 Mpa $\quad\quad fy =$ $\quad$ 240 Mpa
$\gamma_{mo} =$ $\quad$ 1.10 $\quad\quad KL/r =$ $\quad$ 67 $\quad\quad\quad\quad\quad\quad Ae =$ $\quad$ 15400 mm2

| | | | | |
|---|---|---|---|---|
| fcc = | 3.14^2 x 200000 / 67^2 = | 439 MPa | | |
| λ = | (240 / 439)^0.5= | 0.74 | | |
| φ = | 0.5[1 + 0.34 x (0.74 - 0.2) + 0.74^2= | 0.87 | | |
| χ = | 1 /[0.87 + (0.87^2 - 0.74^2)^0.5]= | 0.753 | | |
| fcd = | 0.753 x 240 / 1.1 = | 164 Mpa | <= fy / $\gamma_{mo}$ | <=  240 / 1.1 |
| | | | | <=  218 Mpa |

fcd =     **164** Mpa

**Pd** = Ae . fcd =     15400 x 164 /1000=     **2526 kN**

*(ii) Compression due to buckling about major axis, z-z*

Design compression strength, Pdz = Ae. fcd     where, Ae = effective sectional area
                                                                              fcd = design compressive stress

$$fcd = \chi\, fy / \gamma_{mo} \quad <= fy / \gamma_{mo} \quad\quad (IS\ 800:2007\ 7.1.2.1)$$
$$\chi = 1 / [\,\phi + (\phi^2 - \lambda^2)^{0.5}\,]$$
$$\phi = 0.5\,[\,1 + \alpha(\lambda - 0.2) + \lambda^2\,]$$
$$\lambda = (fy / fcc)^{0.5}$$

$$fcc = \pi^2 E / (KLz/rz)^2 \quad\quad \text{Euler buckling stress}$$

Now, putting following values into above equation

| | | | | | |
|---|---|---|---|---|---|
| α = | 0.34 | E = | 200000 Mpa | fy = | 240 Mpa |
| $\gamma_{m0}$ = | 1.10 | KLz/rz | 23 | Ae = | 15400 mm2 |

| | | | | |
|---|---|---|---|---|
| fcc = | 3.14^2 x 200000 / 23^2 = | 3728 MPa | | |
| λ = | (240 / 3728)^0.5= | 0.25 | | |
| φ = | 0.5[1 + 0.34 x (0.25 - 0.2) + 0.25^2= | 0.54 | | |
| χ = | 1 /[0.54 + (0.54^2 - 0.25^2)^0.5]= | 0.982 | | |
| fcd = | 0.982 x 240 / 1.1 = | 214 Mpa | <= fy / $\gamma_{mo}$ | <=  240 / 1.1 |
| | | | | <=  218 Mpa |

fcd =     **214** Mpa     (=fcdz)

**Pdz** = Ae . fcd =     15400 x 214 /1000=     **3296 kN**

*(iii) Compression due to buckling about minor axis, y-y*

Design compression strength, Pdy = Ae. fcd     where, Ae = effective sectional area
                                                                              fcd = design compressive stress

# Steel Structures

$$f_{cd} = \chi f_y / \gamma_{mo} \quad \leq f_y / \gamma_{mo} \quad \text{(IS 800:2007 7.1.2.1)}$$
$$\chi = 1 / [\phi + (\phi^2 - \lambda^2)^{0.5}]$$
$$\phi = 0.5 [1 + \alpha(\lambda - 0.2) + \lambda^2]$$
$$\lambda = (f_y / f_{cc})^{0.5}$$

$$f_{cc} = \pi^2 E / (KL_z/r_z)^2 \quad \text{Euler buckling stress}$$

Now, putting following values into above equation

| | | | | | | |
|---|---|---|---|---|---|---|
| $\alpha =$ | 0.49 | $E =$ | 200000 Mpa | $f_y =$ | 240 Mpa | |
| $\gamma_{m0} =$ | 1.10 | $KL_y/r_y$ | 52 | $A_e =$ | 15400 mm2 | |

$f_{cc} =$  $3.14^2 \times 200000 / 52^2 =$   729 MPa
$\lambda =$  $(240 / 729)^{0.5} =$   0.57
$\phi =$  $0.5[1 + 0.49 \times (0.57 - 0.2) + 0.57^2] =$   0.75
$\chi =$  $1 / [0.75 + (0.75^2 - 0.57^2)^{0.5}] =$   0.808

$f_{cd} =$  $0.808 \times 240 / 1.1 =$   176 Mpa   $\leq f_y / \gamma_{mo} \leq$   240 / 1.1
                                                    $\leq$   218 Mpa
$f_{cd} =$   176 Mpa   $(= f_{cdy})$

**Pdy** $= A_e \cdot f_{cd} =$   $15400 \times 176 / 1000 =$   **2710 kN**

### *(iv) Flexural strength due to buckling about major axis, z-z*

Design flexural strength about major axis = $M_{dz}$   (IS 800:2007 - 8.2)

$M_{dz}$ shall be the lower value obtained according to the limit state of yielding (plastic moment) and lateral torsional buckling.

### a) Laterally supported beam ( *web not susceptible for shear buckling; design shear < 0.6Vd* )

Nominal flexural strength, $M_{nz} = M_{pz} = f_y Z_{pz} =$ Plastic moment

$f_y =$ Yield stress = 240 Mpa
$Z_p =$ plastic sectional modulus about the x axis $= \Sigma A \bar{y}$
  $= 2 \times 250 \times 20 \times 470/2 + 2 \times (450/2) \times 12 \times (450/4)$
  $= 2957500$ mm4

$M_{nz}$ = 240 x 2957500/1000000 = 710 kNm.    (= Mpz)

Design flexural strength = $M_{dz}$ = β . $M_{nz}$ / $\gamma_{mo}$ < 1.2 $Z_z$ $f_y$ /$\gamma_{mo}$    (IS 800:2007 - 8.2.1.2)

$M_{dz}$ = 1 x 710 / 1.1    =    645 kNm.    where, β = 1 for plastic and compact section

1.2 $Z_z$ $f_y$ /$\gamma_{mo}$ =    1.2 x 2627381 x 240 / 1.1/1000000 =    688 kNm    > Mdz; Okay.

Design flextural strength, **Md** =    **645** kNm

### b) Lateraly unsupported beam ( *susceptible to Lateral-torsional buckling* )

Design flexural strength = $M_{dz}$ = β . $Z_{pz}$ .$f_{bd}$    (IS 800:2007 - 8.2.2)

The expressions are
$f_{bd}$ = $\chi_{LT}$ $f_y$ / $\gamma_{mo}$
$\chi_{LT}$ = 1 / [ $\phi_{LT}$ + ($\phi_{LT}^2$ - $\lambda_{LT}^2$)^0.5] <= 1
$\phi_{LT}$ = 0.5 [ 1 + $\alpha_{LT}$ ($\lambda_{LT}$ - 0.2) + $\lambda_{LT}^2$]
$\lambda_{LT}$ = (β $Z_{pz}$ $f_y$ / $M_{cr}$)^0.5 <= (1.2 $Z_e$ $f_y$ / $M_{cr}$)^0.5
$M_{cr}$ = ( $\pi^2$ $EI_y$ hf / 2 $L_{LT}^2$) . [ 1 + (1/20) . (($L_{LT}$ / $r_y$) / (hf/tf))^2 ]^0.5

Now, putting following values into above equation

| | | | | | | | |
|---|---|---|---|---|---|---|---|
| hf = | 470 mm | $I_y$ = | 52148133 | mm4 | | $L_{LT}$ = | 3900 mm |
| tf = | 20 mm | E = | 200000 | Mpa | | | |
| $r_y$ = | 58 mm | | | | | | |

| | | | | | |
|---|---|---|---|---|---|
| β = | 1 | $\alpha_{LT}$ | 0.49 | | (IS 800:2007 - 8.2.2) |
| $f_y$ = | 240 Mpa | $Z_p$ = | 2957500 | mm3 | |
| $\gamma_{mo}$ = | 1.10 | $Z_z$ = | 2627381 | mm3 | |

we get,
$M_{cr}$ = ( $\pi^2$ $EI_y$ hf / 2 $L_{LT}^2$) . [ 1 + (1/20) . (($L_{LT}$ / $r_y$) / (hf/tf))^2 ]^0.5

$\pi^2$ $EI_y$ hf / 2 $L_{LT}^2$ =    3.14^2 x 200000 x 52148133 x 470 / 2 x 3900^2 =    1588790757

(($L_{LT}$ / $r_y$) / (hf/tf))^2 =    ((3900/58) / (470/20) )^2 =    8.19

# Steel Structures

$M_{cr} = 1588790757 \times [1 + (1/20) \times 8.19]^{0.5} = \quad 1886249965 \text{ Nmm}$

$\lambda_{LT} = (\beta Z_p f_y / M_{cr})^{0.5} <= (1.2 Z_e f_y / M_{cr})^{0.5}$

$(\beta Z_p f_y / M_{cr})^{0.5} = (1 \times 2957500 \times 240 / 1886249965)^{0.5} = \quad 0.6134$

$(1.2 Z_e f_y / M_{cr})^{0.5} = (1.2 \times 2627381 \times 240 / 1886249965)^{0.5} = \quad 0.6334$

$\lambda_{LT} = 0.6134 = \mathbf{0.61}$

$\phi_{LT} = 0.5 [ 1 + \alpha_{LT} (\lambda_{LT} - 0.2) + \lambda_{LT}^2 ] = 0.5[1+0.49(0.61-0.2)+0.61^2] = \quad 0.7865$

$\chi_{LT} = 1 / [ \phi_{LT} + (\phi_{LT}^2 - \lambda_{LT}^2)^{0.5} ] <= 1$
$\quad = 1 / [0.7865 + (0.7865^2 - 0.61^2)^{0.5}]$
$\quad = 0.779$

$f_{bd} = \chi_{LT} f_y / \gamma_{mo} \quad = 0.779 \times 240 / 1.1 = \quad 170 \text{ Mpa}$

Design flexural strength = $M_{dz} = \beta . Z_{pz} . f_{bd} = \quad 1 \times 2957500 \times 170/10^6 = \quad$ **503 kNm**

*If the value of $\lambda_{LT}$ is less than 0.4, Lateral torsional buckling need not be checked and member shall be treated as Laterally supported beam (IS 800:2007 8.2.1.2)*

Hence, the Design flexural strength will be $\quad \mathbf{M_{dz} = \quad 503}$ kNm

## (v) Flexural stength due to buckling about minor axis, y-y

Design flexural strength about minor axis = Mdy $\quad$ (IS 800:2007 - 8.2)

Lateral torsional buckling need NOT be considered, when bending is about minor axis of the section. (IS 800:2007 - 8.2.2)

### a) Laterally supported beam ( *web not susceptible for shear buckling; design shear < 0.6Vd* )

Nominal flexural strength, $M_{ny} = M_{py} = f_y Z_{py}$ = Plastic moment

$f_y$ = Yield stress = $\quad$ 240 Mpa
$Z_p$ = plastic sectional modulus about the x axis = $\Sigma A \bar{y}$
$\quad = \quad 4 \times (250/2) \times 20 \times (250/4)$
$\quad = \quad 625000 \quad mm^4$

$M_{ny} = 240 \times 625000/1000000 = \quad 150$ kNm. $\quad (= M_{py})$

Design flexural strength = $M_{dy} = \beta \cdot M_{ny} / \gamma_{mo}$ < $1.2 Z_y f_y / \gamma_{mo}$ (IS 800:2007 - 8.2.1.2)

$M_{dy}$ = 1 x 150 / 1.1 = 136 kNm.  where, $\beta$ = 1 for plastic and compact section

$1.2 Z_y f_y / \gamma_{mo}$ = 1.2 x 427467 x 240 / 1.1/1000000 = 112 kNm -

Design flexural strength, $M_{dy}$ = 112 kNm

**b) Lateraly unsupported beam (** *susceptible to Lateral-torsional buckling* **)**

*Lateral torsional buckling need NOT be considered, when bending is about minor axis of the section.*
*(IS 800:2007 - 8.2.2)*

*(vi) Slenderness factors for overall member strength against buckling* (IS 800:2007 - 9.3.2.2)

$C_{mz}$ = 1  (IS 800:2007 Table 18)
$C_{my}$ = 1
$C_{mLT}$ = 1    [ 0.6 + 0.4 $\psi$ = 1 for $\psi$ = 1 ]

n = N/Nd       where N = applied compresssion and Nd = design strength in compresion.
ny = Py / Pdy =  469 / 2710 =   0.17
nz = Pz / Pdz =  469 / 3296 =   0.14     (IS 800:2007 - 9.3.2.2)

$\lambda$ = $(f_y / f_{cc})^{0.5}$
$f_{cc}$ = $\pi^2 E / (KL/r)^2$     Euler buckling stress     (IS 800:2007 - 7.1.2.1)

For $\lambda y$ :    Kly /ry =    52
          $f_{cc}$ =  3.14^2 x 200000 / 52^2 =       729 MPa
          $\lambda y$ = $(240 / 729)^{0.5}$=        0.57

For $\lambda z$ :    Klz/rz =    23
          $f_{cc}$ =  3.14^2 x 200000 / 23^2 =       3728 MPa
          $\lambda z$ = $(240 / 3728)^{0.5}$=       0.25

          $\lambda_{LT}$ =   0.61        (IS 800:2007 - 8.2.2)

Ky = 1 + ($\lambda y$ - 0.2)ny <= 1 + 0.8 ny
  = 1 + (0.57 - 0.2)0.17 <= 1 + 0.8 x 0.17
  =   1.063  <=    1.136
  =   1.063

Kz = 1 + ($\lambda z$ - 0.2)nz <= 1 + 0.8 nz
  = 1 + (0.25 - 0.2)0.14 <= 1 + 0.8 x 0.14
  =   1.007  <=    1.112

# Steel Structures

$$= 1.007$$

$$K_{LT} = 1 - [0.1 \lambda_{LT} \, ny / (C_{mLT} - 0.25)] \geq 1 - [0.1 \, ny / (C_{mLT} - 0.25)]$$
$$= 1 - [0.1 \times 0.61 \times 0.17 / (1 - 0.25)] \geq 1 - [0.1 \times 0.17 / (1 - 0.25)]$$
$$= 0.9862 \geq 0.9773$$
$$= 0.986$$

### Nominal Shear strength of web, Vn  (IS 800:2007 - 8.4)

The nominal shear strength of web, Vn is goverered by plastic shear resistance.

$$Vn = Vp = Av \, fyw / (3)^{0.5} \qquad \text{(IS 800:2007 - 8.4.1)}$$

Shear Area, $Av = d \times tw = 450 \times 12 = 5400 \text{ mm}^2$
Yield strength of web, $fyw = fy = 250 \text{ Mpa}$

$$Vn = Vp = 5400 \times 250 / 3^{0.5} / 1000 = 779 \text{ kN}.$$

### Design Shear strength of web, Vd

$$Vd = Vn / \gamma_{m0} \qquad \text{(IS 800:2007 - 8.4)}$$

$$\mathbf{Vd =} \; 779 / 1.1 = \mathbf{708 \text{ kN}}$$

### Strength design for combined axial force and flexure

| property | Limit state method | | | | | Working stress method | | | | |
|---|---|---|---|---|---|---|---|---|---|---|
| | Flange: Semi compact | | | | | Flange: Semi compact | | | | |
| | Web : Plastic. | | | | | Web : Plastic. | | | | |
| Load | DL + LL | Factored load | | | | DL + LL | Unfactored load | | | |
| | Axial | $M_{vertical}$ | $M_{horz}$ | $V_{vertical}$ | $V_{horz}$ | Axial | $M_{vertical}$ | $M_{horz}$ | $V_{vertical}$ | $V_{horz}$ |
| | P | $M_z$ | $M_y$ | $V_z$ | $V_y$ | P | $M_z$ | $M_y$ | $V_z$ | $V_y$ |
| | 469 | 133 | 5 | 0 | 40 | 469 | 133 | 5 | 0 | 40 |
| Load factor | 1.5 | 1.5 | 1.5 | 1.5 | 1.5 | | | | | |
| Factored load | 704 | 200 | 8 | 0 | 60 | 469 | 133 | 5 | 0 | 40 |
| | **Combined Axial compression and bending** | | | | | **Axial Comp.** | (IS 800:2007 - 11.3.1) | | | |
| | *Section Strength* (IS 800:2007 - 9.3.1) | | | | | | | | | |
| | N= 704 kN | Nd = | 2526 kN | | | Ps = 469 kN | Ae = | 15400 mm2 | | |

*[contd]*

| | |
|---|---|
| $M_y$ = 8 kNm  $M_{dy}$= 112 kNm<br>$M_z$ = 200 kNm  $M_{dz}$ = 503 kNm<br><br>$N/N_d + M_y/M_{dy} + M_z/M_{dz} \leq 1$<br><br>=704/2526 + 8/112 + 200/503<br>=0.28 + 0.07 +0.4 =   0.75<br>                      < 1 ; Safe.<br><br>_Overall member strength for bucking failure_<br>(IS 800:2007 - 9.3.2.2)<br><br>$P/P_{dy}+K_y C_{my} M_y/M_{dy} + K_{LT} M_z/M_{dz} \leq 1$<br><br>$P/P_{dz} + 0.6K_y C_{my} M_y/M_{dy} + K_z C_{mz}M_z/M_{dz} \leq 1$<br><br>P=         704 kN     $P_{dy}$ =    2710 kN<br>                     $P_{dz}$ =    3296 kN<br>$M_y$ =    8 kNm    $M_{dy}$ =   112 kNm<br>$M_z$ =    200 kNm  $M_{dz}$ =   503 kNm<br>$K_y$ =    1.063     $C_{my}$ =        1<br>$K_z$ =    1.007     $C_{mz}$ =        1<br>$K_{LT}$ =  0.986<br><br>$P/P_{dy}+K_y C_{my} M_y/M_{dy} + K_{LT} M_z/M_{dz}$<br>=704/2710 + 1.063x1x8/112 + 0.986x200/503<br>  =  0.2598  +   0.0759  +   0.392<br>  =  0.7278      < 1 ; Safe. | Actual comp stress = fc = Ps/Ae<br>                        =   469x1000/15400<br>                        =   30.45 Mpa<br>Permissible comp. stress, fac = 0.60 fcd<br>                        =   0.6 x 164<br>                        =   98 Mpa<br>                        fc<fac; Safe.<br>where, fcd =      164 Mpa ( see calc.above)<br><br>Bending.            (IS 800:2007 - 11.4.1)<br><br>$M_z$ =    133 kNm<br>$M_y$ =    5 kNm<br>$M_s$ =    $(133^2 + 5^2)^{0.5}$=      133.09 kNm<br>$Z_e$ =    2627381    mm3<br><br>Actual bending stress, fbc or fbt = Ms / Ze<br>                        =   133093952 / 2627381<br>                        =   50.66 Mpa<br><br>Permissible bending stress, fabc or fabt =<br>                        =   0.60 Md / Ze<br>                        =   0.6x503x10^6 / 2627381<br>                        =   114.87 Mpa   fbc<fabc; Safe.<br><br>where, Md =     503 kNm   (see calc.above)<br>       (Mdz) |
| $P/P_{dz}+0.6K_y C_{my} M_y/M_{dy} + K_z C_{mz}M_z/M_{dz} \leq 1$<br>=704/3296 + 0.6x1.063x1x8/112 + 1.007x1x200/503<br>  =  0.2136  +   0.0456  +   0.4004<br>  =  0.6595      < 1 ; Safe.<br><br>**Shear check**         (IS 800:2007 - 9.2.1)<br>Factored shear , V =    60 kN<br>Design shear strength, Vd =    708 kN<br>                            Safe.<br><br>60%of Vd =   425 kN<br>_V< 60% of Vd; No reductionin Moment capacity, Md._ | **Combined Axial compression and bending**<br>a) _Member stability requirement:_<br><br>$fc/f_{acy} + 0.6 K_y C_{my}f_{bcy}/ f_{abcy} + K_{LT} f_{bcz}/f_{abcz}$<br>                                   <= 1.0<br>and,<br>$fc/f_{acz} + 0.6 K_y C_{my}f_{bcy}/ f_{abcy} + K_z C_{mz} f_{bcz}/f_{abcz}$<br>                                   <= 1.0<br><br>fc =   30.45 Mpa    fy =      240 Mpa<br>                                       MPa<br>facy = 0.6 fcdy =    0.6x176 =       106<br>facz = 0.6 fcdz =    0.6x214 =       128<br>fbcy = My/Zy =    5000000/427467=   12 |

_[contd]_

$f_{abcy} = 0.6\ M_{dy}/Z_y =$   $0.6 \times 112 \times 10^6/427467 = 157$

$f_{bcz} = M_z/Z_z =$   $133000000/2627381 = 51$

$f_{abcz} = 0.6\ M_{dz}/Z_z =$   $0.6 \times 503 \times 10^6/2627381 = 115$

$K_y =$ 1.063   $C_{my} =$  1
$K_z =$ 1.007   $C_{mz} =$  1
$K_{LT} =$ 0.986

$f_c/f_{acy} + 0.6\ K_y\ C_{my} f_{bcy}/f_{abcy} + K_{LT}\ f_{bcz}/f_{abcz}$
$= 30.45/106 + 0.6 \times 1.063 \times 1 \times 12/157 + 0.986 \times 51 \times 1/115$
=   0.2873  +   0.0487   +   0.4373
=   0.7733    < 1 ; Safe.

and,

$f_c/f_{acz} + 0.6\ K_y\ C_{my} f_{bcy}/f_{abcy} + K_z\ C_{mz}\ f_{bcz}/f_{abcz}$
$= 30.45/128 + 0.6 \times 1.063 \times 1 \times 12/157 + 1.007 \times 1 \times 51/115$
=   0.2379  +   0.0487   +   0.4466
=   0.7332    < 1 ; Safe.

**b) Member strength requirement:**

$f_c/0.6 f_y + f_{bcy}/f_{abcy} + f_{bcz}/f_{abcz} \le 1.0$
$= 30.45/(0.6 \times 240) + 12/157 + 51/115$
=   0.2115  +   0.0764   +   0.4435
=   0.7314    < 1 ; Safe.

**Combined Axial compression, bending and shear**
(IS 800:2007 - 11.5.4) & ( IS 800 -1962  10.5.4)

Shear stress, $\tau_b = V_s/A_v$
=   40000/(450 \times 12)
=   7.41 Mpa

$f_e = (f_c^2 + f_{bc}^2 + f_c.f_{bc} + 3\tau_b^2)^{0.5} < 0.9 f_y$
$= (30.45^2 + 51^2 + 30.45 \times 51 + 3 \times 7.41^2)^{0.5}$
<   $0.9 \times 240$
=   72 Mpa   <   216 Mpa
Safe.

End

**Example 8.14.7 : Design of Axial loaded bracing members (Comp./Tension) - IS 800 - 2007.**

**8.14.7.1 Single angle members with Equal legs**

End connection: Welding or Bolting with a minimum of two bolts and loaded through one leg

**Reference.**
    1.0 IS 800 : 2007   General construction in steel - Code of practice

**Input Data.**

**Material**

| IS 2062 | | | | | | |
|---|---|---|---|---|---|---|
| Grade | E 250 (Fe 410 W) A | | | | | |
| Yield Stress | $f_y$ | 250 | 240 | 230 | MPa | (IS 800:2007 - Table 1) |
| d or t | | <20 | 20-40 | >40 | | |
| Tensile Stress | $f_u$ | 410 | 410 | 410 | MPa | |
| Elasticity | E | | | 200000 | MPa | (IS 800:2007 - 2.2.4.1) |
| Mat. safety fac | $\gamma_{m0}$ | 1.10 | $\gamma_{m1}$ | 1.25 | | (IS 800:2007 - Table 5) |

**Member Loading**      *Unfactored load*

| Load case | Description | Comp kN P | $M_{vertical}$ kN m $M_z$ | $M_{horz}$ kN m $M_y$ | $V_{vertical}$ kN $V_z$ | $V_{horz}$ kN $V_y$ | Tension KN T | Nature of load |
|---|---|---|---|---|---|---|---|---|
| 1 | DL + LL | | | | | | | |
| 2 | DL + LL + WL | 50 | | | | | 50 | |
| 3 | DL + LL + EL | | | | | | | |
| 4 | DL + WL | | | | | | | |
| 5 | DL + EL | | | | | | | |
| | | | | | | | | |

**Methods of design**
    1 Limit State design
    2 Working stress design

**Load, Load factors and Load combinations**

Load description                Load factors
D = Dead load                  As per IS : 800 2007
L = Live load
LLr = Roof live load          Load Combination
WL = Wind load             As per design basis reports and referenced codes of practice

# Steel Structures

EL = Earthquake load     ( IS 875, IS 800)

**Section properties:**     all dimensions are in mm

| Section Provided: | wt/m (kg) | $A_g$ (mm$^2$) | $r_x$ (mm) | $r_y$ (mm) | $r_{vv}$ (mm) | $I_x$ (mm$^4$) | $I_y$ (mm$^4$) | $C_{yy}$ (mm) |
|---|---|---|---|---|---|---|---|---|
| ISA 100 100 10 | 14.9 | 1903 | 30.50 | 30.5 | 19.4 | 1770000 | 1770000 | 28.4 |
| | Leg sizes (mm) | | | Connection bolt | | | | An |
| z — ⌐ z | b | d | t | nos | φdia | φhole | rows | mm2 |
| | 100 | 100 | 10 | 3 | 16 | 18 | 1 | 1723 |

where, An = net area after deduction of bolt holes.
 An =  Ag - (no of rows x φ hole x t)

**Buckling class**

     Buckling class =   c   for buckling about any axis  (IS 800:2007 - Table 10)
     Imperfection factor, α =   0.49          (IS 800:2007 - Table 7)

**Proportioning limit**                    (IS 800:2007 - Table 2)

  ε = (250 / fy )^0.5 =   (250 / 250)^0.5=    1.00

| For single angle (axial comp.) | Ratio | limit | | |
|---|---|---|---|---|
| b / t = 100 / 10 = | 10 | 15.7 ε = | 15.7 | Semi compact |
| d / t = 100 / 10 = | 10 | 15.7 e = | 15.7 | Semi compact |
| (b+d) / t = (100+100) / 10 = | 20 | 25 ε = | 25 | Semi compact |

**Effective Lengths (L) and Slenderness ratio ( L/r ) :**     (IS 800:2007 - Table 11)

| Laterally Unsupported length | Lx = | 4 M | Kx = | 0.85 | KLx = | 3.40 M |
|---|---|---|---|---|---|---|
| | Ly = | 2 M | Ky = | 1 | Kly = | 2.00 M |

 Maxm slenderness ratio (L/r) shall not exceed    250   (IS 800:2007 - Table 3)
 Here,  KLx/rx =  3400/30.5 =   111
    KLy/ry =  2000/30.5 =   66
    KLy/rvv =  2000/19.4 =   103
       L/r =   111   <250, OK.

**Sectional capacity / Design strength:**

**Axial compression**
*(i) Compression due to flexural-torsional buckling (Ly/rvv)*     (IS 800:2007 - 7.5.1.2)
 *( when loaded through one leg)*
Design compression strength, Pd = Ae. fcd

where, $A_e$ = effective sectional area
$f_{cd}$ = design compressive stress

$$f_{cd} = \chi f_y / \gamma_{mo} \le f_y / \gamma_{mo}$$
$$\chi = 1 / [\phi + (\phi^2 - \lambda^2)^{0.5}]$$
$$\phi = 0.5 [1 + \alpha(\lambda - 0.2) + \lambda^2]$$

$\lambda = \lambda e = (k1 + k2 \cdot \lambda_{vv}^2 + k3 \cdot \lambda\phi^2)^{0.5}$ where, $k1 = $ 0.7 (IS 800:2007 Table 12)
$k2 = $ 0.6 partially fixed at end
$k3 = $ 5

$\lambda_{vv} = $ $(L/r_{vv}) / [\epsilon (\pi^2 E /250)]$ $L = Ly = $ 2000 mm
= $(2000/19.4) / [1 \times (3.14^2 \times 200000 / 250)]$
= 0.013

$\lambda_\phi = $ $[(b+d)/2.t] / [\epsilon (\pi^2 E /250)]$
= $((100+100)/(2 \times 10)) / [1(3.14^2 \times 200000 / 250)]$
= 0.001

$\lambda = \lambda e = (k1 + k2 \cdot \lambda_{vv}^2 + k3 \cdot \lambda\phi^2)^{0.5}$
= $(0.7 + 0.6 \times 0.013^2 + 5 \times 0.001^2)^{0.5}$
= **0.837**

Now, putting following values into above equations

$\alpha = $ 0.49 $E = $ 200000 Mpa $f_y = $ 250 Mpa
$\gamma_{m0} = $ 1.10 $KLy/rvv = $ 103 $A_e = $ 1903 mm2

$\lambda = $ $\lambda e = $ 0.837
$\phi = $ $0.5[1 + 0.49 \times (0.837 - 0.2) + 0.837^2 = $ 1.01
$\chi = $ $1 / [1.01 + (1.01^2 - 0.837^2)^{0.5}] = $ 0.635
$f_{cd} = $ $0.635 \times 250 / 1.1 = $ 144 Mpa $\le f_y / \gamma_{mo}$ $\le$ 250 / 1.1
$\le$ 227 Mpa

$f_{cd} = $ **144** Mpa

**Pd** = $A_e \cdot f_{cd}$ = $1903 \times 144 / 1000 = $ **274** kN

<u>(ii) Compression due to buckling about major axis, z-z [Lz / rz (=rx)]</u> (IS 800:2007 7.5.1.1)
<u>(when loaded concentrically to its centroid)</u>
Design compression strength, Pdz = $A_e \cdot f_{cd}$ (IS 800:2007 7.1.2)
where, $A_e = $ effective sectional area
$f_{cd} = $ design compressive stress

# Steel Structures

$$fcd = \chi\, fy / \gamma_{mo} \quad <= fy / \gamma_{mo}$$
$$\chi = 1 / [\phi + (\phi^2 - \lambda^2)^{0.5}]$$
$$\phi = 0.5\,[1 + \alpha(\lambda - 0.2) + \lambda^2]$$
$$\lambda = (fy / fcc)^{0.5}$$

$$fcc = \pi^2 E / (KLz/rz)^2 \quad \text{Euler buckling stress}$$

Now, putting following values into above equation

| | | | | | |
|---|---|---|---|---|---|
| $\alpha =$ | 0.49 | $E =$ | 200000 Mpa | $fy =$ | 250 Mpa |
| $\gamma_{m0} =$ | 1.10 | KLz/rz | 111 | Ae = | 1903 mm2 |

$fcc =$ 3.14^2 x 200000 / 111^2 =      160 MPa
$\lambda =$ (250 / 160)^0.5=      1.25
$\phi =$ 0.5[1 + 0.49 x (1.25 - 0.2) + 1.25^2]=      1.54
$\chi =$ 1 /[1.54 + (1.54^2 - 1.25^2)^0.5]=      0.41

fcd =    0.41 x 250 / 1.1 =      93 Mpa    <= fy / $\gamma_{mo}$    <=    250 / 1.1
                                                                                 <=    227 Mpa

fcd =    93 Mpa    (=fcdz)

**Pdz** = Ae . fcd =    1903 x 93 /1000=    **177 kN**

*(iii) Compression due to buckling about minor axis, y-y (KLy / ry)*      (IS 800:2007 7..5.1.1)
*( when loaded concentrically to its centroid)*

Design compression strength, Pdy = Ae. fcd      where. Ae =    effective sectional area
                                                                                       fcd =    design compressive stress

$$fcd = \chi\, fy / \gamma_{mo} \quad <= fy / \gamma_{mo} \quad \text{(IS 800:2007 7.1.2.1)}$$
$$\chi = 1 / [\phi + (\phi^2 - \lambda^2)^{0.5}]$$
$$\phi = 0.5\,[1 + \alpha(\lambda - 0.2) + \lambda^2]$$
$$\lambda = (fy / fcc)^{0.5}$$

$$fcc = \pi^2 E / (KLz/rz)^2 \quad \text{Euler buckling stress}$$

Now, putting following values into above equation

$\alpha =$    0.49      $E =$    200000 Mpa      $fy =$    250 Mpa

$\gamma_{m0}$ = 1.10   KLy/ry = 66   Ae = 1903 mm2

fcc = 3.14^2 x 200000 / 65.5737704918033^2 = 459 MPa
λ = (250 / 459)^0.5= 0.74
φ = 0.5[1 + 0.49 x (0.74 - 0.2) + 0.74^2= 0.91
χ = 1 /[0.91 + (0.91^2 - 0.74^2)^0.5]= 0.695

fcd = 0.695 x 250 / 1.1 =   158 Mpa   <= fy / $\gamma_{mo}$   <=   250 / 1.1
                                                              <=   227 Mpa

fcd = 158 Mpa   (=fcdy)

**Pdy** = Ae . fcd =   1903 x 158 /1000=   **301 kN**

*The design compressive strength, **Pd** will be lowest of above calculated strengths of the member.*

**Pd** =   **177 kN**


## Axial Tension

The factored design tension, T  <  Design tensile strength of member, Td

<u>(i) Design tensile strength due to yielding of gross section, Tdg</u>        (IS 800:2007 - 6.2)

Tdg = Ag fy / $\gamma_{mo}$ =   1903 x 250 / 1.1/ 1000 =   433 kN

<u>(ii) Design tensile strength due to rupture of critical section, Tdn</u>

Tdn = 0.9 Anc. fu / $\gamma_{m1}$ + β. Ago. fy / $\gamma_{m0}$        (IS 800:2007 - 6.3.3)

β = 1.4 - 0.076 (w/t) (fy/fu) (bs/Lc)  <= 0.9 fu $\gamma_{mo}$ / fy $\gamma_{m1}$ >= 0.7

where,  Anc = net area of connected leg =   100x10-(1x18x10) =   820 mm2
        Ago = gross area of outstanding leg =   100 x 10 =   1000 mm2

        w = width of outstanding leg = d =                        100 mm
        bs = shear lag width = w + w1 (= Cyy) - t =   100+28.4-10=   118 mm
        Lc = length of end connection = (nos of bolt -0.5) x 3 φ hole =   135 mm

        f u =   410 Mpa        $\gamma_{m0}$ =   1.10    t =   10 mm
        fy =    250 Mpa        $\gamma_{m1}$ =   1.25

# Steel Structures 179

$\beta = 1.4 - 0.076 \, (w/t) \, (fy/fu) \, (bs/Lc) <= 0.9 \, fu \, \gamma_{mo} / fy \, \gamma_{m1} >= 0.7$

$= 1.4 - 0.076 \times (100/10) \times (250/410) \times (118/135) <= 0.9 \times 410 \times 1.1 / (250 \times 1.25) >= 0.7$

$= 0.99 \quad <= \quad 1.2989 \quad >= \quad 0.7 \quad$ Ok.

$Tdn = 0.9 \, Anc. \, fu / \gamma_{m1} + \beta . \, Ago. \, fy / \gamma_{mo}$
 $= 0.9 \times 820 \times 410 / 1.25 + 0.99 \times 1000 \times 250 / 1.1$
 $= 467064 \, N$
 $= 467 \, kN$

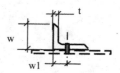

### (iii) Design tensile strength due to block shear, Tdb     (IS 800:2007 - 6.4.1)

a)   $Tdb = [\, Avg. \, fy / (1.732 \, \gamma_{mo}) + 0.9 \, Atn. \, fu / \gamma_{m1} \,]$   OR

b)   $Tdb = (0.9 \, Avn. \, fu / (1.732 \, \gamma_{m1}) + Atg. \, fy / \gamma_{mo})$   whichever is smaller

Avg = gross shear area along bolt line parallel to line (1-2) of external force
  = Lc . t
  = 135 x 10   =   1350 mm2

Avn = net shear area along bolt line parallel to the line (1-2) of external force
  = Lc . t - ( no of bolts x φ hole x thickness)
  = 135 x 10 - (3x18x10) =   810 mm2

Atg = gross area in tension from bolt hole to toe of the angle, perpendicular to line (2-3) of force
  = (d - Cyy) . t
  = (100 - 28.4) x 10 =   716 mm2

Atn = net area in tension from bolt hole to toe of the angle, perpendicular to line (2-3) of force
  = (d - Cyy) . t - ( no of rows x φ hole x thickness)
  = (100 - 28.4)x10 - (1x18x10)=   536 mm2

a)   $Tdb = [\, Avg . \, fy / (1.732 \, \gamma_{mo}) + 0.9 \, Atn . \, fu / \gamma_{m1} \,]$
  = 1350 x 250 /(1.732 x 1.1) + 0.9 x 536 x 410 / 1.25
  = 335374 N
  = 335 kN

OR

b)   $Tdb = (0.9 \, Avn. \, fu / (1.732 \, \gamma_{m1}) + Atg. \, fy / \gamma_{mo})$   whichever is smaller
  = 0.9 x 810 x 410 / (1.732 x 1.25) + 716 x 250 /1.1
  = 300783 N
  = 301 kN

So, Tensile strength due to block shear, Tdb =   301 kN

now, we have   Tdg =   433 kN,   Tdn =   467 kN and   Tdb =   301 kN

*The design tensile strength, **Td** will be lowest of above calculated strengths of the member.*

Td = 301 kN

**Strength design**     **Compression /Tension**

| property | Limit state method | | Working stress method | |
|---|---|---|---|---|
| | Semi compact | | Semi compact | |
| Load | Axial Comp, P = | 50 kN | Axial Comp, P = | 50 kN |
| | Axial Tension, T = | 50 kN | Axial Tension, T = | 50 kN |
| | Load factor = | 1.5 | **Axial compression :** | (IS 800:2007 11.3.1) |
| | Factored load : | | Permissible comp. stress, fac = 0.6 fcd | |
| | P = 1.5 x 50 = | 75 kN | | |
| | T = 1.5 x 50 = | 75 kN | fcd = | 93 Mpa (see calc. above.) |
| | | | Ag = | 1903 mm2 |
| | Design Comp strength, Pd = | 177 kN | | |
| | Factor load , P = | 75 kN | Permissible Comp strength, Pd = 0.6 fcd .Ag | |
| | | P < Pd; Safe. | = | 0.6 x 93 x 1903 |
| | | | = | 106187 N |
| | Design Tensile strength, Td = | 301 kN | = | 106 kN |
| | Factor load , T = | 50 kN | Actual load, P = | 50 kN |
| | | T < Td; Safe. | | P < Pd; Safe. |

*[contd.]*

| | |
|---|---|
| | **Axial Tension:** (IS 800:2007 11.2.1) |
| | Tensile stresses for : |
| | a) yielding  $f_{at} = 0.6 f_y = $  0.6 x 250 |
| | $ = $ 150 MPa |
| | b) rupture  $f_{at} = 0.69 T_{dn}/A_g$ |
| | $ = $ 0.69 x 467000 / 1903 |
| | $ = $ 169 MPa |
| | c) block shear  $f_{at} = 0.69 T_{db}/A_g$ |
| | $ = $ 0.69 x 301000 / 1903 |
| | $ = $ 109 MPa |
| | Permissible value of $f_{at} = $  109 MPa |
| | Permissible Tensile strength, $T_d = f_{at} . A_g$ |
| | $ = $ 109 x 1903 |
| | $ = $ 207427 N |
| | $ = $ 207 kN |
| | Actual load, $T = $  50 kN |
| | $T < T_d$; Safe. |

### *8.14.7.2 Double angle members with Equal legs*

End connection: Welding or Bolting with a minimum of two bolts and loaded through one leg

**Member Loading**  *Unfactored load*

| Load case | Description | Comp kN P | $M_{vertical}$ kN m $M_z$ | $M_{horz}$ kN m $M_y$ | $V_{vertical}$ kN $V_z$ | $V_{horz}$ kN $V_y$ | Tension KN T | Nature of load |
|---|---|---|---|---|---|---|---|---|
| 1 | DL + LL | | | | | | | |
| 2 | DL + LL + WL | 100 | | | | | 100 | |
| 3 | DL + LL + EL | | | | | | | |
| 4 | DL + WL | | | | | | | |
| 5 | DL + EL | | | | | | | |
| | | | | | | | | |
| | | | | | | | | |

| **Section properties:** | | | Built-up member dimensions | | | | | | | |
|---|---|---|---|---|---|---|---|---|---|---|
| Section Provided: | | s | wt/m (kg) | $A_g$ (mm$^2$) | $r_x$ (mm) | $r_y$ (mm) | $r_{vv}$ (mm) | $I_x$ (mm$^4$) | $I_y$ (mm$^4$) | $C_{yy}$ (mm) |
| 2 ISA 100 100 10 | | 12 | mm b/b  29.8 | 3806 | 30.50 | 46.0 | 19.4 | 3540000 | 8043868 | 28.4 |
| | | | Single Leg dimensions | | | | | | | |
| | | | b | d | t | Connection bolt | | | | An |
| | | s | mm | mm | mm | nos | φdia | φhole | rows | mm2 |
| | | | 100 | 100 | 10 | 3 | 16 | 18 | 1 | 3626 |

where, An = net area after deduction of bolt holes.
An = Ag - (no of rows x φ hole x t)

Distance between tacking plate (welded spacer plate) = **a** = $\boxed{300}$ mm (IS 800:2007 - 10.2.5.2)
(the value of ' a ' shall not exceed 32 t or 300 mm, whichever is less)

### Buckling class

| | | |
|---|---|---|
| Buckling class = c | for buckling about any axis | (IS 800:2007 - Table 10) |
| Imperfection factor, α = 0.49 | | (IS 800:2007 - Table 7) |

### Proportioning limit  (IS 800:2007 - Table 2)

$\varepsilon = (250 / f_y)^{0.5}$ = $(250 / 250)^{0.5}$ = 1.00

### Outstanding leg

b / t = 100 / 10 = $\boxed{10}$ Compact (IS 800:2007 - 3.7.2)

*Limiting width to thickness ratio:*

| | | |
|---|---|---|
| Plastic, $\lambda r$ = | 9.4ε = 9.4 x 1 = | 9 |
| Compact, $\lambda p$ = | 10.5e = 10.5 x 1 = | 11 |
| Semi compact, $\lambda sp$ = | 15.7e = 15.7 x 1 = | 16 |

### Effective Lengths (L) and Slenderness ratio ( L/r ) :  (IS 800:2007 - Table 11)

| | | | | | | |
|---|---|---|---|---|---|---|
| Laterally Unsupported length | Lz = | 6 M | Kz = | 0.85 | KLz = | 5.10 M |
| | Ly = | 6 M | Ky = | 1 | Kly = | 6.00 M |

Maxm slenderness ratio (L/r) shall not exceed    **250**    (IS 800:2007 - Table 3)

Here,    KLz/rz =    5100/30.5 =    167
         KLy/ry =    6000/46 =    130
                          L/r =    167    <250, OK.

### Sectional capacity / Design strength:

### Axial compression

*(i) Compression due to buckling about major axis, z-z [Lz / rz (=rx)]*     (IS 800:2007 7.5.2.1)

# Steel Structures

Design compression strength, Pdz = Ae. fcd  (IS 800:2007 7.1.2)

where. Ae = effective sectional area
fcd = design compressive stress

$$fcd = \chi \, fy / \gamma_{mo} \quad <= fy / \gamma_{mo}$$
$$\chi = 1 / [\phi + (\phi^2 - \lambda^2)^{0.5}]$$
$$\phi = 0.5 [1 + \alpha(\lambda - 0.2) + \lambda^2]$$
$$\lambda = (fy / fcc)^{0.5}$$

$$fcc = \pi^2 E / (KLz/rz)^2 \quad \text{Euler buckling stress}$$

Now, putting following values into above equation

| | | | | | | |
|---|---|---|---|---|---|---|
| $\alpha$ = | 0.49 | E = | 200000 Mpa | fy = | 250 Mpa | |
| $\gamma_{m0}$ = | 1.10 | KLz/rz = | 167 | Ae = | 3806 mm2 | =Ag |

fcc = 3.14^2 x 200000 / 167^2 =     71 MPa
$\lambda$ = (250 / 71)^0.5=     1.88
$\phi$ = 0.5[1 + 0.49 x (1.88 - 0.2) + 1.88^2=     2.68
$\chi$ = 1 /[2.68 + (2.68^2 - 1.88^2)^0.5]=     0.218

fcd = 0.218 x 250 / 1.1 =     50 Mpa     <= fy / $\gamma_{mo}$ <=  250 / 1.1
                                                                  <=  227 Mpa

fcd =     50 Mpa     (=fcdz)

**Pdz** = Ae . fcd =     3806 x 50 /1000=     **190 kN**


## (ii) Compression due to buckling about minor axis, y-y (KLy / ryy)  (IS 800:2007 7.5.2.1)

Design compression strength, Pdy = Ae. fcd     where. Ae = effective sectional area
fcd = design compressive stress

$$fcd = \chi \, fy / \gamma_{mo} \quad <= fy / \gamma_{mo} \quad \text{(IS 800:2007 7.1.2.1)}$$
$$\chi = 1 / [\phi + (\phi^2 - \lambda^2)^{0.5}]$$
$$\phi = 0.5 [1 + \alpha(\lambda - 0.2) + \lambda^2]$$
$$\lambda = (fy / fcc)^{0.5}$$

$$fcc = \pi^2 E / (KLz/rz)^2 \quad \text{Euler buckling stress}$$

Now, putting following values into above equation

| | | | | | |
|---|---|---|---|---|---|
| $\alpha =$ | 0.49 | $E =$ | 200000 Mpa | $fy =$ | 250 Mpa |
| $\gamma_{m0} =$ | 1.10 | $KLy/ryy=$ | 130 | $Ae =$ | 3806 mm2 |

$fcc =$ 3.14^2 x 200000 / 130^2 = 117 MPa
$\lambda =$ (250 / 117)^0.5= 1.46
$\phi =$ 0.5[1 + 0.49 x (1.46 - 0.2) + 1.46^2]= 1.87
$\chi =$ 1 /[1.87 + (1.87^2 - 1.46^2)^0.5]= 0.329

$fcd =$ 0.329 x 250 / 1.1 = 75 Mpa $<= fy / \gamma_{mo}$ $<=$ 250 / 1.1
$<=$ 227 Mpa

$fcd =$ 75 Mpa (=fcdy)

**Pdy** = Ae . fcd = 3806 x 75 /1000= **285 kN**

*The design compressive strength,* **Pd** *will be lowest of above calculated strengths of the member.*

**Pd =** **190 kN**


## Axial Tension

The factored design tension, T < Design tensile strength of member, Td

<u>(i) Design tensile strength due to yielding of gross section, Tdg</u>       (IS 800:2007 - 6.2)

Tdg = Ag fy / $\gamma_{mo}$ =   3806 x 250 / 1.1/ 1000 =       865 kN

<u>(ii) Design tensile strength due to rupture of critical section, Tdn</u>

$T_{dn} = 0.9\, A_{nc}\, f_u / \gamma_{m1} + \beta \cdot A_{go}\, f_y / \gamma_{m0}$ \hfill (IS 800:2007 - 6.3.3)

$\beta = 1.4 - 0.076\,(w/t)(f_y/f_u)(b_s/L_c) \le 0.9\, f_u\, \gamma_{mo} / f_y\, \gamma_{m1} \ge 0.7$

where, $A_{nc}$ = net area of connected leg = 2 x [100x10-(1x18x10)] = 1640 mm2
$A_{go}$ = gross area of outstanding leg = 2 x (100 x 10) = 2000 mm2

w = width of outstanding leg = d = 100 mm
bs = shear lag width = w + w1 (= Cyy) - t = 100 + 28.4 - 10 = 118 mm
Lc = length of end connection = (nos of bolt -0.5) x 3 φ hole = 135 mm

| | | | | | |
|---|---|---|---|---|---|
| f u = | 410 Mpa | $\gamma_{m0}$ = | 1.10 | t = | 10 mm |
| fy = | 250 Mpa | $\gamma_{m1}$ = | 1.25 | | |

$\beta = 1.4 - 0.076\,(w/t)(f_y/f_u)(b_s/L_c) \le 0.9\, f_u\, \gamma_{mo} / f_y\, \gamma_{m1} \ge 0.7$
= 1.4-0.076x(100/10)x(250/410)x(118/135) <= 0.9x410x1.1 / (250x1.25) >= 0.7
= 0.99 <= 1.30 >= 0.7 Ok.

$T_{dn} = 0.9\, A_{nc}\, f_u / \gamma_{m1} + \beta \cdot A_{go}\, f_y / \gamma_{m0}$
= 0.9 x 1640x410 / 1.25 + 0.99 x 2000 x 250 / 1.1
= 934128 N
= 934 kN

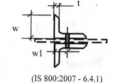

### (iii) Design tensile strength due to block shear, Tdb \hfill (IS 800:2007 - 6.4.1)

a) $T_{db} = [\,A_{vg}\, f_y / (1.732\, \gamma_{mo}) + 0.9\, A_{tn}\, f_u / \gamma_{m1}\,]$ OR

b) $T_{db} = (0.9\, A_{vn}\, f_u / (1.732\, \gamma_{m1}) + A_{tg}\, f_y / \gamma_{mo})$ whichever is smaller

Avg = gross shear area along bolt line parallel to line (1-2) of external force
= 2. Lc . t
= 2 x 135 x 10 = 2700 mm2
Avn = net shear area along bolt line parallel to the line (1-2) of external force
= 2 [Lc . t - (no of bolts x φ hole x thickness)]
= 2 x [135 x 10 - (3x18x10)] = 1620 mm2

Atg = gross area in tension from bolt hole to toe of the angle, perpendicular to line (2-3) of force
= 2 . (d - Cyy) . t
= 2 x (100 - 28.4) x 10 ) = 1432 mm2
Atn = net area in tension from bolt hole to toe of the angle, perpendicular to line (2-3) of force
= 2 [ (d - Cyy) . t - ( no of rows x φ hole x thickness)]
= 2 [ (100 - 28.4) x 10 - (1 x 18 x 10) ] = 1072 mm2

a) $T_{db} = [\,A_{vg} \cdot f_y / (1.732\, \gamma_{mo}) + 0.9\, A_{tn} \cdot f_u / \gamma_{m1}\,]$
= 2700 x 250 /(1.732 x 1.1) + 0.9 x 1072 x 410 / 1.25
= 670748 N
= 671 kN
OR
b) $T_{db} = (0.9\, A_{vn}\, f_u / (1.732\, \gamma_{m1}) + A_{tg}\, f_y / \gamma_{mo})$ whichever is smaller

= 0.9 x 1620 x 410 / (1.732 x 1.25) + 1432 x 250 /1.1
= 601565 N
= 602 kN

So, Tensile strength due to block shear, Tdb = 602 kN

now, we have   Tdg =   865 kN,   Tdn =   934 kN and   Tdb =   602 kN

*The design tensile strength,* **Td** *will be lowest of above calculated strengths of the member.*

Td =   602 kN

**Strength design**     **Compression /Tension**

| property | Limit state method | | Working stress method | |
|---|---|---|---|---|
| | Compact | | Compact | |
| Load | Axial Comp, P = | 100 kN | Axial Comp, P = | 100 kN |
| | Axial Tension, T = | 100 kN | Axial Tension, T = | 100 kN |
| | Load factor = | 1.5 | **Axial compression :** (IS 800:2007 11.3.1) | |
| | Factored load : | | Permissible comp. stress, fac = 0.6 fcd | |
| | P = 1.5 x 100 = | 150 kN | | |
| | T = 1.5 x 100 = | 150 kN | fcd = 50 Mpa (see calc. above.) | |
| | | | Ag = 3806 mm2 | |
| | Design Comp strength, Pd = | 190 kN | Permissible Comp strength, Pd = 0.6 fcd .Ag | |
| | Factor load , P = | 150 kN | = 0.6 x 50 x 3806 | |
| | | P < Pd; Safe. | = 114180 N | |
| | | | = 114 kN | |
| | Design Tensile strength, Td = | 602 kN | Actual load, P = 100 kN | |
| | Factor load , T = | 150 kN | P < Pd; Safe. | |
| | | T < Td; Safe. | | |
| | | | **Axial Tension:** (IS 800:2007 11.2.1) | |

*[contd]*

|  |  |
|--|--|
|  | Tensile stresses for :<br>a) yielding    $f_{at} = 0.6 f_y = $    $0.6 \times 250$<br>                              $=$    150 MPa<br>b) rupture    $f_{at} = 0.69 T_{dn}/A_g$<br>               $=$    $0.69 \times 934000 / 3806$<br>               $=$    169 MPa<br>c) block shear    $f_{at} = 0.69 T_{db}/A_g$<br>               $=$    $0.69 \times 602000 / 3806$<br>               $=$    109 MPa<br><br>Permissible value of $f_{at} = $    109 MPa<br><br>Permissible Tensile strength, $T_d = f_{at} . A_g$<br>                              $=$    $109 \times 3806$<br>                              $=$    414854 N<br>                              $=$    415 kN<br>            Actual load, $T = $    100 kN<br>                                    $T < T_d$; Safe. |

End

**Example 8.14.8 : Design of Beam with Rolled steel section. ( IS 800 - 2007 )**
[ Beams with Equal flanges and Bent about the axis of max. strength x-x axis ]

**Reference.**
    1.0 IS 800 : 2007   General construction in steel - Code of practice

**Input Data.**

**Material**

| IS 2062 | | | | | | |
|---|---|---|---|---|---|---|
| Grade | | E 250 (Fe 410 W) A | | | | |
| Yield Stress | $f_y$ | 250 | 240 | 230 | MPa | (IS 800:2007 - Table 1) |
| d or t | | <20 | 20-40 | >40 | | |
| Tensile Stress | $f_u$ | 410 | 410 | 410 | MPa | |
| Elasticity | E | | | 200000 | MPa | (IS 800:2007 - 2.2.4.1) |
| Mat. safety fac | $\gamma_{m0}$ | | | 1.10 | | (IS 800:2007 - Table 5) |

**Member Loading**     *Unfactored*

| Load case | Description | Axial kN P | $M_{vertical}$ kN m $M_x$ | $M_{horz}$ kN m $M_y$ | $V_{vertical}$ kN $V_x$ | $V_{horz}$ kN $V_y$ | | Nature of load |
|---|---|---|---|---|---|---|---|---|
| 1 | DL + LL | | 203 | | 135 | | | |
| 2 | DL + LL + WL | | | | | | | |
| 3 | DL + LL + EL | | | | | | | |
| 4 | DL + WL | | | | | | | |
| 5 | DL + EL | | | | | | | |
| | | | | | | | | |
| | | | | | | | | |

**Methods of design**
    1 Limit State design
    2 Working stress design

**Load, Load factors and Load combinations**

Load description                Load factors
D = Dead load                    As per IS : 800 2007
L = Live load
LLr = Roof live load            Load Combination
WL = Wind load                As per desgin basis reports and referenced codes of practice
EL = Earthquake load        ( IS 875, IS 800)

**Dimensioning**

           Effective Span    L =    6 m           (IS 800:2007 - 8.1.1)
                   Spacing    s =    3 m
Laterally unsupported length  $L_{LT}$ =    2 m    *(for lateral torsional buckling)*    (IS 800:2007 - 8.3)

# Steel Structures

*Floor load intensity:*
DL = 5 kN/m2    $w_{DL}$ = 15 kN/m
LL = 10 kN/m2   $w_{LL}$ = 30 kN/m
                w       = 45 kN/m

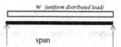
w (uniform distributed load)
span

**Design Load**   *Unfactored*

| Load Case | P  | Mx  | My  | Vx  | Vy |
|-----------|----|-----|-----|-----|----|
|           | kN | kNm | kNm | kN  | kN |
| 1         |    | 203 |     | 135 |    |

Mx = 45 x 6^2/8 = 203 kNm.
Vx = 45 x 6/2 = 135 kN.

**Section properties:**

| DESIGN SECTION | | | |
|---|---|---|---|
| RSJ | **ISMB 450** | | |
| FLANGE PLATE | **200** mm. | wide | |
| [plate per flange] | | **8** mm. | thick |
| Designation | | | |
| ISMB 450 + 200 x 8 TH PPF | | | |

| EQUIVALENT PLATED SECTION | | |
|---|---|---|
| FLANGE PL | **200** mm. | b |
|  | **21.1** mm. | tf |
| WEB PL | **424** mm. | d |
|  | **9.4** mm. | tw |
| WT/M | 97.52 | kg. |
| Area | 12426 | mm2. |
| Ix | 478043294 | mm4. |
| Iy | 19006667 | mm4. |
| rx | 196 | mm. |
| ry | 39 | mm. |
| tw | 9.4 | mm |
| Zx (Ze) | 2050808 | mm3 |
| J (It)  Σ bt3/3 | 1369913 | mm4 |
| $h_f$   d+tf | 445 | mm. |
| For welded section | $\alpha_{LT}$ = | 0.49 |

(IS 800:2007-8.2.2)

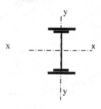

Flange area, Af = 150 x 17.4 + 200 x 8 = 4210 mm²
Web area, Av = 450 x 9.4 = 4230 mm²

**Deflection:**                                                (IS 800:2007 - 5.6.1)
$\delta = 5 w L^4 / 384 EI$ =   5 x 45 x 6000 ^4 / 384 x 200000 x 478043294 =   7.94 mm
                          Allowable deflection limit L /300 =   20.00 mm
                                                                     OK.

**Proportioning limit**                                        (IS 800:2007 - Table 2)

ε = (250 / fy ) ^0.5 =   (250 / 250)^0.5=   1.0

Check for flange

0.5b / tf =   100 / 21.1 =   **4.73**   Compact        (IS 800:2007 - 3.7.2)
*Limiting width to thickness ratio:*
Plastic, λr =           8.4ε = 8.4 x 1 =    8
Compact, λp =           9.4ε = 9.4 x 1 =    9
Semi compact, λsp =    13.6ε = 13.6 x 1 =  14

Check for web

d / tw =   424 / 9.4 =   **45**   Plastic.             (IS 800:2007 - 3.7.2)
*Limiting depth to thickness ratio:*
Plastic, λr =          84ε = 84 x 1 =    84
Compact, λp =         105ε = 105 x 1 =  105
Semi compact, λsp =   126ε = 126 x 1 =  126

*Shear buckling check* (IS 800:2007 - 8.4.2)

Resistance to shear buckling to be checked,
(i)    when   $d/t_w >$    $67\varepsilon_w$    for web without stiffeners
                   = 67 x 1 =    **67**

(ii)    when   $d/t_w >$    $67\varepsilon_w(K_v/5.35)^{0.5}$    for web with stiffeners
                   = 67 x 1(9.35/5.35)^0.5
                   = **88.57**

where $K_v = 5.35 + 4/(c/d)^2$ for $c/d \geq 1$
c = spcg. of stiffener and d is depth of web.
so, let us assume, $K_v$ =   5.35 +4 / (dw/dw)^2 =   5.35+4 =          9.35

In this case,     Shear buckling check not required.

## Nominal Shear strength of web, Vn                 (IS 800:2007 - 8.4)

The nominal shear strength of web, Vn is goverered by plastic shear resistance.

$V_n = V_p = A_v f_{yw} / (3)^{0.5}$                           (IS 800:2007 - 8.4.1)
Shear Area, $A_v = d \times t_w$ =     450 x 9.4 =      4230 mm2
Yield strength of web, $f_{yw} = f_y$ =             250 Mpa

$V_n = V_p$ =    4230 x 250 /3^0.5/1000 =     611 kN.

## Design Shear strength of web, Vd

$V_d = V_n / \gamma_{m0}$                                                     (IS 800:2007 - 8.4)

**$V_d$** =    611 / 1.1=      **555 kN**

## Design flexural strength, Md                        (IS 800:2007 - 8.2)

Md shall be the lower value obtained according to the limit state of yielding (plastic moment) and lateral torsional buckling.

**a) Laterally supported beam** ( *web not susceptible for shear buckling; design shear $\leq 0.6 V_d$* )

Nominal flexural strength, $M_n = M_p = f_y Z_p$ = Plastic moment

$f_y$ = Yield stress =      250 Mpa
$Z_p$ = plastic sectional modulus about the x axis = $\Sigma A y$
   =     2 x 200 x 21.1 x 445.1/ 2 + 2 x (424/2) x 9.4 x (424/4)
   =       2300796     mm4

$M_n$ =    250 x 2300796/1000000 =       575 kNm.    (= Mp)

Design flexural strength = $M_d = \beta . M_n / \gamma_{mo} < 1.2 Z_e f_y /\gamma_{mo}$     (IS 800:2007 - 8.2.1.2)

$M_d$ =    1 x 575 / 1.1     =      523 kNm.    where, $\beta = 1$ for plastic and compact section

$1.2 Z_e f_y /\gamma_{mo}$    =     1.2 x 2050808 x 250 / 1.1/1000000 =     559 kNm    > Md; Okay.

Design flexural strength, **$M_d$** =     **523 kNm**

**b) Lateraly unsupported beam** ( *susceptible to Lateral-torsional buckling* )

$\quad\quad$ Design flexural strength = Md = $\beta \cdot Z_p \cdot f_{bd}$ $\quad\quad\quad\quad$ (IS 800:2007 - 8.2.2)

$\quad\quad$ The expressions are

$f_{bd} = \chi_{LT} \, f_y / \gamma_{mo}$

$\chi_{LT} = 1 / [\, \phi_{LT} + (\phi_{LT}^2 - \lambda_{LT}^2)^{0.5}\,] \; <= 1$

$\phi_{LT} = 0.5\,[\,1 + \alpha_{LT}(\lambda_{LT} - 0.2) + \lambda_{LT}^2\,]$

$\lambda_{LT} = (\beta\, Z_p\, f_y / M_{cr})^{0.5} \; <= \; (1.2\, Z_e\, f_y / M_{cr})^{0.5}$

$M_{cr} = (\,\pi^2\, EI_y\, hf / 2\, L_{LT}^2\,) \cdot [\,1 + (1/20) \cdot ((L_{LT}/r_y)/(hf/tf))^2\,]^{0.5}$

Now, putting following values into above equation

| | | | | | | |
|---|---|---|---|---|---|---|
| hf = | 445 mm | Iy = | 19006667 | mm4 | $L_{LT}$ = | 2000 mm |
| tf = | 21.1 mm | E = | 200000 | Mpa | | |
| ry = | 39 mm | | | | | |

| | | | | | |
|---|---|---|---|---|---|
| $\beta$ = | 1 | $\alpha_{LT}$ | 0.49 | | (IS 800:2007 - 8.2.2) |
| fy = | 250 Mpa | Zp = | 2300796 | mm3 | |
| $\gamma_{m0}$ = | 1.10 | Ze = | 2050808 | mm3 | |

we get,

$M_{cr} = (\,\pi^2\, EI_y\, hf / 2\, L_{LT}^2\,) \cdot [\,1 + (1/20) \cdot ((L_{LT}/r_y)/(hf/tf))^2\,]^{0.5}$

$\pi^2\, EI_y\, hf / 2\, L_{LT}^2 = \;$ 3.14^2 x 200000 x 19006667 x 445 / 2 x 2000^2 = $\quad\quad$ 2084804240

$((L_{LT}/r_y)/(hf/tf))^2 = \quad$ ((2000/39) / (445/21.1) )^2 = $\quad\quad$ 5.91

$M_{cr} = $ 2084804240 x [1 + (1/20) x 5.91]^0.5 = $\quad\quad$ 2372924882 $\quad$ Nmm

$\lambda_{LT} = (\beta\, Z_p\, f_y / M_{cr})^{0.5} \; <= \; (1.2\, Z_e\, f_y / M_{cr})^{0.5}$

$(\beta\, Z_p\, f_y / M_{cr})^{0.5} \; = \;$ (1 x 2300796 x 250 / 2372924882)^0.5 = $\quad\quad$ 0.4923

$(1.2\, Z_e\, f_y / M_{cr})^{0.5} = $ (1.2 x 2050808 x 250 / 2372924882)^0.5= $\quad\quad$ 0.5092

$\lambda_{LT} = \;$ 0.4923 $\;=\;$ **0.49**

$\phi_{LT} = 0.5\,[\,1 + \alpha_{LT}(\lambda_{LT} - 0.2) + \lambda_{LT}^2\,] \quad$ = 0.5[1+0.49(0.49-0.2)+0.49^2] = $\quad\quad$ 0.6911

$\chi_{LT}$ = $1 / [\phi_{LT} + (\phi_{LT}^2 - \lambda_{LT}^2)^{0.5}]$ <= 1
  = $1 / [0.6911 + (0.6911^2 - 0.49^2)^{0.5}]$
  = 0.849

$f_{bd} = \chi_{LT} f_y / \gamma_{mo}$    = 0.849 x 250 / 1.1 =    193 Mpa

Design flexural strength = Md = $\beta$ . Zp . $f_{bd}$ =    1 x 2300796 x 193/10^6 =    444 kNm

*If the value of $\lambda_{LT}$ is less than 0.4, Lateral torsional buckling need not be checked and member shall be treated as Laterally supported beam (IS 800:2007 8.2.1.2)*

Hence, the Design flexural strength will be    **Md =**    **444** kNm

**Intermediate web stiffeners**

d/tw =    45  <=    200 $\varepsilon_w$    =200 x 1 =    200    (IS 800:2007 - 8.6.1.1)

*Intermediate stiffeners not required.*

**Strength design**

|  | Limit state design | Working stress design |
|---|---|---|
|  | (IS 800:2007 - Section 8) | (IS 800:2007 - Section 11) |
|  | Factored design moment = M<br>Design flexural strength of the section = Md<br>Factored design shear = V<br>Design shear strength of the section = Vd<br>M <= Md<br>V <= Vd | Actual design moment = Ms<br>Permissible flexural strength = Ma<br>Actual design shear = Vs<br>Permissible shear strength = Va<br>Ms <= Ma<br>Vs <= Va |
| Load<br>w = | 1.5D + 1.5L<br>= 1.5 x 15 + 1.5 x 30 =    67.5 kN/m | D + L<br>= 15 + 30 =    45 kN/m |
| Span<br>End | L =    6 M<br>simply supported | L =    6 M<br>simply supported |
| Moment | M = 67.5 x 6^2 / 8 =    304    kNm<br>Md =    444    kNm<br>**Md > M; Safe.** | Ms =    45 x 6^2 / 8=    kNm<br>    203    kNm<br>*Allowable flexural strength, Ma :*    (IS 800-11.4.1)<br>*a) For laterally supported beam:*<br>fabc = fabt = Ma/Zx    (Zx top = Zx bot)<br>Ma = fabc. Zx = 0.66 fy Zx    (fabc = fabt)<br>    =    0.66x240x2050808/10^6<br>    =    324.85 kNm<br><br>*b) For laterally Unsupported beam:*<br>fabc = fabt = 0.6 Md/Zx    (Zx top = Zx bot)<br>Md =    444 knM<br>Ma = fabc. Zx = (0.6Md/Zx) Zx    (fabc = fabt)<br>    =    0.6 Md<br>    =    0.6 x 444 =    266 kNm<br><br>Allowable Flexural mom, **Ma =**    **266** kNm.<br>Actual Design Moment, Ms =    203 kNm.<br>**Ma > Ms; Safe.** |
| Shear | V =    67.5 x 6 / 2 =    203 kN<br>Vd =    555 kN<br>**Vd > V; Safe.** | Vs =    45 x 6 / 2 =    135 kN<br>*Shear strength, Vs :*    (IS 800-11.4.2)<br>*a) Pure shear*<br>Va = $\tau_{ab}$ . Av , where Av shear area<br>    =    0.4 fy x Av<br>    =    0.4 x 250 x 424 x 9.4/1000<br>    =    399 kN<br>*b) when subject to shear buckling*<br>Va = $\tau_{ab}$ . Av , where Av shear area<br>    =    (0.7 Vd / Av) x Av<br>    =    0.7 Vn<br>    =    0.7 x 555<br>    =    389 kN<br><br>Allowable Shear, **Va =**    **389** kNm.<br>Actual design shear, Vs =    135 kNm.<br>**Va > Vs; Safe.** |

End

# Steel Structures

*Example 8.14.9: Column base for built-up I section (AISC)*

**Reference.**
    1.0 Steel designers Manual - ELBS

**Input Data.**

**Material**

| Structural Steel : | | ASTM A 36 | | | |
|---|---|---|---|---|---|
| Yield Stress | $F_y$ | 36 | ksi | 250 | MPa |
| Tensile Stress | $F_u$ | 65 | ksi | 450 | MPa |
| Elasticity | Es | | | 200000 | MPa |
| Anchor bolt | fy | 36 | psi | 250 | MPa |
| A36 | Es' | 29000000 | psi | 200000 | MPa |
| Concrete pedestal | | | | | |
| Conc. Strength | fc' | 3600 | psi | 25 | MPa |
| | Ec | 3420000 | psi | 23586 | MPa |
| | m | | | 11 | |

(AISC Manual 14 Table 2-4)

(ACI 318-19 Chapter 19.2)

(ACI 318-19 Chapter 20.2.2.2)

**Member Loading**
**ASD**

| Load case | Description | Axial | $M_{vertical}$ | $M_{horz}$ | $V_{vertical}$ | $V_{horz}$ | | Nature of load | |
|---|---|---|---|---|---|---|---|---|---|
| | | KN | KN m | KN m | KN | KN | | | |
| | | P | $M_x$ | $M_y$ | $V_x$ | $V_y$ | | | |
| 1 | D | | | | | | | | |
| 2 | D + L | 214 | 12 | | 57 | | | | |
| 3 | D + (Lr or S or R) | | | | | | | | |
| 4 | D + 0.75L + 0.75(Lr or S or R) | | | | | | | | |
| 5 | D + (0.6W or 0.7E) | 128 | 192 | | 50 | | | | |
| 6 | D + 0.75L+0.75(0.6W)+0.75(Lr/S/R) | 180 | 71 | | 14 | | | | |
| 7 | D + 0.75L+ 0.75(0.7E) + 0.75 S | | | | | | | | |
| 8 | 0.6D + 0.6W | | | | | | | | |
| 9 | 0.6D + 0.7E | | | | | | | | |

## Dimensions

**Column size:** $D_c$ = 400 mm.
$W_c$ = 150 mm.
**Pedestal :** $D_p$ = 900 mm.
$W_p$ = 600 mm.

**Base plate :**
Width , B = 400 mm.
Length N = 700 mm.
Thickness t = 28 mm.

**Anchor Bolt :**
Nominal Dia. $\phi_{dia}$ = 36 mm.
Nos/ side n = 3
Dist.from Col.face = 75 mm.
Edge distance x = 75 mm.
spacing s = 133 mm.

**Stiffener plate a:**
Height = 600 mm.
Thick = 12 mm.
Nos = 2
$\omega$ = 8 mm. Fillet weld size

**Stiffener plate b:**
Height = 550 mm.
Thick = 12 mm.
Nos = 4

**Shear Lug:**
Depth L = 125 mm.
Wide b = 150 mm.
Thk $t_1$ = 28
Nos = 1
Grout G = 50 mm.

**Load resistant Factors :**
for Moment and axial
ASD $\Omega$ = 1.67
LRFD $\phi$ = 0.9
for bearing
ASD $\Omega$ = 1.67
LRFD $\phi$ = 0.65

| **Design Load** | | Unfactored |
|---|---|---|
| Axial Comp | P = | 128 kN |
| M abt major axis | $M_x$ = | 192 kNm |
| Shear along Y | $F_x$ = | 50 kN |

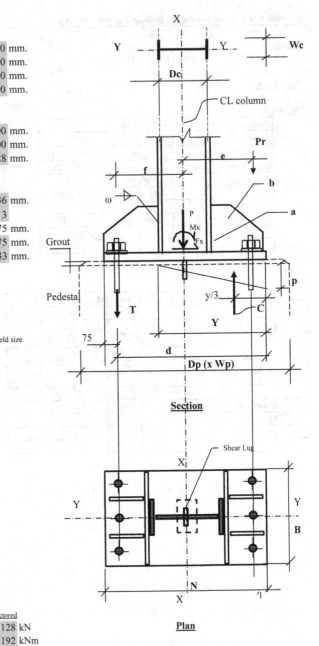

Section

Plan

# Steel Structures

**Strength design**

*Eccentricity check*

| | | | | | | |
|---|---|---|---|---|---|---|
| e = M/P = | 192 / 128 = | 1.5 m | | 1500 mm | | |
| Length of base Plate, N = | | 700 mm | L/6 = | 117 mm | e > L/6; Tension occures. |

If e (=M/P) is less than L/6, i.e. resultant vertical load falls within middle third, there will be no tension in base plate/ anchor bolt. Otherwise, tension occurs between bottom of base plate and top of concrete pedestal.

At this plane, the force transfer from steel column base and pedestal will be similar to a Reinforced concrete beam with reinforcement (here anchor bolt) carrying tensile force and at the opposite end compression load transferred by surface contact between base plate grout.

*N A depth, Y:*

$$Y = m.fc.d / (m fc + fs)$$

let us consider,

$$m = 11$$
$$fc = 0.85 * fc' = 0.85 \times 25 = 21.25 \text{ Mpa}$$
$$fs = 0.6 * fy = 0.6 \times 250 = 150 \text{ Mpa}$$
$$d = N - x = 700 - 75 = 625 \text{ mm}$$
$$Y = 11 \times 21.25 \times 625 / (11 \times 21.25 + 150) = \boxed{381} \text{ mm}$$

*Base Plate sizing*

Taking moment about T,

$$C.(d - y/3) = P.(d - N/2) + Mx$$

$$C = [P.(d - N/2) + Mx] / (d - Y/3)$$
$$= [128000 \times (625 - 0.5*700) + 192000000] / (625 - 381/3)$$
$$= 456225 \text{ N}$$
$$= 456 \text{ kN}$$

Required width of base plate required = Br

$C = 1/2 \cdot Y \cdot fc \cdot Br$     $Br = 2C / (Y \cdot fc)$

$Br = 2 \times 456000 / (381 \times 21.25) = $     113 mm     < B provided

<u>Actual width of base plate shall be determined after selecting nos and spacing of anchor bolt.</u>

<u>Anchor bolt</u>

$T = C - P$     $T = 456 - 128 = $     328 kN

Bolt provided :     3 nos     36 mm dia ($\phi_{dia}$)   fy =     250 Mpa

Tensile capacity of one bolt = $3.14 \times (0.87 \times 36)^2 / 4 \times 250 / 1000 = $     193 kN

Total tension capacity =     $3 \times 193 = $     579 kN

Factor of Safety, $\Omega$ =     579 /328 =     1.77  < 1.67 ; Safe.

Minimum width of base plate to accommodate anchor bolts =     369 mm
$3 \times 3 \times (36+5) = $     369 mm

Width provided , B =     400 mm
< Safe.

<u>Base Plate sizing</u>

Length, N =     700 mm     C =     456 kN
Width, B =     400 mm     Y =     381 mm

Avg base pressure = p =     $456000 / (400 \times 381/2) = $     5.98 Mpa
$C = 1/2 \, Y \cdot B \cdot p$
$p = C / (B \cdot Y/2)$

Max. pressure at edge of base plate, p  =     $5.98 \times 2 = $     **11.96 Mpa**

Taking section at outside face of main gusset welded to column flange

<u>stiffener a</u>
height =     550 mm
thick =     12 mm

<u>base plate</u>
width =     400 mm
thick =     28 mm

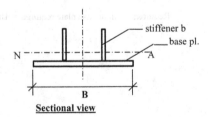

**Sectional view**

NA dist. from bottom of base plate:
y = (2x550x12x(275+28) + 400x28x14) / (2x550x12 + 400x28)
  = 170 mm

$I_{NA}$ = 2x(12x550^3/12+550*12*(275+28-170)^2)+400x28^3/12+400*28*(170-0.5*28)^2
  = 839539733   mm4

Zx top = 839539733/(550+28-170) = 2057695 mm3
Zx bot = 839539733 / 170 = 4938469 mm3

Resulting stress from base pressure, p
Projection outside main gusset = (700-400-2x12)/2= **138** mm
$M_1$ = 11.96*138^2/2=   113883 Nmm  = Upward moment from base pressure

Comp stress at top = 113883 / 2057695 = 0.055 Mpa
Tensile stress at bot = 113883 / 4938469 = 0.023 Mpa
Permissible stress = 0.66 x 250= 165 Mpa

Resulting stress from Anchor bolt tension, T
Projection outside main gusset = 138 - 75 =   **63** mm
T = 328 kN
$M_2$ = 328 x 1000 x 63 = 20664 Nmm  Tension at top
Tensile stress at top = 20664 / 2057695 = 0.010 Mpa
Comp stress at bot = 20664 / 4938469 = 0.004 Mpa
Permissible stress = 0.6 x 250 = 150 Mpa

Local stress from Anchor bolt tension, T
Force per bolt = 328 / 3 = 109 kN
Each bolt panel is having three end fixed and one end free, so the bolt tension will be
resisted by   **3** faces.

Projection from stiffener b = (133 - 12) / 2 = 60.5 mm
$M_{/side}$ = 109000/3 x 60.5 = 2198167 Nmm
eff width = 60.5 x 2 = 121 mm
Z = b $t^2$ / 6 = 121 x 28^2/6= 15811 mm3
Stress developed = 2198167 / 15811= 139 Mpa
Permissible stress = 0.66 x 250= 150 Mpa   **Safe.**

| ASD | Mn = fy Zp / Ω = | **3.55** kNm | Ω = | 1.67 | (AISC Manual 14ed) |
|---|---|---|---|---|---|
| | Zp = b $t^2$/4 = 23716 mm3 | | t = | 28 mm | |
| | | | b = | 121 mm | |
| | Mu = 2.198167 x 1.5 = | **3.30** kNm | fy = | 250 Mpa | |
| | | < Mn; Safe. | LF = | 1.5 | |

*Stiffener - a*
T = 328 kN   Total bolt tension
Projection outside column face = (700-400)/2 = 150 mm
Avg base pressure = 5.98 Mpa   B = 400 mm

C' = Total upward base pressure outside column flange = 400 x 150 x 5.98/1000
  =   **359** kN
Max. Load transferred by main gusset Stiffener mkd a = **359** kN.
weld size, w = 8 mm  Fillet weld capacity = 0.7 x 8 x 110 = 616 N/mm
No of vert weld run =   **2**
Length of weld required/side = 358800/ (2x616)= 291 mm
          Height of stiffener, a = 600 mm   **Safe.**

*Stiffener - b :*    included in base plate design.

*Shear Lug*

| | | | |
|---|---|---|---|
| Shear force, Fx = | **50** kN | fc' = | 25 Mpa |
| Lug: Embedded depth below grout, Le = | | 125 - 50 = 75 mm | (Le = L - G) |
| | | width = 150 mm | |

Following criteria are used to determine embeded length and shear capacity of lug :
(i) bearing strength strength of concrete, and            (ACI 318-19 Table 22.8.3.2)
(ii) Shear strength of concrete in front of lug
(iii) Strength design

(i) Bearing capacity of concrete, $B_n = 0.85 f_c' A_1$    where, $A_1$ = bearing area below grout.
$\phantom{Bearing capacity of concrete, B_n}= 0.85 \times 25 \times 75 \times 150/1000$
$\phantom{Bearing capacity of concrete, B_n}= 239$ kN
$B_u = 239/1.67 = 143$ kN    $> F_x$; Safe.

(ii) Shear strength of concrete in front of lug, $F_r$ :
Projected area of the failure plane at the face of pedestal = a x b

Vertical length =    75+450=    525 mm
Horj width = $W_p$ = pedestal width =    600 mm

Effective stress area =    525 x 600 - 75 x 150 =    303750 mm2
Uniform tensile stress on eff. area= $4 \phi (f_c)^{0.5}$ =    15 Mpa
$4 \times 0.75 \times (25)^{0.5}$
$F_r = $    303750 x 15 / 1000 =    4556 kN
$F_r / \Omega = 4556 / 1.67 =$    2728.1 kN    $> F_x$; Safe.

*[ Note: The concrete design shear strength for the lugs shall be determined based
on a uniform tensile stress of $4 \phi (f_c')^{0.5}$ acting on an effective stress area defined
by projecting a 45 degree plane from the bearing edge of the shear lug to the free surface.
The bearing area of the lug is to be excluded from the projected area.
Use $\phi = 0.75$ .    Refer ACI Steel design guide 1 for base plate design/ App B of ACI 349-01]*

(iii) Strength design :

$M_u = F_x ( G + L_e/2)$  kNm
$\phantom{M_u}= 50 \times (50 + 75/2)/1000 = $    4.375 kNm

$M_n = F_y Z_p / \Omega = F_y (b t_1^2 /4)/ \Omega = M_u$

$t_1 = [ 4 M \Omega / ( F_y . b) ]^{0.5}$
$\phantom{t_1}= [4 \times 4375000 \times 1.67 / (250 \times 150) ]^{0.5}$
$\phantom{t_1}= 27.92$ mm
Thickness provided =    28 mm    Safe.

[ Alternately, shear key with two flats making a cross pattern may be provided to reduce plate thickness.]

End

# Steel Structures

*Example 8.14.10 : Wing plate type Column base for built-up I section (IS 800 - 2007)*

**Reference.**
    1.0 Design of welded structure - O J Blodgett

**Input Data.**

**Materials**
| | | | |
|---|---|---|---|
| Concrete grade = | M25 | | ( IS : 456 : 2000 ) |
| Characteristic strength, fck = | 25 N/mm2. | | |
| Bending stress in comp. σcbc = | 8.5 N/mm2 | | |
| Modulus of Elasticity, Ec = | 28500 N/mm2 | | |

| | | | |
|---|---|---|---|
| Structural steel and anchor bolt : | IS 2062 | Grade E 250 (Fe 410 W) A | ( IS : 800 : 2007 ) |
| Yield stress    fy = | 240 N/mm2. | | |
| Ultimate tensile stress   fu = | 410 N/mm2. | | |
| Modulus of Elasticity, Es = | 200000 N/mm2. | | |

**1.0 Design force**

**Column mkd :**    A -1

**Column base load:**

| Load case | Axial P KN | Mom M KNm |
|---|---|---|
| 1 | 1226 | 1558 |

**2.0 Dimensions:**

| | | | |
|---|---|---|---|
| **Column size:** | Dc = | 1200 | mm. |
| | Wc = | 900 | mm. |
| **Pedestal :** | Dp = | 3000 | mm. |
| | Wp = | 2500 | mm. |
| **Base plate :** | | | |
| Length | L = | 1800 | mm. |
| Width | B = | 1200 | mm. |
| Thickness | t = | 25 | mm. |
| **Anchor Bolt :** | | | |
| Nominal Dia. | φdia = | 36 | mm. |
| Nos. per side | n | 5 | |
| Dist.from Col.face | | 500 | mm. |
| Edge distance | x | 100 | mm. |

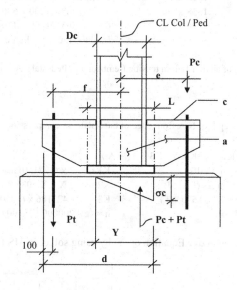

**Stiffener plate a :**   **Cover plate c :**          SECTIONAL VIEW.
Depth   600 mm.    Depth   600 mm.
Thick   12 mm.     Thick   25 mm.

### 3.0 Design base pressure:

| Load case | Axial | Mom | Base Plate | | Base area | | Design Base pr. | | Permissible base pr.(*) | | Remarks |
|---|---|---|---|---|---|---|---|---|---|---|---|
| | P | M | L | B | A | Zxx | Max | Min | a | σ bp | |
| | KN | KNm | mm | mm | mm2 | mm3 | N/mm2 | N/mm2 | | N/mm2 | |
| 1 | 1226 | 1558 | 1800 | 1200 | 2E+06 | 6E+08 | 3.0 | -1.84 | 1.84 | 11.47 | Safe. |

( * ) Permissible bearing pressure, σbp = 0.25 * fck * a        (IS 456 2000 - 34.2)
where, a = (A1/A2)^0.5 but not greater than 2
A1 = Area of supporting pedestal.    A2 = loaded area on supporting pedestal.

### 4.0 Design of Anchor bolt :

*To find out N-A depth*                  *[Design of welded structure - O J Blodgett, Section 3.3-8]*

P    1226 kN.        e = M/P =    1558000 / 1226 =         1271 mm.
M    1558 kNm.       f =    0.5 x 1200 + 500 =              1100 mm.
                     d =    0.5 x 1800 + 1100 +100=         2100 mm.
                     Modular ratio, n = Es/Ec =                 7

Area of anchor bolt in tension ( embeded in Pedestal), As =    5 x 0.25 x 3.14 x (1 x 36)^2
                                                 As =    5087 mm2

Y = Depth of N.A from edge of base plate on compression face

$Y^3 + K1 \cdot Y^2 + K2 \cdot Y + K3 = 0 \, (= A)$
K1 = 3 ( e - d/2)        K1 =   3 x (1271-1050)=                    663
K2 = 6 n As (f+e) / B    K2 =   6 X 7 X 5086.8 x (1100 + 1271)/1200=  423186
K3 = - K2 (0.5*d + f )   K3 =   - 423186 x (1050 + 1100) =         -909849900
n = Es/Ec                n =    200000 / 28500 =                      7

Solving above Equation of Y by using solver in MS Excel, we get  Y =    572 mm

| A | Y | |
|---|---|---|
| -2.6E+08 | 5.72E+02 |  *[Excel what if analysis, Goal seek : Set cell A =0 by changing Y]* |

*Graphical representaion of Y value:*

| Y(mm) | A |
|---|---|
| -210 | -9.8E+08 |
| 0 | -9.1E+08 |
| 21 | -9E+08 |
| 105 | -8.6E+08 |
| 168 | -8.2E+08 |
| 210 | -7.8E+08 |
| 525 | -3.6E+08 |
| 630 | -1.3E+08 |
| 840 | 5.06E+08 |
| 1050 | 1.42E+09 |
| 1260 | 2.68E+09 |

From the above chart, we get    Y = 572 mm

Depth of N-A considered for design,    Y = 572 mm    *conservative approach*

*Anchor bolt*

Taking moment about Pt,   $P*f + M - \sigma c*0.5*Y.B.(d-x-Y/3) = 0$
$\sigma c = P*f + M / [0.5* Y B (d - x - Y/3)]$
= 1226000 x 1100 + 1558000000 / [0.5 x 572 x 1200(2100 - 100 -572/3)]
= 4.68 N/mm2.    **< 0.85 fck; Safe.**

Pc = 0.5* σc * Y * B = 0.5 x 4.68 x 572 x 1200/1000 = 1605 kN
Pt = Pc - P = 1605 - 1226 = 379 kN

Net area of bolts/side, As net = 5 x 0.25x3.14 x (0.85x36)^2 =    3675 mm2
Tensile stress in bolt = Pt / As net =    379000 /3675=    103 Mpa.
Allowable tension ,fat = 0.6*fy =    0.6 x 240 =    120 Mpa.    (IS800 2007 -11.2.1)
**Safe.**

## 5.0 Design of base plate.

| | | | |
|---|---|---|---|
| Width of base plate, B = | 1200 mm. | Length of base plate, L = | 1800 mm |
| No of Anchor bolt per side = | 5 | Depth of column, Dc = | 1200 mm |
| No of panel per side = | 5 | | |
| No of vertical stiffeners = | 6 | | |

Let us calculate bending stress of base plate using Roark's formula of Stress and Strain - W.C.Young for uniform load over the entire panel.

| Intensity | q = | 4.68 Kn/m2 = | 0.0047 N/mm2. |
|---|---|---|---|
| Panel size | a = | 1200 / 5 = | 240 mm. |
| | b = | (1800 - 1200) / 2 = | 300 mm. |
| | t = | | 25 mm. |
| | E = | 204700 N/mm2. | |

Simply supported three sides and one edge free

a/b = 0.80
$\alpha$ = 0.106
$\beta$ = 0.45

| a/b | $\beta$ | $\alpha$ |
|---|---|---|
| 0.5 | 0.36 | 0.08 |
| 0.7 | 0.45 | 0.106 |
| 1 | 0.67 | 0.14 |
| 1.5 | 0.77 | 0.16 |
| 2 | 0.79 | 0.165 |
| 4 | 0.8 | 0.167 |

Bending stress, $\sigma = \beta q b^2 / t^2$
   = 0.45 x 0.0047 x 300^2 / 25^2   =   0.3046 Mpa
Permissible bending stress = 0.75 fy =   0.75 x 240 =       180 Mpa.   ( IS : 800 : 2007 - 11.3.3)
**Safe.**

Max, deflexn,   y = $\alpha q b^4 / (E.t^3)$ =
   = 0.106 x 0.0047 x 300^4 / (204700 x 25^3) =     0.0013 mm
Permissible limit = L /325 =   240 / 325 =       0.7385 mm.
**Safe.**

## 6.0 Design of Stiffener plate (a) and Cover plate (c) :

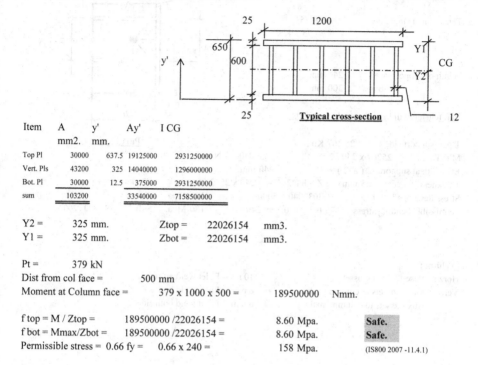

| Item | A mm2. | y' mm. | Ay' | I CG |
|---|---|---|---|---|
| Top Pl | 30000 | 637.5 | 19125000 | 2931250000 |
| Vert. Pls | 43200 | 325 | 14040000 | 1296000000 |
| Bot. Pl | 30000 | 12.5 | 375000 | 2931250000 |
| sum | 103200 | | 33540000 | 7158500000 |

Y2 = 325 mm.    Ztop = 22026154   mm3.
Y1 = 325 mm.    Zbot = 22026154   mm3.

Pt = 379 kN
Dist from col face = 500 mm
Moment at Column face = 379 x 1000 x 500 = 189500000 Nmm.

f top = M / Ztop =  189500000 /22026154 =   8.60 Mpa.   **Safe.**
f bot = Mmax/Zbot = 189500000 /22026154 =   8.60 Mpa.   **Safe.**
Permissible stress = 0.66 fy =  0.66 x 240 =   158 Mpa.   (IS800 2007 -11.4.1)

<u>Provide Fillet weld of 6 mm for vertical stiffener on both sides.</u>

## 7.0 Design of Cover plate.

Tension in a single bolt = 379 /5
= 75.8 KN

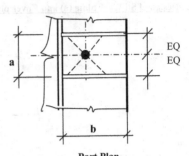

Part Plan

Width of support, a = 240 mm
b = 300 mm

<u>Bolt tension will have three edge supports.</u>

T per /support side = 25.267 Kn
M = T . a/2 = 25267 x 240 / 2 = 3032040 Nmm
Eff width at support = 2 ( a /2 ) = 240 mm
Thickness = 25 mm  Z = b t^2 /6 = 240 x 25^2 / 6 = 25000 mm3
Stress developed = M / Z = 3032040 /25000 = 121 Mpa
Permissible bending stress = 0.75 fy = 0.75 x 240 = 180 Mpa.  **Safe.** ( IS : 800 : 2007 - 11.3.3)

## Welding:

Horz : Base Plate to Gusset -            10 mm  Fillet weld all round
Vert :  Main gussets to column shaft -   8 mm   Fillet weld both side
        Aux gussets to column / main gusset -  6 mm  Fillet weld both side

## 8.0 Shear Lug

Method of design same as in Example 8.14.9

End

# Steel Structures

## Example 8.14.11 : Column base for Twin legged built-up I section (IS 800 - 2007)

### Reference.
    1.0 IS 800 : 2007    Indian standard code for general construction of steel.
    2.0 SP 6-3 : 1962   Hand book for Steel Columns and Struts

### Input Data.

#### Materials
Concrete grade =    M25                                                       (IS : 456 : 2000)
Characteristic strength, $f_{ck}$ =      25 N/mm2.
Bearing pressre on concrete, $\sigma_{cbc}$ =   8.5 N/mm2
Modulus of Elasticity, $E_c$ =       28500 N/mm2

Structural steel and anchor bolt :    IS 2062     Grade E 250 (Fe 410 W) A
Yield stress         $f_y$ =        230 N/mm2.
Ultimate tensile stress   $f_u$ =      410 N/mm2.
Modulus of Elasticity, $E_s$ =     200000 N/mm2

### Design force

#### Load at Column Base

| Load Case | Description | Axial P kN | Momen Mx kNm | Shear Sy kN |  |  |  |  | Load factor |
|---|---|---|---|---|---|---|---|---|---|
| 1 |  | 3421 | 823 | 501 |  |  |  |  | 1 |
| 2 |  | 4818 | 1050 | 553 |  |  |  |  | 1 |
| 3 |  | 1693 | 1553 | 635 |  |  |  |  | 1 |
| 4 |  | 4342 | 1359 | 756 |  |  |  |  | 1 |
| 5 |  | 1226 | 1588 | 603 |  |  |  |  | 1 |

**Dimensions**
**Column size:** *Twin columm*
Dist betn legs          d =      1500 mm
Depth of Col leg        D =      750 mm
Width of Col leg        W =      400 mm
Flange thickness        tf =     25 mm
Web thickness           tw =     16 mm
**Base plate :**
Length                  L =      2100 mm
Width                   B =      1050 mm
Thickness               ts =     28 mm

**Anchor Bolt :**
Nos. per side           n =      4
Nominal Dia.            ϕdia =   28 mm
Net area (0.7 gross area)  ϕarea = 431 mm2
(0.7. π.ϕdia^2 / 4)
Bolt pretension stress  fpt =    126 N/mm2

Edge distance of bolt holes   e =   100 mm
*Lever arm for Moment Capacity*  Y =   2000 mm
*( Y = L - e )*
**Gusset plate:**
Main gusset             h1 =     1500 mm    (IS800 2007 -7.7.2.3)
                        t1 =     20 mm
Aux gusset              h1 =     300 mm
                        t2 =     12 mm

Pt = Pretension on bolt
T = Tension in bolt

*Case I*

*Case II*

*Stress diagram*

## N-A depth and Moment capacity:
### Case I
Pretension in anchor bolts, $Pt = n \cdot fpt \cdot \phi area =$      $4 \times 126 \times 431 =$      217 KN

To find out NA depth, a1 with the conventional triangular distribution, we get
$$2 \cdot Pt + P = B \cdot \sigma cbc \cdot a1 / 2$$
$$a1 = 2Pt + P / (0.5B \times \sigma cbc)$$

Example:    Load Case    [ 1 ]    
- P = 3421 kN      Mx = 823 kNm
- B = 1050 mm      L = 2100 mm
- $\sigma cbc$ = 8.5 N/mm2      e = 100 mm

a1 = $2 \times 217000 + 3421000 / (0.5 \times 1050 \times 8.5) =$    864 mm
Pc = 0.5. B. $\sigma cbc$. a1 =    $0.5 \times 1050 \times 8.5 \times 864 / 1000 =$    3856 kN
Y = L - e =    2100 - 100 =    2000 mm
Lever arm, X = Y - a1/3 =    2000 - 864/3 =    1712 mm
Moment Capacity = Pc. Lever arm (X) =    $3856 \times 1712/1000 =$    6601 kNM
                                                                                                > Mx; Safe.,

Max. Base pr, $pmax$ = P/LB + Mx / (BL^2/6)
     = $3421000/(2100 \times 1050) + 823000000/(1050 \times 2100^2 / 6)$
     = 1.5515 +    1.0664 =    2.618 MPa      < 8.5 MPa ; Safe.,

Permisible bearing pr.on pedestal =    8.5 MPa.

### Case II
Pretension in anchor bolts, $Pt = n \cdot fpt \cdot \phi area =$      217 KN

To find out NA depth, a2, assuming bearing on concrete on rectangular stress block,
$$2 \cdot Pt + P = B \cdot \sigma cbc \cdot a2$$
$$a2 = 2Pt + P / (B \times \sigma cbc)$$

Example:    Load Case    [ 1 ]   
- P = 3421 kN      Mx = 823 kNm
- B = 1050 mm      L = 2100 mm
- $\sigma cbc$ = 8.5 N/mm2      e = 100 mm

a2 = $2 \times 217000 + 3421000 / (1050 \times 8.5) =$    432 mm
Pc = B. $\sigma cbc$. a2 =    $1050 \times 8.5 \times 432 / 1000 =$    3856 kN

Y = L - e =    2100 - 100 =    2000 mm
Lever arm, X = Y - a2/2 =    2000 - 432 / 2 =    1784 mm

Moment Capacity = Pc. Lever arm (X) =    $3856 \times 1784/1000 =$    6879 kNM
                                                                                                > Mx; Safe.,

Max. Base pr, $pmax$ = P/LB + Mx / (BL^2/6)
     = $3421000/(2100 \times 1050) + 823000000/(1050 \times 2100^2 / 6)$
     = 1.5515 +    1.0664 =    2.618 MPa      < 8.5 MPa ; Safe.,
Permisible bearing pr.on pedestal =    8.5 MPa.

**Base Pressure & Moment capacity :**

| Load Case | Axial P kN | Moment Mx kNm | Base pressure pmax MPa | Base pressure pmin MPa | Rectangular Stress block a2 mm | Rectangular Stress block X mm | Rectangular Stress block Mcap kNm | Triangular Stress block a1 mm | Triangular Stress block X mm | Triangular Stress block Mcap kNm | Observation Rect. block | Observation Trian block |
|---|---|---|---|---|---|---|---|---|---|---|---|---|
| 1 | 3421 | 823 | 2.62 | 0.49 | 432 | 1784 | 6878 | 864 | 1712 | 6601 | Safe. | Safe. |
| 2 | 4818 | 1050 | 3.55 | 0.82 | 589 | 1706 | 8959 | 1177 | 1608 | 8444 | Safe. | Safe. |
| 3 | 1693 | 1553 | 2.78 | -1.24 | 238 | 1881 | 4001 | 477 | 1841 | 3917 | Safe. | Safe. |
| 4 | 4342 | 1359 | 3.73 | 0.21 | 535 | 1732 | 8275 | 1070 | 1643 | 7849 | Safe. | Safe. |
| 5 | 1226 | 1588 | 2.61 | -1.50 | 186 | 1907 | 3166 | 372 | 1876 | 3115 | Safe. | Safe. |

*(Note: The above Loads are factored load.)*

**Base Plate thickness, ts :**

$$t_s = (2.5 \, w \, c^2 \, \gamma_{mo} / f_y)^{0.5} > t_f \qquad \text{(IS800 : 2007 -7.4.1.1)}$$

where,

Uniform base pr, $w = p_{max}/2 =$  1.87 MPa    $p_{max} =$  **3.73** MPa
projection beyond main gusset, $c =$  130 mm    $c =$ (1050-750-2x20)/2
partial safety factor for material, $\gamma_{mo} =$  1.10    (IS800 : 2007 -Table 5)
$f_y =$  230 MPa

$t_s =$  (2.5 x 1.87 x 130^2 x 1.1 / 230)^0.5 =  20 mm    > tf
ts provided =  28 mm    **Safe**

Tension in a single bolt, Pt =  54 KN
(431x126/1000)

panel dimension :
a =  1050/4 =     263 mm
b =  (2100-1500-16)/2 =    292 mm

Bolt tension will have three edge supports.

T per /support side =  18.000 Kn
M = T*a/2 =  18000 x 263 / 2 =   2367000  Nmm    **Part Plan**
Eff width at support = 2 ( a /2 ) =   263 mm
Thickness =  28 mm  Z = b t^2 /6 =  263 x 28^2 / 6 =    34365 mm3

# Steel Structures

Stress developed = M / Z = 2367000 /34365 = 69 Mpa
Permissible bending stress = 0.75 fy = 0.75 x 230 = 172.5 Mpa.       ( IS : 800 : 2007 - 11.3.3)
**Safe.**

## Gusset Plate :

Max. pressure at edge of base plate, pmax =   **3.73** Mpa

Taking section through main gusset at outer edge of column flange

Main gusset
height= 1500 mm
thick = 20 mm
[ Aux gussets ignored]
Base plate
width = 1050 mm
thick = 28 mm

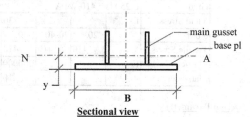

Sectional view

NA dist. from bottom of base plate:
y = 527 mm

$I_{NA}$ = 22769149400 mm4
Zx top = 22769149400/(1500+28-527) = 22746403 mm3
Zx bot = 22769149400 / 527 = 43205217 mm3

Projection outside column flange = (2100-1500-400)/2= **100** mm
$M_1$ = 3.73*100^2/2= 18650 Nmm = Upward moment from base pressure

Comp stress at top = 18650 / 22746403 = 0.001 Mpa
Tensile stress at bot = 18650 / 43205217 = 0.000 Mpa
Permissible stress = 0.75 x 230= 173 Mpa        ( IS : 800 : 2007 - 11.3.3)
**Safe.**

## Welding:

Horz : Base Plate to Gusset -         10 mm  Fillet weld all round
Vert:  Main gussets to column shaft -   8 mm   Fillet weld both side
       Aux gussets to column / main gusset -  6 mm  Fillet weld both side

## Shear Lug

Method of design same as in Example 8.14.9

End

## Example 8.14.12  Design of Steel Crane Gantry girder.  (IS 800 - 2007)
[ Section with Unequal flanges and Bent about the axis of max. strength x-x axis ]

**Reference.**
    1.0 IS 800 : 2007    General construction in steel - Code of practice

**Material**

| IS 2062 | | | | | |
|---|---|---|---|---|---|
| Grade | E 250 (Fe 410 W) A | | | | |
| Yield Stress | $f_y$ | 250 | 240 | 230 | MPa |
| d or t | | <20 | 20-40 | >40 | |
| Tensile Stress | $f_u$ | 410 | 410 | 410 | MPa |
| Elasticity | E | | | 200000 | MPa |
| Mat. safety fac | $\gamma_{m0}$ | | | 1.10 | |

(IS 800:2007 - Table 1)

(IS 800:2007 - 2.2.4.1)
(IS 800:2007 - Table 5)

**Design Parameter:**

Crane girder Span    L=    6 m.

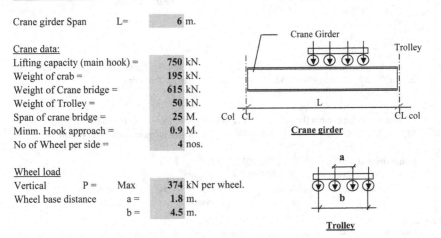

Crane data:
Lifting capacity (main hook) =    750 kN.
Weight of crab =    195 kN.
Weight of Crane bridge =    615 kN.
Weight of Trolley =    50 kN.
Span of crane bridge =    25 M.
Minm. Hook approach =    0.9 M.
No of Wheel per side =    4 nos.

Wheel load
Vertical    P =    Max    374 kN per wheel.
Wheel base distance    a =    1.8 m.
    b =    4.5 m.

[ Crane data and wheel load to be checked with manufacturer's load data and input information]

Live load on walk way =    5 kN/m2

Impact factor :    (IS 875: 1987 part 2 - 6.3)
a. Vertical wheel load    25%
b. Horizontal surge per track (any one)    10% weight of crab + lifting load
c. Longitudinal surge per track    5% of static wheel load.

# Steel Structures

## Calculation of wheel load

| | |
|---|---|
| Lifting capacity (main hook), W = | 750 kN. |
| Weight of crab = | 195 kN. |
| Weight of Crane bridge = | 615 kN. |
| Weight of Trolley = | 50 kN. |
| Span of crane bridge, S = | 25 M. |
| Minm. Hook approach = | 0.9 M. |
| No of Wheel per side = | 4 nos. |

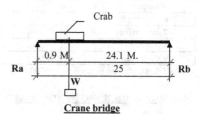

**Crane bridge**

### Wheel load :

| | Ra : | Rb : | |
|---|---|---|---|
| Lift | 723 | 27 kN | |
| Bridge | 308 | 308 kN | |
| Crab | 188 | 7 kN | |
| Trolley | 50 | 50 kN | |
| 25% | 228 | 9 kN | impact (crab+lift) |
| Sum | 1496 | 401 kN | |

Wheel load maxm. = 1496 /4 = **374** kN

## Crane girder member section

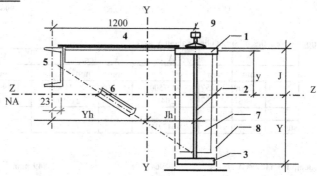

| Item | Description | nos | W mm | Thk mm | Len m. | Section | Spcg. m. | Wt/m kN | Remarks |
|---|---|---|---|---|---|---|---|---|---|
| 1 | Top flng. | 1 | 300 | 25 | | | | 0.589 | |
| 2 | Web plate | 1 | 1150 | 12 | | | | 1.083 | |
| 3 | Bot. Flange | 1 | 250 | 25 | | | | 0.491 | |
| 4 | Surge girder | 1 | vert | | 1.2 | ISA 65 65 6 | 1 | 0.070 | |
| | lacing bars | 1 | diag | | 1.7 | ISA 65 65 6 | 1 | 0.098 | |
| | W/way chqd pl | 1 | 1200 | 6 | | | | 0.565 | |

*[contd]*

| | | A | h | A h | | | ISMC 250 | | | 0.304 |
|---|---|---|---|---|---|---|---|---|---|---|
| 5 | Surge beam | 1 | | | | | ISMC 250 | | | 0.304 |
| 6 | Stay angle | 1 | | | | 1.70 | ISA 65 65 6 | | 2 | 0.049 |
| 7 | Int. stiffner | 2 | 100 | 8 | 1.10 | | | | 0.9 | 0.154 |
| 8 | Bearing stiff. | 2 | 300 | 25 | 1.22 | | | | | 1.437 |
| 9 | Crane rail | 1 | CR 60 | | | | | | | 0.450 |

Overall depth of main girder = 1200 mm             sum   **5.289** kN/m

Sectional properties of Main Girder about major axis :

| item | A | h | A h | y | $I_{CG}$ | $A y^2$ | Izz |
|---|---|---|---|---|---|---|---|
| | $mm^2$ | mm | $mm^3$ | mm | $mm^4$ | $mm^4$ | $mm^4$ |
| 1 | 7500 | 1187.5 | 8906250 | 560.5 | 390625 | 2356201875 | 2356592500 |
| 2 | 13800 | 600 | 8280000 | 27 | 1520875000 | 10060200 | 1530935200 |
| 3 | 6250 | 12.5 | 78125 | 614.5 | 325521 | 2360064063 | 2360389583 |
| sum | 27550 | | 17264375 | | | Izz = | 6247917283 |

h = CG dist of items from bottom line      $I_{zz} = I_{CG} + A y^2$
y = CG dist of items from N.A ( Z - Z )

NA depth from bottom, $Y = \Sigma A \cdot h / \Sigma A =$   17264375 / 27550 =   **627** mm.
**J =**   1200 - 627 =   573 mm.
$Zz^{top}$ = Izz / J =   6247917283 / 573 =   10903870   $mm^3$
$Zz^{bot}$ = Izz / Y =   6247917283 / 627 =   9964780   $mm^3$

Sectional properties of Surge Girder about Y axis :

| item | A | h' | A h' | y' | $I_{CG}$ | $A.y'^2$ | Iyy |
|---|---|---|---|---|---|---|---|
| | $mm^2$ | mm | | mm | $mm^4$ | $mm^4$ | $mm^4$ |
| 1 | 7500 | 1257 | 9427500 | 408 | 56250000 | 1249914711 | 1306164711 |
| 5 | 3867 | 57 | 220419 | 792 | 2191000 | 2424194552 | 2426385552 |
| | | | | | | | |
| sum | 11367 | | 9647919 | | | Iyy = | 3732550264 |

NA depth from left = A . h' / A =   9647919 / 11367 =   **849** mm.
**Jh =**   1257 - 849 =   408 mm.   $Zy^{Right}$ = Iyy / Jh =   9143156   $mm^3$
**Yh =**   849 - 57 =   792 mm.   $Zx^{Left}$ = Iyy / Yh =   4714211   $mm^3$
**y' =**   CG dist from N.A. (Z - Z )
**Iyy =**   $I_{CG}' + A'.y'^2$
**h' =**   CG dist from left

# Steel Structures

**Load on Crane girder**

*Vertical load*

(i) DL : Self weigt including surge girder and walk way cheqd plate and rail, $w_{DL}$ =     5.29 kN/m

(ii) LL :     on walkway, $w_{LL}$ = 1.2 x 5 = 6 kN/m

(iii) CR : Wheel load from crane

    P =     374 kN

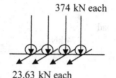

23.63 kN each

*Horizontal Load*

(iv)    Horizontal surge per track = ( Weight of Crab + Lifting load ) x Impact factor
                             = (195 + 750) x 0.1
                             = 94.5 kN
       No of wheels per side =     4       Horz load per wheel =   23.63 kN

(v)    Longitudinal surge per track is 5% of static wheel load, which is required for column design.

**To calculate actual moment and shear for crane girder member section**
*Bending moment:*

a) for wheel load : Resultant and nearest wheel position is bisected by CL of girder
          *[This positon does not apply when b exceeds 0.586L]*

P =     374 kN
a =     1.8 m
b =     4.50 m
L =     6 m
*Deflexion :*
$\delta_{max} = 63PL^3/1000EI + 5wL^4/384EI$
     = 4.23 mm
     < L/1000; Safe.
$\delta_{allow}$ = L/1000 =    6     mm
( IS 800 2007 Table 6 )

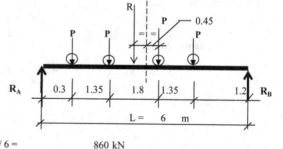

$R_A$ =    374 x (1.2 + 2.55 + 4.35 + 5.7) / 6 =        860 kN
$R_B$ =    4 x 374 - 860 =        636 kN
Mv 1 =     860 x 3 - 374 x 2.7 - 374 x 1.35 =     **1065 kNm**
similarly,
for horizonal surge, load per wheel =     23.63 kN
Mhorz =     1065 x 23.63 / 374 =     **67 kNm**

b) For self weight          Mv 2 = 5.289 x 6^2/8 =    **23.80 kNm**
c) For LL on walkway      Mv 3 = 6 x 6^2/8 =       **27.00 kNm**
<u>for load combination, DL + CR + LL</u>

               **Actual moment, M :**      Mv =    1065 + 23.8 + 27 =       **1116 kNm**
                                           Mh =         **67 kNm**

<u>Shear force:</u>
a) for wheel load
       P =       374 kN

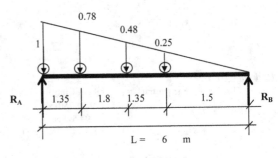

<u>Influence Line dia for shear</u>

$R_{A1}$ =    374 x (1 + 0.78 + 0.48 + 0.25) =      939 kN.

b) for self weight       $R_{A2}$ =    5.289 x 6 x 0.5 =        16 kN
c) for LL on w/way      $R_{A3}$ =    6 x 6 x 0.5 =            18 kN

<u>for load combination, DL + CR + LL</u>

              **Actual Shear,**      V =     16 + 939 + 18 =        **973 kN**

                   Fh for crane surge =      939 x 23.63 / 374 =       **59 kN**

# Steel Structures

## Design strength of main girder Md and Vd.

*Limiting width to thickness ratio* (IS 800:2007 - Table 2)

Top flange:  $0.5b/t_f = 150 / 25 =$    6.00    Plastic
Bot flange:  $0.5b/t_f = 125 / 25 =$    5.00    Plastic

$\varepsilon = (250 / f_y)^{0.5} = (250 / 240)^{0.5} = 1.02$

Plastic, $\lambda_r =$     $8.4\varepsilon = 8.4 \times 1.02 =$    9
Compact, $\lambda_p =$    $9.4\varepsilon = 9.4 \times 1.02 =$    10
Semi compact, $\lambda_{sp} =$    $13.6\varepsilon = 13.6 \times 1.02 =$    14

Web :      $d/t_w = 1150 / 12 =$    **95.83**
           $67\varepsilon = 67 \times 1.02 =$    68.34

**Web is susceptible to shear buckling & design bending strength will governed by elastic critical moment as per 8.2.2.**    (IS 800:2007 - 8.2)

## Shear buckling check    (IS 800:2007 - 8.4.2)

Resistance to shear buckling to be checked,
(i)   when   $d/t_w >$    $67\varepsilon_w$    for web without stiffeners
                         $= 67 \times 1.02 =$    **68**

(ii)  when   $d/t_w >$    $67\varepsilon_w(K_v/5.35)^{0.5}$    for web with stiffeners
                         $= 67 \times 1.02(12.74/5.35)^{0.5}$
                         $=$    **105**

where $K_v = 5.35 + 4/(c/d)^2$ for $c/d \geq 1$
c = spcg. of stiffener and d is depth of web.                $c/d =$  $900/1150 =$    **0.78**
so, let us assume, $K_v =$  $5.35 + 4/(900/1150)^2 =$    11.88
OR,  $K_v = 4 + 5.35/(c/d)^2$ for $c/d < 1$
             $K_v =$  $4 + 5.35/(900/1150)^2 =$    12.74
                                              $K_v =$  **12.74**

In this case,   **Provide web stiffener; Shear buckling check not necessary.**

## Design Shear strength of web, Vd    (IS 800:2007 - 8.4)

The nominal shear strength of web, Vn is goverered by plastic shear resistance.

$V_n = V_p = A_v f_{yw} / (3)^{0.5}$    (IS 800:2007 - 8.4.1)

Shear Area, $A_v = d \times t_w =$    $1150 \times 12 =$    13800 mm2
Yield strength of web, $f_{yw} = f_y =$                      250 Mpa

$V_n = V_p =$  13800 x 250 /3^0.5/1000 =   1992 kN.

$V_d = V_n / \gamma_{mo}$    (IS 800:2007 - 8.4)

$V_d =$  1992 / 1.1=   **1811** kN

Actual design shear , V =   973 kN <   0.6Vd  = 0.6 x 1811 =   1086.6 KN

## Design flexural strength, Md      (IS 800:2007 - 8.2.1.2)

The top chord is connected to surge girder lacing system and the web not susceptible for shear buckling (design shear < 0.6Vd); hence the flexural moment will be detrmined as laterally supported beam.

Nominal flexural strength, $M_n = M_p = f_y Z_p =$ Plastic moment
$f_y$ = Yield stress =    240 Mpa
$Z_p$ = plastic sectional modulus about the Z axis = $\Sigma A y$
  =   300 x 25 x ( 573- 12.5) + 250 x 25 x ( 627- 12.5) + 2 x (1150/2) x 12 x (1150/4)
  =   12011875  $mm^3$
$Z_e$ = elastic sectional modulus about the Z axis
  =   10903870  $mm^3$

$M_n =$  240 x 12011875/1000000 =    2883 kNm.  (= $M_p$)

Design flexural strength, $M_d = \beta \cdot M_n / \gamma_{mo} < 1.2 Z_e f_y / \gamma_{mo}$   (IS 800:2007 - 8.2.1.2)
$M_d =$ 1 x 2883 / 1.1 =    **2620.9** kNm.

$1.2 Z_e f_y / \gamma_{mo}$   =   1.2 x 10903870 x 240 / 1.1/1000000 =    2855 kNm   > Md; Okay.

where , $\beta =$    1    for plastic and compact section.
and   $\beta = Z_e / Z_p =$   12011875 / 10903870 =    0.91 for semicompact section.
here, the flange is    Plastic     and so,  $\beta =$    1

## Strength design of main girder

| Limit state design | Working stress design |
|---|---|
| (IS 800:2007 - Section 8) | (IS 800:2007 - Section 11) |
| Factored design moment = M | Actual design moment = Ms |
| Design flexural strength of the section = Md | Permissible flexural strength = Ma |
| Factored design shear = V | Actual design shear = Vs |
| Design shear strength of the section = Vd | Permissible shear strength = Va |
| M <= Md<br>V <= Vd | Ms <= Ma<br>Vs <= Va |

*[contd]*

| Load | 1.2DL + 1.2CR + 1.2 LL | DL + CR + LL |
|---|---|---|
| Span<br>End | L = 6 M<br>simply supported | L = 6 M<br>simply supported |
| Moment | *Factored load:*<br>$Mv = 1.2 \times 1116 = 1339$ kNm<br>$Mh = 1.2 \times 67 = 80$ kNm<br>$M = (Mv^2 + Mh^2)^{0.5}$<br>$= (1339^2 + 80^2)^{0.5}$<br>$= 1341$ kNm<br><br>$Md = 2621$ kNm<br><br>**Md > M; Safe.** | *Unfactored load:*<br>$Ms = (Mv^2 + Mh^2)^{0.5}$<br>$= (1116^2 + 67^2)^{0.5}$<br>$= 1118$ kNm<br><br>**Allowable flexural strength, Ma :** (IS 800-11.4.1)<br>*For laterally supported beam:*<br>$fabc = Ma/Zz^{top}$<br>$Ma = fabc. Zz^{top} = 0.66 \, fy \, Zz^{top}$<br>$= 0.66 \times 240 \times 10903870 / 10^6$<br>$= 1727$ kNm<br><br>$fabt = Ma/Zz^{bot}$<br>$Ma = fabt. Zz^{bot} = 0.66 \, fy \, Zz^{bot}$<br>$= 0.66 \times 240 \times 9964780 / 10^6$<br>$= 1578$ kNm<br><br>Allowable Flexural mom, Ma = 1578 kNm.<br>Actual Design Moment, Ms = 1118 kNm.<br>**Ma > Ms; Safe.** |
| Shear | *Factored shear*<br>$V = 1.2 \times 973 = 1167.6$ kN<br>$Vd = 1811$ kN<br>**Vd > V; Safe.**<br><br>Horz shear will be transferred to column by surge girder. | *Unfactored shear*<br>$Vs = 973$ kN<br><br>**Shear strength, Va :** (IS 800-11.4.2)<br>Pure shear<br>$Va = \tau_{ab} . Av$ , where Av shear area<br>$= 0.4 \, fy \times Av$<br>$= 0.4 \times 250 \times 1150 \times 12 / 1000$<br>$= 1380$ kN<br><br>Allowable Shear, Va = 1380 kNm.<br>Actual design shear, Vs = 973 kNm.<br>**Va > Vs; Safe.** |

*[contd]*

|   | | |
|---|---|---|
| weld betn flange & web | **Weld between flange to web:**<br><br>Required Shear at joint, F = V Q / Iz<br><br>Q = A y =    300x25x560.5=   2101875 mm3<br>Iz =     6247917283    mm4<br>F =    1167600 x 2101875 / 6247917283<br>   =      393 N/mm<br><br>Force per mm run =    393/2 =    197 N<br>                      < w; OK.<br>Capacity of         6 mm fillet weld, w :<br>w =0.7x0.6x1100=    462 kg/cm =    462 N/mm<br>weld stress = 110( Mpa<br><br>*Provide*<br>Top weld :    Full penetration but weld.<br>Bott weld :    Fillet weld 6 mm per side. | **Weld between flange to web:**<br><br>Required Shear, F = Vu Q/Ix<br><br>Q = A y =         2101875 mm3<br>Ix =     6247917283    mm4<br>F =    973000 x 2101875 / 6247917283<br>   =      327 N/mm<br><br>Force per mm run =    327/2 =    164 N<br>                      < w; OK.<br>Capacity of         6 mm fillet weld, w :<br>w =0.7x0.6x1100=    462 kg/cm =    462 N/mm<br>weld stress = 110( Mpa<br><br>*Provide*<br>Top weld :    Full penetration but weld.<br>Bott weld :    Fillet weld 6 mm per side. |

**Strength design of stiffeners**

**Intermediate web stiffeners**      (on both sides of web plate)      (IS 800:2007 - 8.7.2)

| size | | | | | | |
|---|---|---|---|---|---|---|
| | wide | tb | 100 mm | $\varepsilon = (250/fy)^{0.5} =$ | 1.0 | |
| | thick | tq | 8 mm | depth of web, d = | 1150 mm | |
| | spcg | c | 900 mm | web thickness, tw = | 12 mm | |

design        Max allowable width =    $20 \cdot tq \cdot \varepsilon$            (IS 800:2007 - 8.7.1.2)
                                  =    20 x 8 x 1
                                  =    160 mm     tb < allowable limit; OK.

           Moment of area required, $Is_{rqd} >= 1.5 \, d^3 \, tw^3 / c^2$        (IS 800:2007 - 8.7.2.4)

             $Is_{rqd} =$    $1.5 \times 1150^3 \times 12^3 / 900^2 =$     4866800     mm4

# Steel Structures

Is provided = 8 x (2x100)^3 / 12 =     5333333   mm4

Is prov > Is reqd.; OK.

The stiffeners will be welded to web and underside of top flange by 6mm fillet weld on both sides. The bottom of intermediate stiffeners will be separated from tension flange of girder by a gap of 50 mm.

## End bearing stiffeners                                             (IS 800:2007 - 8.7.4)

The bearing stiffener is welded to end of girder and shall transfer the end shear on center of column keeping a gap of 6 mm (3 + 3) for erection clearance for next girder sitting on the same column cap. This gap will be filled by pack plate.

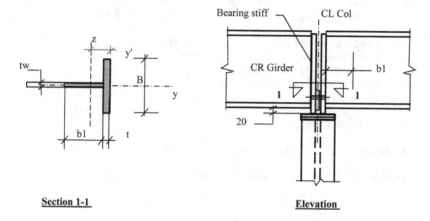

**Section 1-1**                                                    **Elevation**

The bearing stiffener will be designed as column.

    The dimension of bearing stiffener

| | | | | | |
|---|---|---|---|---|---|
| B = | 300 mm | bi = 20 tw = | 240 mm | Length, L = | 1207.5 mm |
| t = | 25 mm | tw = | 12 mm | A = | 10380 mm2 |

$L =$   0.5 x 25 + 1150 + 25 + 20 =     1207.5 mm
$y' =$   300 x 25 x 12.5 + 240 x 12 x 145 / (300 x 25 + 240 x 12)  =        46 mm
$I_z =$   300 x 25^3/12 + 300 x 25 x (46-12.5)^2 + 12 x 240^3/ 12 + 240 x 12 x (25 + 120-46)^2
 =    50858380    mm4
$r_z =$   (50858380 / 10380)^0.5  =       70 mm
$I_y =$   25 x 300^3/12 + 240 x 12^3/12=     56284560    mm4
$r_y =$   (56284560 / 10380)^0.5 =    73.64 mm

$KL/r_{min}$    $0.7 \times 1207.5 / 70 =$    12

*Axial Compression Strength*                                                          (IS 800:2007 - 7.1)

Design compression strength, $Pd = Ae \cdot fcd$    where, $Ae =$ effective sectional area
$fcd =$ design compressive stress

$$fcd = \chi\, fy / \gamma_{mo} \quad <= fy / \gamma_{mo}$$    (IS 800:2007 7.1.2.1)
$$\chi = 1 / [\phi + (\phi^2 - \lambda^2)^{0.5}]$$
$$\phi = 0.5\,[\,1 + \alpha(\lambda - 0.2) + \lambda^2\,]$$
$$\lambda = (fy / fcc)^{0.5}$$
$$fcc = \pi^2 E / (KL/r)^2 \quad \text{Euler buckling stress}$$

Now, putting following values into above equation

$\alpha =$  0.49    $E =$  200000 Mpa    $fy =$  240 Mpa
$\gamma_{m0} =$  1.10    $KL/r =$  12    $Ae =$  10380 mm2

$fcc =$    $3.14^2 \times 200000 / 12^2 =$    13694 MPa
$\lambda =$    $(240 / 13694)^{0.5} =$    0.13
$\phi =$    $0.5[1 + 0.49 \times (0.13 - 0.2) + 0.13^2] =$    0.49
$\chi =$    $1 / [0.49 + (0.49^2 - 0.13^2)^{0.5}] =$    1.039
$fcd =$    $1.039 \times 240 / 1.1 =$    227 Mpa    $<= fy / \gamma_{mo}$    $<=$ 240 / 1.1
                                                                $<=$ 218 Mpa

$fcd =$    **218** Mpa

**Pd** $= Ae \cdot fcd =$    $10380 \times 218 / 1000 =$    **2356** kN

Load on Bearing Stiffener. $P =$    End shear of Cr girder $=$    973 kN

| Limit state design | Working stress design |
|---|---|
| (IS 800:2007 - Section 8) | (IS 800:2007 - Section 11) |
| Factored Axial Load = P<br>Design comp. strength of the section = Pd<br><br>P <= Pd | Actual Load = Ps<br>Permissible comp. strength = Pa<br><br>Ps <= Pa |

*[contd]*

# Steel Structures

| Load | 1.2DL + 1.2CR + 1.2 LL | DL + CR + LL |
|------|------------------------|--------------|
| $P =$ <br> $P_d =$ | $1.2 \times 973 =$    1168 kN <br>                 2356 kN <br> **$P_d > P$; Safe.** | $P_s =$   973 kN    $A_e =$   10380 mm2 <br> Actual comp stress = $f_c = P_s/A_e$ <br>                     = 973×1000/10380 <br>                     = 94 Mpa <br> Permissible comp. stress, $f_{ac} = 0.60 f_{cd}$ <br>                     = 0.6 × 218 <br>                     = 131 Mpa <br> **$f_c < f_{ac}$; Safe.** <br> where, $f_{cd} =$   218 Mpa. |

## Surge girder

Refer the skecth showing crane girder member section.

The surge girder will transfer the horizontal load and moment through its lacing system (item 4) and Chord members (item 1 and 5) to building column.

### Load :

| Horz surge from crane | Vert. load from W/way |
|---|---|
| $M_h =$   67 kNm | $M_v =$   23.8 + 27 =   50.80 kNm |
| $F_h =$   59 kN. | $V =$   (5.289 + 6) × 6 /2 =   33.87 kN |

### Member 5

This member will span between columns carrying vertical load from W/way and axial force being a chord of surge girder.

Axial load = $M_h / I_{yy}$ (of surger girder) = 67000000 / 3732550264 =    0.018 N
                                                     ( Negligible; hence ignored for strength design.)

$M =$   50.8 / 2 =    25.4 kNm
$V =$   33.87 / 2 =    16.94 kNm

Section provided :    **ISMC 250**      $Z_x =$    305300 mm3
                                               $A_v =$    1775 mm2

Allowable moment
$M_a = 0.6 f_y Z_x =$    0.6 × 250 × 305300 /1000000        (IS 800:2007 -11.4.1)
                    =    45.80 kNm
                      **> M ; Safe,**

Allowable Shear
Va = 0.6 fy Av =   0.6 x 250 x 1775 /1000            (IS 800:2007 -11.4.2)
              =   266.25 kNm
                  > V ; Safe,

*Lacing Member 4*

L =         1200 mm        Diagonal length =    1697 mm
Axial force, Ps =    59 x 1697 / 1200 =    83 kN

Section provided :    ISA 65 65 6        Ae =    744 mm2

Allowable comp. load
Pa = 0.6 fy Ae =    0.6 x 250 x 744 /1000             (IS 800:2007 -11.3.1)
              =    112 kN
                   > Ps ; Safe,

*Stay angle Member 6*

L =    1697 mm        Spacing =    2 m

UDL on w/way =    (5.289 + 6) /2 =    5.6445 kN/m

Axial force, Ps =    5.6445 x 2 x (1697 / 1200) =    16 kN

Section provided :    ISA 65 65 6

| | | |
|---|---|---|
| Ae = | 744 | mm2 |
| rvv = | 12.6 | mm |
| L = | 1697 | mm |
| $\alpha$ = | 0.49 | |
| E = | 200000 | Mpa |
| fy = | 250 | Mpa |
| $\gamma_{m0}$ = | 1.10 | |
| KL/r = | 114 | |
| fcc = | 150 | MPa |
| $\lambda$ = | 1.291 | |
| $\phi$ = | 1.6 | |
| $\chi$ = | 0.393 | |
| fcd = | 89.32 | Mpa |

(IS 800:2007  7.1.2.1)

Allowable comp load
Pa = 0.6 fcd Ae =    0.6 x 89.32 x 744 /1000    =    40 kN    (IS 800:2007 -11.3.1)
                                                     > Ps ; Safe,

End

# 9 Reinforced Concrete and Associated Work

## 9.1 GENERAL

In this chapter, we will go through the design procedures of elementary reinforced concrete members and the limitations that are followed in industrial practices. We will also discuss several types of foundation systems in reinforced cement concrete.

Three methods of strength design are followed: the working stress method, the limit state method, and the ultimate load method.

The working stress method is old and simple in concept. It follows the elastic theory of reinforced concrete following Hooke's law. In this method, tensile stress is only taken by steel. The strain in steel is equal to the strain in concrete. The compressive stress in steel is equal to the compressive stress in concrete multiplied by a factor called the modular ratio of the concrete: the modular ratio is m = Es/Ec (the ratio of modulus of elasticity of steel to that of concrete). But this value varies with the grade of concrete mix. So, it is considered as, m = 280 / σcbc, where σcbc is the permissible stress in bending compression for different grades of concrete. The permissible stress in steel and concrete should not exceed the value obtained by dividing the characteristic strength of material by the factor of safety (three for concrete and 1.8 for steel). The working stress method gives a safe design and less deflection or cracks, but the section sizes are higher than other methods of design.

The ultimate load method considers ultimate strengths at ultimate load.

The limit state method is currently chosen as the best method in strength design. It allows the design load up to its characteristic load multiplied by the *partial safety factor appropriate to the nature of the loading and the limit state* being considered. The design strength of the material is characteristic of the material strength divided by the *partial safety factor appropriate to the material* (1.5 for concrete and 1.15 for steel) *and the limit state* being considered.

There are two limit state concepts: one is the limit state of collapse and another is the limit state of serviceability.

The limit state of collapse of a structure could be assessed from the rupture or buckling due to elastic or plastic instability. At this limit state, the structure will undergo collapse or become unstable under any of the combinations of expected overload.

DOI: 10.1201/9781003211754-9

The limit state of serviceability ensures that the structure will not exceed the permissible value of deflection and the crack width should be within acceptable limits for all combinations of loads (dead load, live load, wind load/seismic load, and others) multiplied by the partial safety factor of the limit state of serviceability.

Member strength design examples are furnished at the end of this chapter.

## 9.2 ROOF SLABS

The design of roof slabs is done to withstand loads from self-weight and finishes at soffits and the top including the weight of roof treatment work. The intensity of the roof life load for accessible and inaccessible conditions will be selected as per load data shown in the design basis report. Equipment like cooling towers, air conditions, ventilation units, roof exhaust fans, and so on that are installed on roofs. The estimation of the load from screed concrete on the top should be done considering additional thickness that may be required to maintain the slope. The design drawing should mention the thickness of the screed in the plan showing the slope direction. Unless specified otherwise, local codes of practice should be followed for the design intensity of the live load on the roof.

The designer should respect the design limits for minimum slab thickness, deflection, span by effective depth ratio, and minimum reinforcements as per applicable codes (ACI 318/IS: 456/BS 8110) or equivalent.

The control of temperature crack as a consequence of temperature variation at the top (outside) and bottom (inside) should be considered in the design of buildings in severe environments. Openings in roof slabs for pipes or duct entries, wherever necessary, should be provided with protective concrete curbs around the opening to prevent rainwater flow. Pipes and ducts should be fitted with a suitably designed hood or overhead rain shed.

## 9.3 FLOOR SLABS

Design limitation for minimum reinforcement, span/depth ratio for deflection, and crack control as stated for roof slabs is also applicable for floor slabs.

Floor slab panels in industrial buildings are designed to withstand equipment and superimposed live loads. There are pedestals, openings, or cutouts for equipment and piping. Additional reinforcements are provided around the openings to replace cutoff re-bars and to resist diagonal cracking. Openings of more than 300 mm should have protective curbs around them. Large openings for cable shafts, pipe hatches, or ventilation ducts and stairwells should have handrails above the curbs.

Equipment bases or pedestals carrying loads higher than the design floor intensity should be provided with crossbeams. Similarly, edge protection beams are necessary for long cutouts. Edge protection angles are necessary for openings around the chequered plate or grating floor. The details are standardized for uniform patterns and leveling. Suitable drainage outlet points connected to downcomer pipes are provided for firewater and emergencies. Expansion joints in floor slabs for large buildings are designed with gaps (50 mm) with edge angles as per standard details.

Levels of slabs should be uniform as per floor finish detailing.

## 9.4 GROUND FLOOR SLABS

Ground floor slabs rest on compacted fill after the completion of backfill around the foundations, and the plinth area is compacted up to the desired level. There are trenches, buried pipes and pits, and equipment foundations closely placed on the ground floor, so it is difficult to achieve the desired compaction unless the plinth filing is done by sand or granular soil.

The unloading bay of the ground floor is subjected to heavy loading. The backfill soil has to be rolled and compacted to get maximum dry density (90% modified proctor density) at optimum moisture content. It is advisable to use crushed stone, morrum, or small-sized boulders or stones mixed with sand and gravel as subgrade in this zone.

Contraction and expansion joints are provided to control cracking due to shrinkage and seasonal temperature variations. The ground floor slab is isolated from pedestals bearing vibratory machines by isolation joints.

For strength design, this slab can be considered as spanning over a soft spot of 3 m and is reinforced at both layers. The grade of concrete should not be less than M25 with a bottom cover of 75 mm if cast directly on compacted earth. Otherwise, this slab is cast on 50 mm thick leveling concrete underlain by compacted subgrade or rolled over soil. Polythene sheets or membrane is also laid on the subgrade to prevent subsoil corrosion.

Paving in the outdoors other than roads, such as slabs in car parking areas, which are subject to uniform loading of 5 kN–30 kN per square meter, can be defined under this category.

## 9.5 WALLS

Reinforced concrete wall structures are widely seen in industrial construction. They are generally used in basements, water retaining structures, large tank foundations, silos, bunkers, chimney wind shields, cooling towers, mining and coal handling substructures, earth retaining walls, riverfront structures, cable vaults, tunnels, trenches and manholes, crossing culverts, and many other cases.

The design of the wall section is done for withstanding flexure stress combined with axial forces in accordance with applicable design codes ACI 318/IS: 456/BS 8110. Design limits like minimum thickness and reinforcement should be as per code. The construction of long walls is done in alternate panels to control shrinkage. Contraction or expansion joints, isolation joints, and crack control joints are provided as per standard details.

The design of walls in some control room buildings is done as blast-resistant walls. Fire barrier walls in transformer yards are designed for limited time (2 h) fire resistance. For deep underground construction, temporary or permanent type diaphragm wall structures are found to be effective and useful.

Precast and prestressed units are used in floors, walls, the lining of reservoirs and canals, covers of trenches or culverts, and so on.

## 9.6 BEAMS

Beams are basically known as flexural members. The designer should look into the design limitations stated in codes and standards for the selection of effective spans, slenderness limits for lateral stability, i.e., allowable clear distance between lateral restraints, cover to concrete and reinforcement bar sizes, minimum reinforcements, and span by effective depth ratio to limit deflection before the selection of beam sections.

Vertical stiffness of beam sections in a bay should be planned according to distribution of load finally transferred to the column section. For example, Beam A is supported by Beam B, here the vertical stiffness of Beam B should be higher than or equal to Beam A. The array of floor beams should be governed by equipment layout and floor loading.

Examples of design calculations are shown at the end of this chapter.

## 9.7 COLUMNS

The column is an axially loaded member with moments about major axis or minor axis or about both axes.

A column is considered as a *short column* when both the *slenderness ratios* [the ratios of effective heights in respect of major axes (Lx)/depth (D) and effective heights in respect of minor axes (Ly)/width (B)] are less than:

a) **12** as per IS: 456: 2000 – 25.1.2
b) **10** (unbraced) and 15 (unbraced) as per BS 8110. Part 1 – 3.8.1.3
c) **22** (unbraced) and 34 + 12(M1/M2) as per ACI 318–19 – 6.2.5

It should otherwise be considered as *slender*.

The columns sections have design limits for sizing and effective length as stated above and minimum and maximum reinforcement. The required strength should be calculated in accordance with the factored load combinations and analysis procedures.

Sample design calculations are shown at the end of this chapter.

## 9.8 FOUNDATIONS INCLUDING PILE AND CAISSON

In this section, we will discuss all types of foundation systems commonly used in industrial projects and how the load is transmitted to the soil. It is assumed that the reader has studied textbooks on soil mechanics.

# Reinforced Concrete and Associated Work

## 9.8.1 FOUNDATION TYPES AND SELECTION CRITERIA

To select an appropriate type of foundation, we need to see the loading pattern, soil bearing capacity, and selection criteria, for example, settlement limitation by a superstructure framing system.

The role of a foundation is to carry the applied load and transfer it to the supporting soil system without crossing the settlement limits. The shape and size of the foundation structure, i.e., pedestal, footing, pile and pile cap, raft, wall, and grade beams, are determined to ensure proper load transmission and then structurally designed to meet the requirements of the strength design of the concrete without cracking beyond the limits.

## 9.8.2 LOADING PATTERN

Here, the pattern of loading means the nature of the forces that are applied on the foundation structure and then how they are transferred to the soil mass below the foundation. It is static when the magnitude of the load does not change over time. Dynamic load is a load that is not static. An example of a static load is the dead load of a building or structure, and dynamic loads are wind loads, moving live loads, and the transient load of a machine at operating conditions. The weight of a slab or pavement of a runway or road is a dead load but the wheel loads over it are dynamic. The factor of safety of foundations for a dynamic load will be more than that of a static load.

The type of load that we get for the design of foundation systems are as follows:

a) Concentrated or point load
b) Uniform distributed load
c) Line load
d) Combination of the above loads

After the analysis of the frame structure resting on column pedestals, a wall bearing strip footing, or a machine foundation block, the forces according to combinations of static, dynamic, and other load cases are resolved in three-dimensional reactions as stated below:

a) Vertical ($F_x$, $F_y$, and $F_z$)
b) Moments ($M_x$, $M_y$, and $M_z$)
c) Shear ($S_x$, $S_y$, and $S_z$)

The role of the foundation engineer is to provide a foundation system or substructure that can transform the applied load in the form of uniform distributed compression and shearing load on the soil below the foundation. Because the soil structure can only take vertical compression and shear, it cannot take tension.

For example, an isolated footing transforms the concentrated load applied on a column pedestal into a distributed load spread on the soil below the footing. Similarly, a combined footing or raft foundation converts the point loads into a distributed load spread onto the bearing soil stratum.

Pile and well foundations are called deep-seated foundations because they transfer the load into the deep underground soil stratum. This applied load is resisted by shaft friction and tip bearing compression of piles. The capacity of a single pile is the sum of its shaft friction (perimeter × soil friction) and bearing capacity at the tip (area at tip × soil bearing capacity at that level). Piles are driven in groups. Load/reaction per any pile in the group due to applied load onto the pedestal should not exceed the allowable capacity of the pile.

Figure 9.1 shows the load transformation into the foundation system.

The picture only shows the distribution of vertical forces. But in practical cases, the forces on the surface of the soil just below the foundation bottom are not always vertical. They appear trapezoidal for moments.

Figure 9.2 shows the distribution for vertical loads and moments and shear on the supporting soil.

**Resistance to shear forces** below ordinary footings is given by frictional resistance or sliding resistance of soil over the contact surface. This resistance is a kind

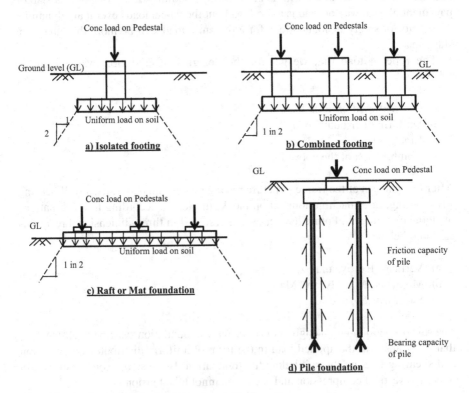

**FIGURE 9.1** Load distribution from foundation to soil.

# Reinforced Concrete and Associated Work

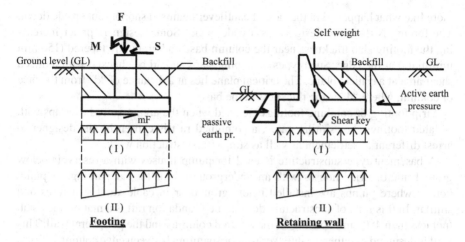

**FIGURE 9.2** Reaction from vertical load and moment.

of adhesion or friction dependent on the roughness of the concrete surface below the footing and soil property (angle of internal friction). This resisting force is equal to the vertical weight of the foundation multiplied by a factor, $\mu$. The value of $\mu$ depends on the soil or rock surface below the foundation (0.3 for ordinary soil to 0.5 coarse-grained soil to rock). Footing bases subjected to high lateral forces, like earth retaining walls, have enhanced sliding resistance by providing shear keys below the footing slab. Passive pressure of earth is also considered as a part of this resisting force.

This sliding check is an important part in foundation design and should not be neglected.

### 9.8.3 Type of Foundations

All kinds of foundations can be broadly divided into two types. One is directly resting on earth and the other type is resting on pile or caisson, i.e., deep-seated foundation.

#### 9.8.3.1 Foundations Resting on Soil

The net contact area of the foundation upon the soil should be computed by the designer, keeping in mind that the maximum load applied on soil should not be more than the safe bearing capacity of soil. The shape and flexibility of footing or raft are designed accordingly.

Building column foundations are usually designed with isolated spread footing. The size of the foundation depends on the load and effective area requirements. If vertical load is predominant, the footing can be made square in shape. For moment carrying foundations, the shape may be rectangular in plan, having a longer side placed along the direction of moment. The slab and column joints at the base are subjected to rotation during moment distribution. The force transfer at this joint is more like what happened in the case of cantilever beams at support but upside down. The footing is therefore designed as a wide beam. Some designers prefer increasing the footing slab thickness near the column base and gradually tapered (150 mm minimum) at the edge. Such types of sloping slabs should be checked with reduced shear area at a critical plane. The critical plane lies at a distance of "d" from the face of column, where "d" is the thickness at the base.

Strip footing or combine footings are used when the area required overlaps with isolated footings. For similar cases, a combined raft is chosen by the designer to arrest differential settlement as well as simplify construction work.

A basement-type substructure is used for pump houses with a reservoir below ground, multistoried houses with an underground floor for car parking, or pump houses where pumps are installed below ground or in cable vault structures and similar. In this type of substructure design, the foundation raft is a mat of thick slab (not less than 450 mm) housing all the internal columns and the peripheral wall. This wall is designed as either a cantilever wall or spanning between edge columns. These columns are monolithically cast with the wall and act like a buttress for the retaining wall. The minimum thickness of these walls should not be less than 300 mm, considering the ease of the placement of reinforcement bars in both faces and the minimum space for pouring concrete. The cover outside in contact with soil should not be less than 50 mm. This basement structure should be designed as a water-retaining structure.

There are other types of below-ground structures like concrete tunnels for carrying cables, conveyors pipes, and transportation systems. These structures should be checked for stability against buoyancy, water tightness, and differential settlement as well as strength design requirements. Box tunnels and open-top concrete ducts resting on soil are used for carrying water in industrial projects.

The foundation for large diameter tanks is designed as resting on soil where the quality of soil is not poor.

### 9.8.3.2 Foundations Resting on Pile or Caisson

When the safe bearing capacity of soil at the founding layer is less than the computed design pressure, pile or caisson foundation are used. This type of foundation is a called deep-seated foundation. Like isolated spread footings, the column pedestal is supported on a thick footing slab, termed as a pile cap. Pile caps rest on a group of piles, which are placed in a regular grid pattern in a spacing not less than 2.5 to 3 D, where D is the diameter of a pile. In the case of a square pile, D is the side of pile in the section. Piles are like long columns carrying the force from the pile cap and transfer it to the soil by friction along the shaft length and the bearing on the tip.

Design construction and testing of the pile foundation are available in standard codes of practice.

Indian standards: IS: 2911 Code of Practice for Design and Construction of Concrete Pile.

American standard: ACI 543 R Design Manufacture and Installation of Concrete Piles.

# Reinforced Concrete and Associated Work

Large diameter piles are termed caisson in the United States. The design is similar to piles as stated above.

### 9.8.3.3 Well Foundations

This type of structure is generally used for intake pump houses and bridge piers.

## 9.8.4 SELECTION OF FOUNDATION TYPES

The type of foundation selected is on the magnitude of load, i.e., design load intensity, soil bearing capacity of the site, and settlement criteria specified for the superstructures.

For example, turbine building columns in a power plant bear an axial load of 5000 kN to 12,000 kN together with moment and shear forces. The soil bearing pressure with standard sizes of foundation comes to around 250–400 $KN/m^2$.

In this case, soft to medium-hard soil will not be able to carry this load but a site with hard soil or rocks can bear the load. So, the turbine building, located at a soft soil site, needs a pile foundation. Piles can be designed to carry this high-bearing pressure at a deep level by friction and bearing. Settlement of the pile group can also be limited by proper design to meet the settlement criteria of the superstructure.

Now the foundation engineer has to select the type of piles that will be cost-effective, technically suitable, and easy for construction. The piles commonly used are as follows:

a) Bored cast-in-place
b) Driven pile – steel pile or precast concrete pile
c) Driven steel casing and then cast-in-concrete
d) Driven – prestressed precast pile

While choosing the appropriate type of pile, the engineer should consider the method of construction, availability of pile driving equipment, and the facility for manufacture and supply. The locally available type and the constructor having sufficient experience in working with that particular type should be an economic and safe choice.

Bores cast-in pile generally need more time in construction, close monitoring on-site for quality control, and muck-management in soft soil. But if the piling work has to be done within 30 m of any existing structure, driven pile cannot be allowed for construction. Bore pile should be a safe solution in that case to save the nearby structure from the effect of ground vibration.

Bore pile installed by mechanical boring tools saves time in construction. In soft soil, precast piles are installed by a mechanical device, which pushes the pile into the soil by high-power hydraulic jacks instead of conventional methods of driving by a hammer. These methods are faster and noiseless.

Let us see a set of constructed projects as an example, showing soils of a different nature and what kind of foundation systems have been adopted there. This is for general guidance only (Table 9.1).

## TABLE 9.1
### Foundation System in Different Soil

| Project | Location | Soil type | Foundation System |
|---|---|---|---|
| A | Indonesia, West Java. (Site 1) | Silty sand and gravel deposits. | Precast prestressed hollow spun pile. |
| B | Indonesia, West Java. (Site 2) | Stiff clay and sand deposits at lower strata. | Bored cast-in-situ pile. |
| C | Vietnam, south of Ho Chi Minh City. | Soft lean clay (improved by ground improvement work). | Precast prestressed solid square pile. |
| D | Bangladesh, Chittagong. | Medium to dense silty fine sand. | Bored cast-in-situ pile. |
| E | India, Orissa. | Highly weathered and decomposed rock. | Large based foundation resting on soil. |
| F | India, Madhya Pradesh. | Highly fractured granitic gneiss rock. | Isolated foundation resting on rock. |
| G | Kuwait, south | Medium to very dense silty sand | Bored cast-in-situ pile. |

The selection of foundation type depends on the soil characteristics and the settlement criteria of the superstructure. Factor of safety against overturning, sliding, and uplift should be within the permissible limit.

Field testing confirmation, for example, tests pile results, are to be carefully considered while recommending allowable design loads for foundation design.

Availability of material and equipment, experience of constructor, management of ground water dewatering, precautions for deep excavation, effect of subsoil dewatering on nearby structures, transportation and handling facilities in the case of precast pile, type of cement and corrosion protection of reinforcement, application of paint or protective coating on-site with contaminated ground water, and so on should be looked into by the foundation engineer.

### 9.8.5 Settlement

Settlement is one of the major criteria that govern the selection of a foundation.

Methods of settlement calculation for foundations resting on soil or pile groups are available in standard textbooks and codes of practice. Permissible limits of total settlement and differential settlement are also shown in codes of standards.

As a general guideline, the settlement of foundations and structures should be within the allowable limits as stated in Table 9.2.

The following practices should be followed to avoid damages in buildings and structures for differential settlement.

a) Building and structures should be provided with expansion joints, preferably at 30 m distance.
b) Building foundations resting on pile and those that are not on pile should not be connected by rigid jointed members. If it is necessary for some

**TABLE 9.2**
**Settlement Limits**

| Item | Allowable Limit |
|---|---|
| Foundations for general building and structures | Total settlement will not exceed 25 mm. Differential settlement will not exceed 0.2% between adjacent columns. |
| Foundation for mechanical and electrical equipment | Overall settlement will not exceed 25 mm. |
| Flat bottom steel tank foundations | Total vertical settlement will not exceed 25 mm at the perimeter. Differential settlement of the center of the tank with respect to the edge will not be more than diameter/90. |

walkways or extensions of floors to connect pile and non-pile structures, the same has to be designed with flexible end connection joints that allow differential movement. Joints should be designed with brackets that can bear the vertical load and allows movement.

c) Tank-supporting structures and foundations should be designed on rigid foundations keeping total settlement within the permissible limit. Full load tests should be done prior to the connection of pipes with flanges. Flexible nozzle connections should be used.

### 9.8.6 SHALLOW DEPTH FOUNDATIONS

The following types of shallow depth foundations are commonly used:

a) Isolated or spread footing
b) Combined footing and strip footing
c) Mat or raft foundation
d) Slab-on-grade

#### 9.8.6.1 Isolated Footing

The footing is resting on soil (Figure 9.3).

*9.8.6.1.1 Construction*

The construction of this foundation is done in the following stages:

a) Excavation
b) Leveling concrete
c) Footing slab: laying a protective layer on top of the lean concrete, formwork placing reinforcement, and casting
d) Pedestal: reinforcement, form work, anchor bolt placement, and casting

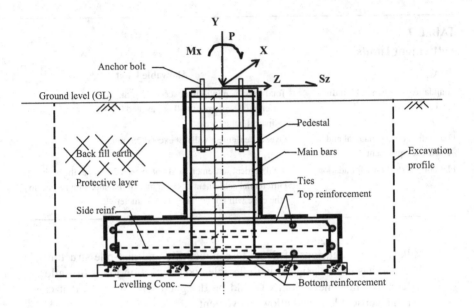

**FIGURE 9.3** Isolated footing.

e) Curing
f) Applying protective paint at the sides
g) Backfilling and compaction

The excavation work should begin after the layout and grid location are fixed. Excavation of pits should be 300–600 mm extra on each side to provide working space. The sides may be vertical or in a slope (1 horizontal: 2 vertical) depending on the stability of the soil and depth of pit. When the pits for adjacent foundations are coming close, the excavated pit should be made common. Excavation of the soil may be done manually or using a mechanical excavator. The final level should be dressed and not exposed for a long time in order to prevent the bed level from losing natural moisture content and getting dry. The leveling concrete should be laid on it as soon as possible. If the work is done below the groundwater table, the bottom of the pit should be kept free from water ingress by a proper dewatering system during construction. For construction in a wet pit, the constructor has to remove muck from the bottom of the pit and the surface should be filled with sand or sand with coarse aggregate, for example, broken brickbats or crushed stones. The thickness of such a subbase should not be less than 150 mm.

The protection of the sides is done by shoring the formwork made of thick wooden planks. Steel sheet pile and joist framework supports are used for deep excavations.

Leveling concrete shall be done with M10 grade of concrete. Brick flat soling gaps filled with sand are not a preferable alternative. If the depth of excavation done is more than the desired level, the additional depth should be filled with sand or mass concrete fill (1:4:8). The engineer shall check the bed level before it is cleared for casting leveling concrete.

For contaminated subsoil, the foundation concrete is protected by a layer of waterproof protection like bitumen emulsion paint or a membrane system. This is recommended by geotechnical specialists in the soil report. Some basement structures require additional multilayer protection for water tightness. This is called tanking.

Tanking membranes are flexible multilayered composite sheets that are polythene sheets bonded to bitumen/polymer adhesive. Geotextile waterproof liners using self-sealing sodium bentonite are also used as an alternative solution. The application of this paint or membrane system should be done on a dry surface of the concrete structure.

Reinforcement at the footing slab should be provided as per the design drawing. In some project works, separate drawings and bar bending schedules for reinforcement work is prepared for fabrication of re-bars, transportation in the work site, and the placement of bars at the site as per the drawings.

Bottom reinforcements are main reinforcements that are a flexural reinforcement of the footing slab bending upward. The footing slab is designed as a wide beam supported at the column. Reinforcement at the top layer is necessary for slabs subjected to uplift or rotation at slab-column joints and thus generating flexural stress at the top. For a footing slab more than 600 mm, reinforcement at the top layers is provided to prevent shrinkage stress. Side face re-bars are necessary for large depth foundation slabs. Diameters of bars in the footing slab should not be less than 12 mm. A crack width check is necessary to ensure protection against subsoil corrosion.

Column bars or dowels are to be placed from the top, keeping adequate anchorage length (development length) for the full strength of the reinforcement bar. The anchorage length is the portion embedded in the footing, which equals the length of the L-shaped bar starting from the top of the footing slab. Again, the projected length of column bars above the top of the footing slab should be the full length of the column or pedestal bars. If the projected length is found to be too long to keep it vertical, the length can be reduced to its full-strength anchor length plus a minimum 500 mm extra for staggered lapping.

The depth of the footing at the column junction should be adequate to transfer the compressive force of reinforcement bars as per design requirements.

The casting of footing slabs is done within closed formwork at the sides after the reinforcement work of the footing slabs is complete. The curing of the slabs will begin from the next day.

The joining surface should be thoroughly cleaned, and loose aggregates removed before starting the formwork for pedestal or column casting. Ties for pedestals or columns should be suitably designed so that all the main bars are tied. After the placement of column vertical bars, ties, and the formwork for the pedestal, the anchor bolts are installed. Wooden or steel frame templates are used to set the anchor bolt group in position while holding it level above the concrete surface until casting is complete. Stripping off the formwork may be done after the concrete is set for freestanding. The pedestal is then allowed for curing work wrapped in Hessian.

The engineer should have a check of the foundations at this stage and advise for any repairs for surface irregularities, if any. The surface will be allowed to dry and is coated with corrosion protection paint or a membrane as per the drawing.

After drying the protection paint, the backfilling work around the foundation is allowed to start. The earth used for backfill should be granular and the selected type from excavated earth. The filling should be done in layers and compacted simultaneously.

### 9.8.6.1.2 Design

i. Methods of geotechnical design for safe-bearing capacity and settlement calculation; concrete design of footings, pedestals, or columns and anchor bolts are available in standard textbooks and codes of practice. So, the following steps are for guidance only.
ii. Load input. To prepare a table of load applied on top of a pedestal from analysis output (STAAD or other) selecting the worst load combinations.
iii. Draw a sketch showing plan dimensions and elevations of the foundation that indicates finish ground level (FGL), top level of pedestal (TOP), bottom level of footing (BOF), groundwater level (GWL), column/pedestal dimensions, plan dimensions and thickness of footing slab, center line axes, orientation of the foundation, and loading.
iv. Select design parameters. Allowable gross safe bearing capacity at founding level, density of backfill soil, groundwater level, grade of concrete, yield stress of reinforcement steel, concrete cover, and Load factor.
v. Transform the load at the base of the foundation. Vertical load, $W$ = applied load on top of pedestal (P) + own weight of footing slab and pedestal + weight of backfill earth. Effect of ground water level should be considered in density and weight calculations. Moment at base, $M = Mx + Sz. H$, where H is the depth of footing from the top of the pedestal to the bottom of the foundation. The above moment expression is for uniaxial load cases only.
vi. Computation of area (A) and section modulus ($Zx$, $Zy$) of the footing base in contact with soil.
vii. Compute the soil bearing pressure and compare the same with allowable gross bearing pressure at the founding level. Resize footing dimensions and depth below ground, if necessary, to keep the developed soil bearing pressure within allowable gross safe bearing pressure. If tension occurs, the size of footing may be revised to avoid corner uplift; otherwise, the sectional properties (A, $Zx$, and $Zy$) should be computed for the actual contact area.
viii. For uniaxial bending, the soil bearing pressure at the base:
ix. $p_{max} = (W / A) + (M / Zx)$ and
x. $p_{min} = (W / A) - (M / Zx)$
xi. Now, check for the factor of safety against sliding, overturning, and uplift, if any.
xii. Once the dimensioning of the footing is finalized, strength design of the footing slab and the pedestal should be done as per permissible strength of concrete grade and reinforcement bars.

xiii. The capacity of anchor bolts should be the minimum of the tensile capacity of a steel rod and the concrete breakout strength over the embedded length inside the concrete pedestal. ACI 318 shows the guidelines for the calculation of concrete breakout strength.
xiv. Compute the settlement of foundation for DL + 0.5 % LL and compare the value with allowable total settlement and differential settlement with respect to the adjacent column foundation. The values should be within permissible settlement criteria (DL means dead load and LL is live load or movable superimposed load]).

### 9.8.6.2 Combined Footing and Strip Footing

When the load influence area of an individual column foundation overlaps with another area, the designer chooses combined footing and/or strip footing or a common raft foundation holding more than one column.

The method of excavation, construction, and surface protection are the same as for isolated foundation.

In the design of a combined or strip foundation, the designer has to select the shape of the footing slab in such a way so that the load distribution is uniform as far as practicable. The shape can be trapezoidal in the plan for a two-column footing having considerable differences in column load.

In a strip footing with multiple columns in a row, the slab connecting columns may be stiffened with a deep long beam monolithically cast with the footing slab. This beam will be designed as a continuous T-beam spanning over the columns. The strength design is done with moment and shear obtained by a method of sections so that it behaves like a long rigid member. This is termed rigid footing. In the rigid type of footing, the pressure distribution is in a straight line. The deflections of the foundation part between columns are negligible. If the centroid of the footing base and the resultant of the column load coincide, there will be no moment and the footing slab will experience uniform pressure distribution. For cases, if the CG of load and pressure areas have a difference of e (eccentricity), the strip will have plane but trapezoidal soil pressure distribution for moment due to eccentricity, $M = W \times e$.

For columns with axial loads only, the magnitude of soil bearing pressure is as follows:

$$p_{max} = (W/A) + (M/Z) \text{ and}$$

$$p_{min} = (W/A) - (M/Z).$$

However, in all practical cases, the column moments are algebraically summed up to get a net resultant moment. The length of footing is adjusted to reduce eccentricity and avoid tension at the corners and ends.

In case of a strip or combined footing resting on soft soil or sand deposits and the footing slab is not thickened enough or not stiffened with a deep beam like a rigid foundation, the upward reaction of soil pressure will not be in a straight line as stated above. The strip will behave like an elastic mat resting on springs. The springs may be assumed as vertical support and the spring constant is equal to subgrade modulus or the coefficient of subgrade reaction of soil. The curvature will be downward at column locations (settlement of soil) and convex upward in the span between columns. The nature of bending depends on load, subgrade reaction, and flexural strength of the footing slab. Analysis for moments and shear for the slab can be done by method of finite element using any standard analysis program (STAAD Pro or equivalent).

### 9.8.6.3 Mat or Raft Foundation

Construction of raft or mat foundation is simple and less time-consuming because the placement of reinforcement bars and concrete is much easier than isolated and strip footing. It is commonly used for residential buildings, multistoried houses, and industrial structures in foreign countries.

Safe-bearing capacity and settlement of mat foundation are different from isolated and strip foundation on the same soil. It is a favorable choice to control settlement.

The construction planning of large mats should be done by keeping in mind the contractor's resources in the supply of cement concrete per hour, available workers, and equipment for pumping and pouring concrete. The area of concrete placement should be divided into zones and the number of construction joints should be minimized to avoid cold joints as far as practicable. Quality control monitoring including observation of cast-in-concrete temperature is very important for this type of construction. The temperature of the concrete mix at the time of placement should be kept below 23°C unless stated otherwise in the design and construction specification. The placement of concrete at each layer should not be more than 300 to 400 mm thick based upon its slump and stability. The pouring sequence should be in forward and backward movements within a defined segment or zone so that the concrete layer below will not pass the initial setting time before the next layer is poured upon it. Construction joints should be vertical and dowels are to be provided as per the drawing.

For rafts more than 1 m thick, volumetric reinforcement (say 16 mm dia. bars @ 600 mm spacing both ways) should be provided to control shrinkage.

The plant and machinery along with contingencies should be adequate to provide continuous pouring and meet the requirements of the initial setting time of cement.

The design of the raft can be done as a rigid mat of uniform thickness housing all the columns over the slab. The thickness of the slab should meet rigidity check requirements (relative stiffness factor) as per IS: 2950 or equivalent American standards.

Gross bearing pressure should be checked against the allowable safe-bearing capacity of the raft. All column loads, weight of backfill earth, surcharge (if any), own weight of the raft, and moments due to eccentricity between load COG (center of gravity) and the center of the contact area with soil below the foundation should be considered in calculating gross bearing pressure. Corner pressures should be as follows:

$$P_1 = W/A + Mx/Zx + My/Zy$$

$$P_2 = W/A + Mx/Zx - My/Zy$$

$$P_3 = W/A - Mx/Zx + My/Zy$$

$$P_4 = W/A - Mx/Zx - My/Zy$$

Uplift at the corners should be avoided by increasing dimensions of the raft as far as practicable, otherwise A, Zx, and Zy should be computed for the actual contact area.

For strength design, net upward pressure should be computed, deducting uniform pressure on the soil due to the weight of backfill earth and surcharge (if any) and own weight of the raft, from gross pressure. Method of section may be used to get the design moments at the center span and below columns (at support) and shear forces at critical sections. Volumetric reinforcement should be provided to arrest shrinkage strain.

For raft over a large area, it is recommended to analyze the raft as resting on spring supports (subgrade modulus of soil) by the finite element method using standard computer analysis programs (STAAD Pro). Minimum reinforcement should not be less than that specified in codes of practice.

Embedment like columns anchors, insert plates, conduits, and depression for trenches are to be designed and installed at the time of reinforcement placement. If there is a machine foundation block or pits that need to be separated from the raft, the area should be defined in the drawing as a design cutout. The analysis and reinforcement detailing should be done considering this planned opening.

## 9.8.7 Deep-Seated Foundations Including Caisson or Well Foundations

When it is observed that the intensity of foundation load is much higher than the safe-bearing capacity of soil in the upper stratum, we look for deep-seated foundations. Deep-seated foundations are made of steel sections or concrete members that transfer the foundation load to a suitable soil stratum deep underground. The load is transferred to the soil by end bearing and frictional resistance capacity of members.

Commonly used deep-seated foundations are pile foundations, caisson, or well foundations.

There are some other types of foundation systems that fall into this category. They are compaction piles, steel sheet pile walls, and concrete diaphragm walls.

Let us discuss some of the most commonly used items in seriatim.

### 9.8.7.1 Pile Foundations

Piles are long slender columns made in structural steel rolled sections, reinforced concrete sections of cast-in-situ, or precast and prestressed precast concrete. These members are installed into the ground by the process of deep boring or driving by a mechanically operated drop hammer. In very soft soil, there is a method of pushing it into the ground by equipment.

Pile foundation consists of a set of long piles installed up to a target depth in the soil strata. Each pile has its capacity in carrying loads in three directions, say x, y, and z directions. After installation, the top of the piles is leveled in a predetermined design elevation, which is termed the cutoff level. Piles are installed at a fixed spacing and in a regular pattern, which is called a group of pile. A group is formed by various numbers of piles, i.e., two-pile group, three-pile group, four-pile group, n-pile group.

Each group is capped by a thick cast-in slab resting on top of piles, i.e., cast over cutoff level. This is like a footing slab but named as a pile cap. Dowel bars are projected above the pile top which gets embedded in the pile cap. The column pedestal is cast monolithically with pile caps resting on top of the cap.

Like a footing, the pile cap carries load from superstructure columns or walls and transfers it into soil through piles. The types of pile cap are isolated, strip, or raft similar to footing on soil.

The difference is that an ordinary footing rests on a uniform support system (soil or equivalent soil spring) but a pile cap rests on point supports (the top of individual pile or stiffened spring equivalent to elastic compression of pile and deformation of soil strata).

The dowels on top of the pile head should be the same as the pile shaft reinforcement at the top unless designed otherwise. So, pile heads are supposed to carry moments that can be generated by the array and strength of bars embedded in the pile cap bottom. The pile caps are connected by grade beams to transfer shear to the adjacent pile group as per the design requirement. The bottom elevation of these grade beams should be the same as the pile cap. If the pile cap is restrained against rotation and lateral sway due to such interconnections, the pile groups may be considered as fixed-head piles/columns at the top. Otherwise, it will be designed as a free-headed pile group.

Normally, pile tops are considered pin-headed connections in the design for simplification in analysis.

The strength design of pile caps and pedestals is similar to the process followed in the design of footing and pedestal, except that the distance of the critical shear plane is d/2 (d is the effective depth of pile cap at support) away from the face of the pedestal. Pile heads are considered pin joints, so forces applied from the pile cap over the group of piles are like transferring it on an imaginary plane connected to the pile head.

The soil under the pile cap will not take any load.

A typical sketch of pile cap foundation is in Figure 9.4 and Figure 9.5.

Assuming the pile cap is rigid. The reaction or force per pile shall be as per the following elastic equation:

$$F_m = P/N + /- Mz.x/\sum x^2 +/- Mx.z/Sz^2$$

# Reinforced Concrete and Associated Work

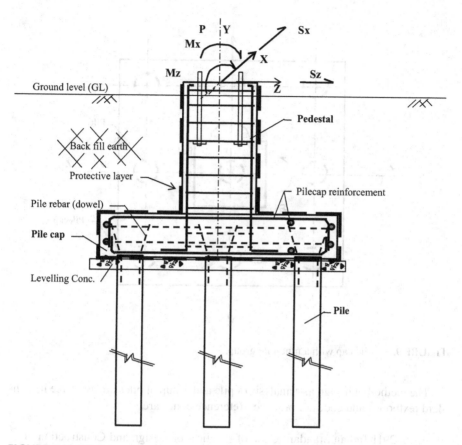

**FIGURE 9.4** Pile cap.

Where, $F_m$ = axial load in any pile, m
P = total vertical load (column load + own weight of pile cap + back fill earth)
N = no of piles in the group
Mx = moment about X-axis
Mz = moment about Z-axis
Sx, Sz = shear force along X-axis and Z-axis

For large mat and combine pile caps, the pile caps are analyzed and designed by using finite element methods by E Tab or STAAD Pro or equivalent programs.

In a manual design, the reactions on piles are calculated using elastic equations as stated above.

The magnitude of reactive forces is compared with the allowable capacity of individual pile. The pile spacing and the number of piles are changed until the pile reactions are within allowable limits. Once the reactions are found within the permissible limit, the strength design of the pile cap and pedestals is done.

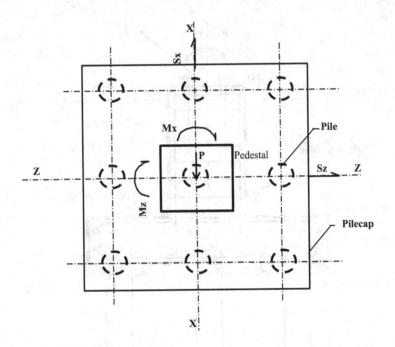

**FIGURE 9.5** Pile cap with a nine-pile group.

The method of design and analysis of pile and group of piles are available in standard textbooks and codes of practice. Reference codes are:

1. IS: 2911 Indian Standard Code of Practice for Design and Construction of Pile Foundation
2. ACI 543 R – 12: Guide to Design, Manufacture and Installation of Concrete Piles
3. IBC: International Building Code

**Design of a pile** can be divided into the following steps:

  a. Geotechnical capacity calculation
  b. Structural strength design

*9.8.7.1.1  Geotechnical Capacity Calculation of Pile*
It includes the determination of axial compression, lateral and uplift capacity of a single pile on the basis of engineering properties of soil strata holding the full length of pile, and the settlement of strata below the toe of pile.

Ultimate resistance of an individual pile in compression is equal to the sum of its shaft friction (perimeter of pile × effective frictional resistance of soil) and bearing capacity of soil at the toe (cross-sectional area of pile at toe × bearing capacity of soil at that level divided by appropriate factor of safety).

The load-bearing capacity of a group of pile may or may not be the same as the capacity of an individual pile multiplied by the number of piles in that group.

Table 9.3 is an example of group efficiency of pile groups in clay layers.

If piles are installed in compressible cohesive soil or in dense cohesion-less soil underlain by compressible soil, the ultimate axial compression capacity of a pile group may be less than that of the sum of the ultimate axial capacity of the individual pile. In this case, the pile group has a group efficiency of less than one.

In cohesion-less soils, the ultimate axial compression capacity of a pile group is generally greater than the sum of the ultimate axial compression capacity of the individual piles comprising the group. In this case, the pile group has group efficiency greater than one.

Settlement of a pile group is also a predominant factor in determining vertical capacity of a pile group. The settlement of a pile group is many times greater than the settlement of an individual pile carrying the same load per pile as each pile in the pile group. In case of a single pile, only a small zone of soil around and below

### TABLE 9.3
### Group Efficiency of Piles in Clay [Pile dia – 550 mm; Spacing 1.5 m]

| Group No (No of Piles) | Converse Labarre Formula | Seiler Keeney Formula | FELD'S RULE |
|---|---|---|---|
| 1 | 1 | ~1 | 1 |
| 2 | 0.88 | 0.935 | 0.938 |
| 3 | 0.88 | 0.935 | 0.875 |
| 4 | 0.776 | 0.854 | 0.813 |
| 5 | 0.776 | 0.854 | 0.800 |
| 6 | 0.739 | 0.812 | 0.771 |
| 7 | 0.739 | 0.812 | 0.786 |
| 8 | 0.702 | 0.785 | 0.797 |
| 9 | 0.702 | 0.785 | 0.666 |
| 10 | 0.702 | 0.785 | 0.763 |
| 11 | 0.683 | 0.767 | 0.750 |
| 12 | 0.683 | 0.767 | 0.698 |
| 13 | 0.672 | 0.754 | 0.692 |
| 14 | 0.664 | 0.744 | 0.714 |
| 15 | 0.650 | 0.730 | 0.683 |
| 16 | 0.664 | 0.754 | 0.672 |
| 17 | 0.659 | 0.736 | 0.721 |
| 18 | 0.659 | 0.736 | 0.688 |
| 19 | 0.653 | 0.744 | 0.645 |
| 20 | 0.653 | 0.744 | 0.656 |

the pile toe is subjected to vertical stress. For a pile group, a considerable depth of soil around and below the pile group is stressed. The settlement of the pile group may be large depending on the compressibility of the soil within the stressed zone. If the settlement criterion given in the project site is less than the computed group settlement, the length and other parameters of piles are to be changed to meet the allowable settlement.

Hence, for the determination of allowable axial pile capacity in compression, the designer has to consider the group efficiency factor and settlement of a group of piles as stated above.

The lateral capacity of piles can be computed by the determination of depth of fixity, lateral deflection at the pile head, and maximum moment of laterally loaded piles as per the method stated in standard codes. Knowing the depth of fixity, which is the length of the equivalent cantilever (L), the deflection of pile head (Y) can be estimated using the following equations:

$$Y = QL^3/3EI \, (cm) \text{ for free head pile, and}$$

$$Y = QL^3/12EI \, (cm) \text{ for fixed head pile.}$$

Where Q is lateral load in kg; E is Young's modulus of pile material in kg/cm$^2$ and I is the moment of inertia of pile section in cm$^4$.

Allowable load Q for a deflection of 5 to 7 mm is usually considered for general building design unless noted otherwise.

There are other methods in determining lateral capacity. Using a computer model of a single pile giving subgrade modulus of soil as lateral support and applying a load (Q) at the top, deflection (Y) can be obtained.

Uplift capacity of a tension should not be considered more than 50% of the frictional resistance of pile.

However, all these theoretical values are to be established by field testing of piles. Design capacities or working load of piles are finalized accordingly.

Negative friction is another phenomenon that reduces the axial load capacity from skin friction. It occurs when piles are installed in an area of deep fill, which is not compressed or consolidated or filled over a compressive layer. There will be downward movement of the fill as it compresses under its own weight or from surcharge over ground. The effect of this downward movement results in drag-down forces, which is called negative friction. How to calculate the negative friction over the shaft of the pile and the effective depth of this compression is available in geotechnical textbooks. The magnitude of the forces should be deducted from the design pile capacity in axial compression.

*9.8.7.1.2 Structural Design*

Structural design of pile is the strength design of the pile section, which may be the steel section of wide flanged, cast-in-concrete circular sections, precast member in square solid sections, or hollow circular sections. The pile shaft is designed as a column section, which can carry the working load specified in the geotechnical capacity calculation, i.e., for axial compression, axial tension, moment at top, tension + lateral, and all other critical load combinations. The structural design can be done by the limit state method of design. The top of the pile will be additionally reinforced as per requirements for transmitting lateral and bending moments above fixity depth. Pile sections lower than this depth should be designed for axial compression and tension. Entire shafts should be reinforced satisfying design forces or minimum steel requirement as per code.

### 9.8.7.2 Pile Load Tests

Testing of piles is conducted at locations of importance to the structure. This testing is done with test piles that are made as per design length and strength design with an adequate factor of safety. Test piles should be identical to working pile.

Static pile tests are done to verify design length and to confirm the safe working load capacity of piles. The tests piles are tested to 2.5 × safe working load. The minimum number of test piles shall be:

a) Three for axial compression (ASTM D1143/IS: 2911)
b) Two for axial tension (ASTM D3689/IS: 2911)
c) Two for lateral loads (ASTM D3966/IS: 2911)

Working piles are installed after the confirmation of safe working loads and tested to two times the safe working load. Dynamic pile testing (high-strain method, ASTM D4945; low strain integrity test, ASTM D5882-00) may also be carried out on working pile. The results of dynamic tests are correlated with the results of static tests on working piles.

### 9.8.7.3 Pile Types

The following types are commonly used in construction:

1. Bored pile
2. Precast concrete pile
3. Steel pile
4. Micro pile
5. Screw pile
6. Compaction pile
7. Sheet pile

*9.8.7.3.1 Bored Pile*

Cast-in-situ bore pile is formed by boring or drilling a hole in the ground and then filling it with concrete with or without reinforcement steel as per the design. In case of drilling holes in a soil, where the side of the borehole cannot stand by itself, the hole is stabilized with the aid of bentonite slurry drilling mud.

A bore pile is formed in the ground to transmit the design load into the soil by the resistance developed at its toe bearing and surface friction between soil and pile surface along its length.

The drilling hole is done by piling rigs with a rotary drill. In case of unstable soil, a steel casing tube (inside diameter same as the pile diameter), is inserted at the top (about 4–6 m) for guiding auger. The augers used are helical screw types.

**Under-reamed bore piles** are formed with an enlarged bulb at the toe. The formation of this bulb is done by inserting an attachment that can extend its arm to enlarge the diameter of the drilling hole at the bottom. An under-reamed pile can give more axial load-bearing capacity for this enlarged toe.

The time of construction and work progress depend on the expertise field management in planning rig movements, providing slurry and mud drainage, and placement of the concrete pump in a suitable place for timely supply.

In a site of medium to loose density sand at an upper layer at Kuwait, a bore pile of 600 mm diameter × 16 m long was installed complete with boring, lowering reinforcement cage, and filling in concrete by pumps in less than an hour. Here, the drilling was done by rigs fitted with screw-type long auger and a full-length casing tube. The casing was withdrawn during the placement and formation of pile. The groundwater table was about 2 m below grade level.

Whereas in Indonesia, 10–12 nos, 600 diameter x 12 m-long under rimmed piles were installed per day using mechanical augurs operated by pile rigs. The subsoil was stiff clay and the groundwater table was much below the pile toe.

### 9.8.7.3.2 Precast Concrete Pile

Precast concrete pile is constructed in reinforced concrete in a factory or casting yard. These piles are made with high-strength concrete and prestressed (pretensioned). It is also made of ordinary reinforced concrete.

The precast piles are generally made in solid rectangular cross-sections or circular hollow sections. Precast prestressed cylindrical hollow section piles are also named **spun pile**. The spun piles are very popular in Indonesia, Vietnam, Japan, and other countries. These pile sections are manufactured at factories using high-strength concrete and a pretensioned prestressing process.

All of these piles are installed by driving into the ground by rigs with a dropping hammer operating in the guides or pushed into soft soil by a hydraulic pile pusher. Installation of piles by a hydraulic pusher is silent and noise free. Precast piles are also installed by boring a hole into the ground and inserting the same into the hole up to the predetermined depth.

The load from the structure is transmitted to the soil by the resistance developed at its toe together with the frictional resistance of the soil around the pile shaft.

Precast piles are cast-in segments, suitable for transportation (maximum 14 m length) and delivery to the site. For long piles, extension of pile is done by jointing one segment on top of the other on-site on an erected position. After the first segment is driven into the ground, keeping a projection of about 1–1.5 m above ground, the next matching segment is aligned on top of it. The jointing planes are provided with steel end connection plates for full-strength butt-weld joints.

The assembly and joining procedures have to be completed in a short time. During pile driving, excessive pore pressure builds up around the pile at such a high level that hydraulic fracturing in the soil may take place around the pile. This causes the soil to consolidate much faster during the pile extension period. As a result, after the pile joining process is completed, the soil strength recovers to some extent and the driving resistance increases considerably. It makes restarting the pile driving exceedingly difficult and even causes refusal.

Precast piles are also installed by the boring method instead of driving. In this process, a hole is dug into the soil by screw-type auger the same as done in the case of bore piles. After the boring is complete, cement slurry grout milk is fed into the hole. Precast pile is then inserted in the hole up to the grouted base. Installation of precast pile by the method of boring can be described in steps as below:

a) Excavation of the hole using screw augur up to a predetermined length
b) Forming grouted shaft at the bottom of the hole (50% of total shaft length) as a mixing area
c) Fill in mixing area with cement slurry grout
d) Forming a grouted base at the bottom of the hole; minimum depth (L)1.5 m or 20 times pile diameter, whichever is more
e) Insert the precast pile (length, Dp) up to the bottom and set on the grouted base, the tip of the pile shall be partly embedded in the grout base
f) Total length of bored hole is Dp + L; grout shaft length is 50% of (Dp + L)

Installation of the preboring method is applicable for all kinds of soil, but the cost and time of construction are more than the driving method. The construction by driving is faster and simpler and hence most popular in Vietnam and other countries. Where the insertion depth and top elevation of piles installed by the boring method are seen to be at a fairly accurate position, it varies a lot in the case of driven piles. Because driving resistance in soil differs from pile to pile, it is not practicable to drive harder in order to bring the heads at the same elevation without breaking. So, the pile heads are seen popping up and down after they are exposed in the foundation pit. Hence, cutting and rebuilding heads for extension is common work before the foundation bed is prepared for reinforcement work.

### 9.8.7.2.3 Steel Pile

Steel piling is very popular in the United States. Rolled steel sections of wide flanges (H sections) are widely used as steel pile sections. It is easy to handle and capable of carrying high compressive and lateral resistance, although the cost is high compared with precast concrete piles. The steel pile can be driven hard to reach a deep stratum to achieve the required frictional resistance in the soil or to get a hard bearing stratum. Extension of length is done by welding. Hollow tube pile is also used in waterfront and marine structures. The inside is filled with concrete.

Corrosion resistance in normal soil is taken care of by its thickness during the working life of superstructures or plant life. However, suitable painting coats and cathodic protection is provided for marine structures and contaminated soil.

### 9.8.7.2.4 Micro Pile

These are mini piles made of steel tubes having a diameter of 150 mm or less. These are short-length tubes inserted into the ground by hammering. The tube piles have a welded shoe, and the inside of the hollow tube is filled in with concrete after driving.

Load-carrying capacity is small but adequate for lightly loaded structures, where regular piling cannot be done for low headroom.

### 9.8.7.2.5 Screw Pile

These piles are steel tubes with helical blades welded at the end. It looks like a screw. These piles are of short length and highly effective for use as anchors in soil.

Most effectively used for supporting legs of solar panel supporting structures in the Middle East. The installation is fast and done by vehicle-mounted rig.

### 9.8.7.2.6 Compaction Pile

Compaction piling is basically a ground improvement system used to increase bearing capacity in weak, loose, and soft soil.

Wooden pile is most popular in India for small construction.

Formation of stone columns and sand compaction pile increase stability and liquefaction potential in weak soils.

Cement deep mixing (CDM) pile is a type of soil improvement used to develop bearing strength of large areas in soft and weak soil. It is a deep soil mixing method done by inserting a rig-mounted long shaft fitted with spiral augers and a cement injection system. The cement binder is injected as slurry and mixed with the soil to form a hardened column or pile. The penetration in the ground to form CDM pile is done in a grid pattern at regular spacing. A reinforced slab pavement cast over a grid of CDM pile is capable of withstanding a high stack of coal and minerals. This method of soil improvement compaction piling is also used for aircraft landing strips in soft soil sites.

### 9.8.7.2.7 Sheet Pile

This piling system is used to resist lateral pressure. Popular sections are rolled steel sections of trapezoidal shaped tough in cross-section and length that are suitable for transportation (below 12 m). Section properties are adequate to withstand driving stress and buckling. Each pile has an interlocking facility at the sides so that they can be driven side by side to form a continuous wall barrier.

Typical sections can be seen in a manufacturer's product catalog.

The basic design is like a cantilever retaining wall partly driven into the ground and part projected above, which retains earth and surcharge load at one side. The cantilever moment and shear generated by the retaining earth and surcharge are resisted by passive pressure of soil acting over the embedded length and strength of sheet piles. Twin walls spaced apart with a reinforced concrete cap at the top will generate more resistance. Sometimes, a wall is tied with anchored ropes as per the structural design to reduce moment and work like a propped cantilever wall.

### 9.8.8 Caisson or Well Foundation

Well foundation is another kind of deep-seated foundation generally used as pier support for a bridge structure crossing rivers and waterways. It is also used as an intake structure for a pump house situated on the riverbed or a collection well in an ash pond, a transfer tower, or a deep underground substructure in plants. The common shapes are cylindrical structure, D-shaped, or square in plan.

The advantage of a well structure is that it can be installed in the ground by sinking, i.e., slow penetration by its own weight and some construction surcharge, if necessary. The construction of the well structure is done by constructors specialized in this field. The construction of a well and sinking it up to design length is a challenge to the constructor.

The method of design of well foundation structures in soil or riverfronts is available in textbooks and relevant design codes of practice.

The location of intake structures and the well is selected on the riverbed after a detailed hydraulic study. Geotechnical investigation of subsoil stratum at the location of the caisson is done to get the engineering properties for design and construction. Dimensioning in plans, wall thickness, and depth of penetration are selected according to the general arrangement drawing of the building or structure to be supported by the well.

## 9.9 STEEL TANK FOUNDATIONS

Foundations for steel tanks are provided to give non-yielding support to the tank base that will not settle or tilt during its functional period. Oil storage tanks generally have a large diameter base compared with their height. The center of the base is cambered upward, which compensates for deflection at the center and the ponding inside. These tanks are anchored to the base when uplift due to wind and the seismic force cause tension at the base. But these lateral forces are not the governing force in foundation design. Foundations are basically loaded by the weight of filling liquid because it is a uniformly distributed load on a circular base.

The following types of foundations are commonly used for tank bases:

a) Sand pad
b) RCC ring wall foundation
c) RCC mat resting on pile

### 9.9.1 Sand Pad

This type of foundation is used for medium to hard soil stratum, where computed settlement is expected to be within a permissible limit. Tanks of base 30-m diameter or more are erected on sand pad foundations. The selected area of this foundation is prepared by clearing, grabbing, excavation, leveling, and compaction by rollers. The patches of bad soil or protruding rocks, if any, should be removed and replaced by compacted granular fill.

The structure of this foundation consists of a short ring-type embankment of crushed stone, a compacted sand fill pad inside, and a topping with sand bitumen bedding layer underlain by a crushed stone subbase.

The embankment is made with crushed stone (63 to 90 mm) or boulders, gaps filled with stone dust/sand, and screenings. The soil below the base of the embankment is strengthened by rolling and compaction of ground soil and then covered with a thick layer of coarse sand fill. The top of this stone embankment should have sufficient width to provide a walkway around after placing the tank wall at the center of the embankment. The embankment will be constructed like a ring wall of centerline diameter same as the diameter of tank so that the tank wall can be placed on the center of this embankment. Construction of these embankments is done in layers to achieve the required degree of compaction according to the design requirement.

The center plot is then filled with selected sand in layers and compacted to achieve a relative density not less than 80%. The top of the sand fill is finally leveled and be graded with a camber of 1 in 90 (or as required by computed settlements).

The steel tank bedding layer at the top consists of a sand bitumen carpet laid over a compacted crushed stone layer of about 250 mm thickness. The sand bitumen carpet consists of coarse sand blended with 80/100 bitumen (8–10% by volume).

Since large dimeter tanks are seated on this type of foundation, holding down bolts will not be necessary for stability requirements against wind or seismic load.

A typical section of sand pad foundation is shown in Figure 9.6.

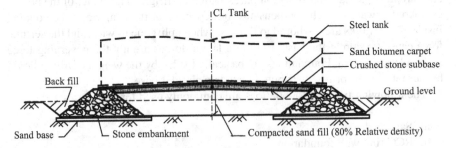

**FIGURE 9.6** Typical section of sand pad type foundation for a steel tank.

## 9.9.2 RCC Ring Wall Foundation

This type of foundation is very popular for all types of soil including soft to medium types, where it is necessary to transfer the load on a good bearing stratum at a low level.

The ring wall foundation is a circular-shaped cylinder-type wall containment made with reinforced cement concrete. The wall thickness generally varies from 200 to 600 mm depending on design requirements. The top of the wall is generally 300 to 600 mm above ground level and the bottom of the wall rests on a short-width concrete base like a footing designed to suit the allowable soil bearing load. Inside of this ring wall, sand is filled or granular fill is mixed with sand. This filling is done in layers and hydraulically compacted to ensure that the weight of the tank is transferred below without any settlement inside the ring wall containment.

This filling is capped with 200–250 mm crushed stone subbase and a sand bitumen layer of about 75 mm thickness laid on top of it. The slope of this bedding layer should be with a crown at the center at a camber of 1 in 90, or as necessary to compensate the computed settlement.

The centerline diameter of the ring wall is the same as the diameter of the steel tank so that the vertical weight of the tank wall passes through the centerline of the RCC ring wall.

Anchor bolts are to be provided to hold the base of the tank against overturning forces due to wind or seismic force, as necessary.

A typical sketch of a ring wall-type foundation is shown in Figure 9.7.

## 9.9.3 RCC Mat Resting on Pile

When the soil stratum is very poor and cannot bear the unit pressure load of a tank within permissible limits of settlement, pile foundation is recommended.

This is a circular mat resting on piles. The thickness of the mat varies from 450 mm and above depending on the spacing of piles and the strength design requirements.

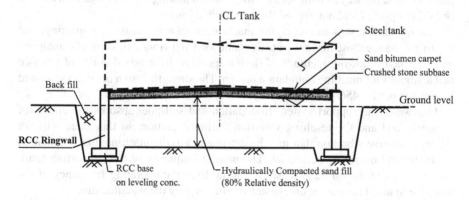

**FIGURE 9.7** Typical section of RCC ring wall foundation for a steel tank.

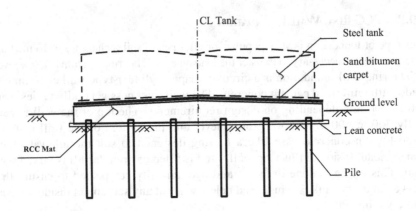

**FIGURE 9.8** RCC mat foundation on pile for steel tank.

There will be a sand bitumen layer of about 50 mm thickness laid on this raft to prevent corrosion in the underside of the tank bottom plate, similar to the other types described previously.

The top elevation of the mat should be 200 to 600 mm above ground depending on the plinth level decided for the plant.

A typical sketch is given for reference (Figure 9.8).

## 9.10 MACHINE FOUNDATION

Machine foundation is a block or framed structure, which gives support to a machine generating vibration and dynamic load when it works. Therefore, the foundation has to bear the static weight of the machine together with the dynamic load and weight of the foundation structure. The resultant load is transferred to the soil bed or group of pile below the foundation.

The pressure on the soil or load on pile and settlement of foundation will be computed in the same way as done for building column footings. The pressure on the soil or load per pile should not exceed the permissible limits.

The foundation structure carrying machines with reciprocating, impacting, and rotating mass are those that generate dynamic load and vibration, should be analyzed and designed following methods of design specified in standard codes of practice and textbooks for machine foundation design. The strength design of concrete should be done as per IS: 456/ACI 318.

In a word, the support system (foundation and soil/pile) absorbs the effects of dynamic load and the resulting vibration during operation, so that there will not be any excessive vibration that may be detrimental to the machine unit, foundation structure, and operating personnel. The natural frequency of the supporting foundation structure should be at least 20% away from the operating frequency of the machine to avoid resonance unless stated otherwise by the manufacturer.

# Reinforced Concrete and Associated Work

Since these foundations carry vibratory equipment, they are to be isolated from adjacent building structures and ground floor slabs by vibration isolation joints, which prevent vibration transmission.

### 9.10.1 Types of Machines and Foundations

In general, machines can be classified into the following types considering their vibration characteristics:

a) Rotating machines
b) Reciprocating machines
c) Impacting machines
d) Other types

Rotating type machines are those that have a shaft attached with a rotating part like a fan with blades. Pumps and motors, centrifuges, and fans are medium-weight rotating machines, in general. Heavyweight machines are turbine-generators, boiler feed pumps, large fans, turbine-driven pumps, and compressors. The vibration in a rotating machine is minimum when it runs in a well-balanced condition like in the manufactured state. If for any reason, there develops an eccentricity (e) between the mass centroid of rotating parts and the center of rotation, the machine will vibrate more than its design condition. The unbalanced force is a function of the mass of the shaft, speed of rotation, and magnitude of the eccentricity.

Reciprocating machines are piston moving in a cylinder or driven by a crankshaft. Compressors and diesel engines are reciprocating machines.

Impacting machines operate with regulated impact. Forging hammers and metal forming press machines belong to this type.

Other types of machines that generate dynamic loads during operation are crushers, mills, and so on.

**Types of foundations** used to support these machines are as follows:

a) Block foundation
b) Framed foundation
c) Inertia block on a vibration isolator
d) Spring mounted equipment
e) Rubber pad mounted equipment
f) Pile foundation

Typical sketches and descriptions of the above types of foundations are shown in Figure 9.9.

**Block foundations** are a solid concrete pedestal holding the machine on top of a pedestal anchored by foundation bolts. The shape of the block above ground level is as per the manufacturer's general arrangement/civil information drawing. This drawing shows loading data, details of foundation bolts/pockets, and so on to be provided on top of a block foundation for fixing the machine base. This information

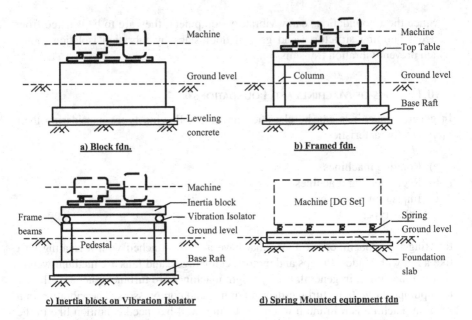

**FIGURE 9.9** Types of machine foundation.

includes layout and details of pockets for fixing anchor bolts, top elevation of the pedestal, and grouting details. The base area of block foundation and depth below ground are decided by the foundation engineer to satisfy the safe bearing capacity and dynamic stability requirements of the foundation.

As a rule of thumb, the mass of the foundation should not be less than two and half times that of the total weight of the machine. Some experts say the weight of the foundation block should not be less than three (3) times the static weight of the machine plus five (5) times the weight of the rotating parts.

However, the foundation engineer has to compute the natural frequency and amplitude/velocity to check with the permissible limits specified in the relevant codes of practice as well as the manufacturer's recommendation, whichever is worst.

**Frame foundation** consists of a top table, supporting columns, and base raft. The top table may be a solid slab or framework of beams and thick slab. There may be raised pedestals to support individual machine units, bearing block, and so on. The dimensioning and shape of the top table and column pedestals should be according to the general arrangement drawing and the recommendation of machine manufacturers. Civil assignment drawings showing general arrangement, equipment layout, anchor bolt fixing provisions, and loading data are given by the manufacturer.

For heavy machinery like turbine-generator foundations in a power plant, information given by manufacturers includes the limits of dynamic stability requirements and deflection criteria, which are to be strictly followed by the foundation designers. The analysis of the frame foundation includes static and dynamic analysis.

Computation of soil bearing pressure/load on pile group and settlement is done in a similar way to other foundations. Dynamic parameters of soil or pile group are necessary for the analysis of these special foundations.

The foundation top table is generally accessible from the operating floor/platform.

A separation gap of 50 mm is kept on all sides to isolate the machine foundation structure from the adjacent floor/operating platform.

**Inertia block and vibration isolator** type foundations are commonly used for coal crushers, screening, and similar vibratory machines. Turbine generator units are also supported in some power plants using this system.

In this method, the machine is installed on a rigid concrete mat, which is termed an inertia block. This block is supported over a set of springs within a casing. The entire system is called a vibration isolation system (VIS). These spring blocks are heavy-duty springs and designed by specialized agencies. It can take loads and absorbs vibration. The reaction from these vibration isolators is furnished by the designer of the specialized agency in the civil assignment drawing. The building designer or foundation engineer has to carry out the design of supporting pedestals and base raft as a normally static loaded structure. No further dynamic analysis is necessary in this case. Sometimes, the spring blocks are designed to rest directly over the floor slab. The floor beams and slab structure are designed for this load as done for the static load. No isolation gaps are necessary for this system.

**Spring mounted equipment** foundations are very simple for design and construction. Equipment, such as a diesel generator, which comes in a container-type unit is fitted with a base frame and spring foot mountings. This kind of support is also named a *cushy foot mount*. It absorbs vibration and only transfers vertical loads. So, the foundation designer has to design the base slab as slab-on-grade. No dynamic analysis or separation joints are required in this case.

**Rubber pad mounted equipment foundation** is also similar to cushy foot mount, but it comes as a cube of about six-inch blocks. These blocks are manufactured to bear the load without transmitting any vibration on the supporting floors or slab-on-grade. The foundation system consists of equipment bearing concrete pedestal/thick slabs and sets of rubber cubes. The concrete pedestal rests on the set of rubber pads/cubes as per the manufacture's drawing. The construction is done keeping rubber blocks in position and embedded in a sand pad enclosed by formwork at the sides. The concrete pedestal is cast on the sand pad and rubber cubes; after it is hardened, the sand is removed from the bottom. The pedestal rests on rubber cubes. This type of foundation is used in paper plants.

**Pile foundation** under a machine foundation block is designed for static and dynamic loading. It is similar to foundation blocks resting on soil except that the block is supported on stiff piles (springs). Pile stiffness is computed according to the properties of the soil stratum and its elastic stiffness.

Pile reactions and settlement are computed after static analysis and then verified with permissible limits. The number of piles and group array is changed to suit the load versus capacity requirement.

### 9.10.2 Codes and Standards

The following codes and standards are used for the design of machine foundations:
IS: 2974 Code of Practice for Design and Construction of Machine Foundation

  i. Part I – Foundation for Reciprocating Type Machines
 ii. Part II – Foundations for Impact Type Machines (Hammer foundations)
iii. Part III – Foundation for Rotary Type Machines (Medium and High Frequency)
 iv. Part IV – Foundation for Rotary Type Machines of Low Frequency
  v. ACI 351.3R-04 Foundations for Dynamic Equipment
 vi. DIN 4024 and VDI 2056 standards
vii. CP 2012

### 9.10.3 Materials Used

The material used is a concrete mix of a higher grade (M25 and above) in general.

For some exceptional cases, steel framework or a platform structure is used to support small machineries. Composite construction, such as steel framing beams and RCC slab floor, is also used to support machines resting on a spring foot or VIS (vibration isolation system) metal springs.

**Grouting** is a part of machine foundation. After the final alignment and leveling, the gap between machine base plate or base frame and the top of the concrete pedestal is filled and packed with cement-based grout. Pockets housing anchor bolts are also filled up with the same grout mix. The mix is prepared like ordinary cement concrete (cement: sand: coarse aggregate) with an admixture that adds a non-shrink property into the grout mix. Free flow ready-mix grout mix products are available that contain all the ingredients. Ready-mix grout products from reputed manufacturers are preferable to site mix grout to ensure strength and quality control requirements.

Grout thickness of more than 75 mm should be done with a mix having coarse aggregates of 6 mm and below. Grades of concrete should be higher than the grade of pedestal concrete. Curing of grout should be done in the same way followed by ordinary cement concrete to resist shrinkage crack.

### 9.10.4 Vibration Isolation

As stated earlier, machine foundations are vibratory units and are separated from building structures and foundations by providing isolation or separation joints. A

# Reinforced Concrete and Associated Work

clear gap of 25–50 mm is also a method of isolation to prevent vibration propagation. Sand filling around vibrating block foundation also reduces transmission of vibration.

### 9.10.5 General Guidelines

Some of the salient points are shown below as designer's checkpoints.

a) Natural frequency should be ± 20% away from the operating frequency of the machine.
b) Vertical and horizontal amplitude/velocity should be within the permissible limit stipulated in the manufacture's recommendation. In absence of manufacturer's data, it should satisfy the criteria recommended in DIN 4024 and VDI 2056. Amplitude/velocity should be computed at the rated speed and for 20% over speed.
c) The base area of the foundation or dimensions in contact with soil should be such that the eccentricity of the combined centroid of the foundation and static weight of the machine from the center of the foundation base should remain within 5% of the linear dimension of the foundation base.
d) Dynamic analysis may not be carried out for small machines weighing less than 10 kN or if the mass of the rotating part is less than one-hundredth of the mass of the foundation.
e) The mass of the foundation should not be less than 2.5 times the weight of the machine.
f) The maximum load or pressure on the soil should not exceed 50% of allowable bearing capacity during normal operating conditions and should not exceed 85% at extreme conditions.
g) The minimum re-bars shall not be less than 30 kg per $m^3$ of concrete unless specified otherwise in relevant design codes as well as specifications given by the owner. Volumetric reinforcement to be provided to control shrinkage crack.
h) Concreting shall be completed in a single operation without any construction joints as far as practicable. Construction joints for unavoidable cases should be decided by the designer and engineer in charge.
i) Temperature monitoring and control concreting should be done in extreme weather conditions.
j) The foundation should be protected with anticorrosive bituminous paint and coatings against contaminated subsoil.

## 9.11 MISCELLANEOUS STRUCTURES

### 9.11.1 RCC Reservoir

These reservoirs are generally used for storing raw water, firewater, and clear water associated with pump houses used in industrial plants. These reservoirs are rectangular-shaped structures with partition walls parallel to the width that divides the storage area into parts, so that chambers can be emptied for cleaning and so on. A pump house runs parallel to the length of the reservoir, which may or may not be integrated with a common wall. The base level of the storage reservoir and the pump house basement is normally at the same elevation based on hydraulic calculation requirements.

The foundation used for such reservoirs is generally a thick RCC raft resting on soil or on pile. The stability against uplift or buoyancy checking is an important factor in designing a foundation system for the site with GWL (groundwater level) near to finish grade level. Semi-underground types of reservoir are found to be most economical in low-bearing capacity soil because the volume of earth replaced by the tank portion below ground level counteracts the gross weight of the tank. Hence, the gross pressure on soil at the foundation level is much less than what it would have been for the reservoir base on the ground. The civil foundation engineer has to coordinate with the hydraulic designer at this stage to arrive at the most economical solution.

A settlement study should be done to ensure total and differential settlement within permissible limits. For sites having poor soil at the upper stratum, the topsoils are replaced by hydraulically compacted sand fill, which is retained by a closed RCC retaining wall underground. The retaining wall can be placed away from the edge of the base slab, keeping a distance so that the effect of surcharge does not fall upon the wall stem. In this case, the design section of the retaining wall will be lighter. The top of the sand fill between the reservoir wall and the retaining wall inner face should be blinded with a plain concrete slab to protect the fill-in sand. There are large tanks of plan dimensions 50 m × 40 m and more that have been designed and constructed on soft to hard soil following the above type of design.

The buoyancy check for open-top reservoirs is important. It was seen that water deposited inside the reservoir during heavy rainfall is not adequate to resist uplift due to subsoil water rising from the collection of surface runoff outside. The runoff water from the plant area flows into the excavated pit around the reservoir wall (when backfilling is not done or loosely compacted) and goes into the bottom of the base slab resulting in the upheaval of the slab at mid span. Such types of phenomenon are seen in the case of open-top forebays of pump houses and neutralization pits.

### 9.11.2 Pipe and Cable Trenches

Pipe and cable trenches are thin-walled reinforced concrete substructures used to carry service pipes or cables. The tops of trenches are covered with removable precast planks spanning over the full width of the part according to the general arrangement drawing. The bottom of a trench has a transverse slope with screed concrete and longitudinal bed slope for drainage of subsoil seepage or rainwater, if any. There are sump pits at intervals for pumping out accumulated water. Cables are laid on racks with steel supports anchored to a wall. There are Uni-strut holds or embedded plates fixed on the surface of an inner wall at intervals to hold the steel supports. Similarly, pipes are supported on cross-steel members welded to fixed embedded plates or by end plates with anchor fasteners.

Cables are also carried through multiple layers of PVC conduits arranged in rows with gaps of 200–250 mm at all sides being separated by spacers. The gaps between conduits are filled with mass concrete to form a concrete duct bank of rectangular or square-shaped cross-sections. Manholes are provided at corners and intervals to facilitate cabling work. These ducts are reinforced at crossing roads.

### 9.11.3 RCC Culverts at Road Crossings

These are rectangular-shaped box structures made with cast-in-situ concrete or precast units. These structures are used for carrying pipes, cables, ducts, and for cross drainage work. The top of the culvert may be the same as the road level or below the subgrade depending on levels and route layout. The clear height of culverts should be adequate for operating persons to access; otherwise, the top cover should be made of a removable type of heavy-duty precast concrete slab of steel deck. Both ends of the crossing culverts should be provided with wing walls to retain earth.

## 9.12 RETAINING WALLS

A retaining wall is used to retain earth or materials at different elevations on either side of it. The material retained or supported by the retaining wall structure is generally termed as backfill. The top surface of backfill may be horizontal or inclined. The portion of material lying above the horizontal plane at the elevation of the top of the wall is called surcharge. The surcharge may be a passing load of vehicles that run on a horizontal ground surface or road/railway embankment protected by the retaining wall structure.

The design of retaining walls has two parts: geotechnical design and then strength design. In geotechnical design, we compute the lateral pressure from backfill and surcharge. Then check the stability against overturning and sliding before stepping into structural strength design.

Two types of retaining wall are commonly used. They are cantilever and vertical walls with buttresses at intervals.

The cantilever type is generally found economical for walls of up to 6 m above the base. The cantilever type has a vertical wall usually with a tapered section with a wide base. The wall surface has its vertical face in contact with backfill. The base raft is projected more inside the material holding area to get the weight of backfill on top of it and thus to gain counteracting moment in stability computation.

Shear keys are provided below the base raft to resist sliding forces.

The counterfort types have vertical wall panels of uniform thickness spanning over a buttress. These buttress members are long depth columns and the structure acts as a cantilever member fixed at the base and free at the top. These buttresses are generally tapered sections with higher depth at the bottom. The wall panels and buttresses are cast monolithically with base slabs.

The retaining wall is a very reliable structure as it has been used for a long time. Retaining walls in railway embankments with a culvert wing wall were made in brick masonry. Those structures are still functioning with minimum maintenance.

Weep holes are provided in earth-retaining structures in order to release water pressure from the backfill inside.

## 9.13 SHEET PILE AND DIAPHRAGM WALLS

Sheet piling and concrete diaphragm walls are used for retaining earth in the case of deep excavation and shore protection work.

Steel sheet pile sections are rolled steel products, which are driven underground and keyed together at the sides to form a continuous steel wall. Sheet pile wall, when driven into the ground, acts as a cantilever wall to protect the sides of an excavated pit. For deep excavations, steel side runners and cross ties or compression members are provided to brace the wall structure horizontally across the pit because cantilever-type construction has a depth limitation. For riverbank protection, sheet pile wall is usually designed and driven as a cantilever structure. It can also be used as a propped cantilever wall by adding a horizontal runner at the top of the wall and then tied with tension members of guy ropes (steel wires) anchored to the shore at intervals.

Sheet pile wall is a costly item, and it is generally used as temporary work for construction and then pulled out for restoration and further uses. It can be used as a permanent earth retaining structure with corrosive protection.

Concrete diaphragm walls are made by excavating a long narrow but deep pit with the help of a crane operating bucket. The sides of the excavated pit are kept stable by filling in bentonite solution. Once the digging is complete, a wall reinforcement cage is lowered into the pit with cover blocks that help the cage into position. Concrete

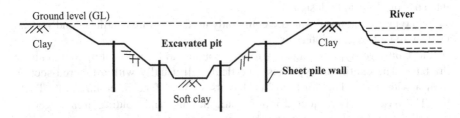

**FIGURE 9.10** Typical section of sheet pile protection for deep excavation.

mix is poured in by the tremie process. These diaphragm walls have mild steel plate inserts (welded to reinforcement cage) that are used for fixing steel compression members across the width of the excavated pit having a diaphragm wall at both sides. Once the diaphragm walls are hardened to reach design strength, excavation work begins from the top. Steel cross members are fixed in position with the progress of excavation. These diaphragm walls serve as an earth retaining watertight barrier, so that the main structure can be built inside without any trouble of subsidence outside the wall. This diaphragm wall may also be used as part of the main construction.

For critical construction sites, for example, onshore work, concrete bored piles are installed in a row close to each other to form a diaphragm wall.

Construction of these works is done by specialized agencies who are experts at such work.

The major uses are for temporary shoring work in deep excavations and building riverfront barriers. The sheet pile wall is a permanent solution for roads and buildings built by the riverside. This wall can control groundwater seepage.

Figure 9.10 shows the use of a sheet pile diaphragm wall protecting a deep excavation area free from seepage, near the bank of a river. In this case, the final depth of the excavation was more than what a standard length of sheet pile wall can reach to control groundwater seepage. To overcome this critical situation, the excavation was planned in steps and subsequently the standard length of sheet piles was driven so that the cumulative depth was more than the target excavation.

## 9.14 EMBANKMENT LINING AND SHORE PROTECTION WORK

### 9.14.1 Embankment Lining

An embankment is a wide-based wall of earthwork. It is built by filling and compacting earth in layers and raised up to the desired level with stable side slopes. Uses of embankments in industrial plants are for the construction of railway tracks and roads, flood protection barriers, large water storage reservoirs, river training, and canal work.

Materials used for an embankment should be good soil that can give a stable slope as steep as possible without compromising safety aspects. The factor of safety depends on cohesion and shearing resistance of the soil. Hence, if a soil that can give a steep side slope is available for construction, the embankment section will be economical as it would occupy less base area.

Water reservoirs or flood protection embankments should be designed and constructed with cohesive soil to minimize seepage loss.

The design of embankments is furnished in standard codes and design manuals. The computation of slope stability analysis and evaluation of seepage loss for water reservoir embankments are major points in the design.

Reservoir and canal embankments are lined with polythene sheets (HDPE/LDPE) at the sides and base to prevent loss of water by seepage. This polyethylene sheet is protected from exposure to sun, wind, and damage by covering with precast concrete tile lining. Protection of the slope on the other side of the embankment (not in contact with water) can be done by stone riprap, cement concrete lining, or by turfing. Toe drains outside the embankment should be provided to collect seepage loss, if any.

The lining may be of cast-in-situ concrete with regular joints for crack control. Temperature or shrinkage crack reinforcement may be provided where necessary. Precast concrete blocks of liftable size are generally found to be efficient for laying and replacements. Concrete lining is expensive but preferred because of its durability and low maintenance.

### 9.14.2 Shore Protection Work

Shore protection is required for the protection of plant boundary walls situated along riverbanks or canal sides or in river water intake pump houses on the shore. It prevents erosion of soil due to the variation of water levels and surface runoff.

Protection of riverbanks or canals is done by stone boulder pitching or concrete blocks/stone riprap or geosynthetic bags filled in sand. The sizes of stone boulders and geosynthetic bags filled with sand are calculated based on the velocity of river/canal flow and empirical guidelines in Indian standard codes (IS: 14262) and hydraulic design guidelines by the US Army Corps of Engineering and Design. Geosynthetic bags are filled with sand and stitched weighing around 500 kg to form a replacement for stone boulders. They should be laid on a prepared slope protected by geofiber mesh. Single or double layers of bags are placed based on estimated design thickness.

Stone boulders ripraps are placed on a prepared slope and gaps are filled with sand or cement mortar at the sides of large water reservoirs. At riverbanks, stones filled in wired mesh like sausage links are placed on the bank slope.

A worked-out design example of riverbed protection work at a plant water discharge pipe is shown in Chapter 12.

## 9.15 SILOS

Silos are tall cylindrical-shaped storage bins made of steel or concrete. The use of a reinforced concrete silo is found in cement plants for storage and the discharge of raw materials and cement. In power plants, silo structures are used for dry ash storage before disposal. Concrete silos and steel bins are extensively used in chemical and food industries, fertilizer plants, and mining industries.

# Reinforced Concrete and Associated Work

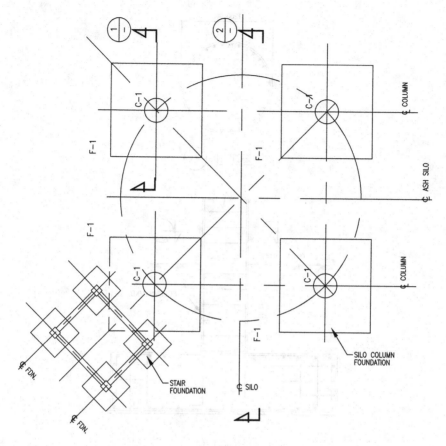

**FIGURE 9.11** Foundation plan.

264 Design of Industrial Structures

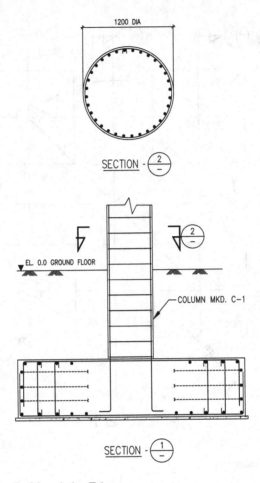

FIGURE 9.12  Detail of foundation F-1.

# Reinforced Concrete and Associated Work

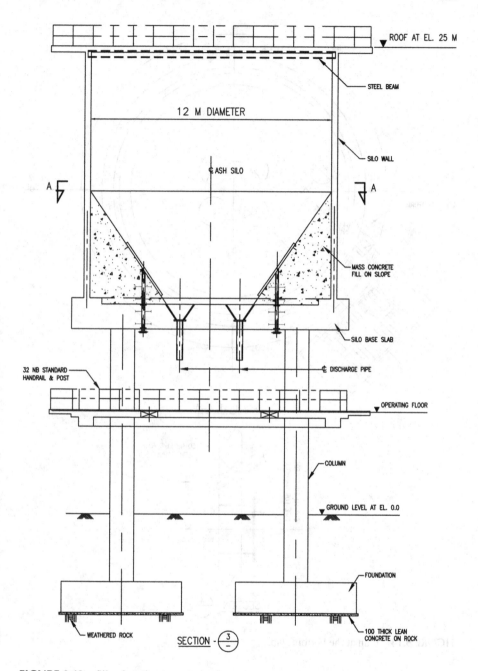

**FIGURE 9.13** Silo elevation.

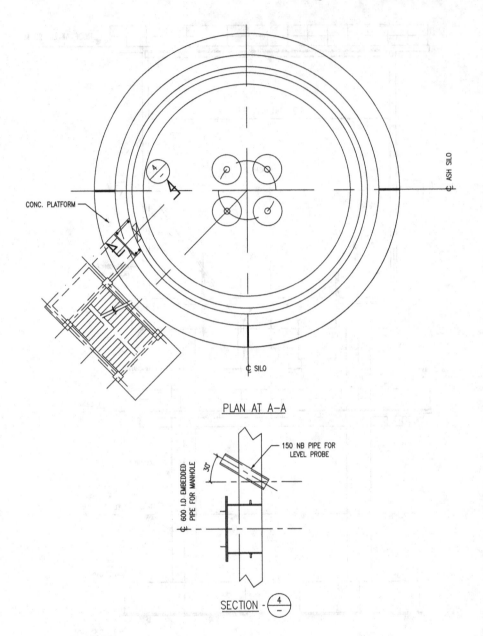

**FIGURE 9.14** Plan at the bottom slab.

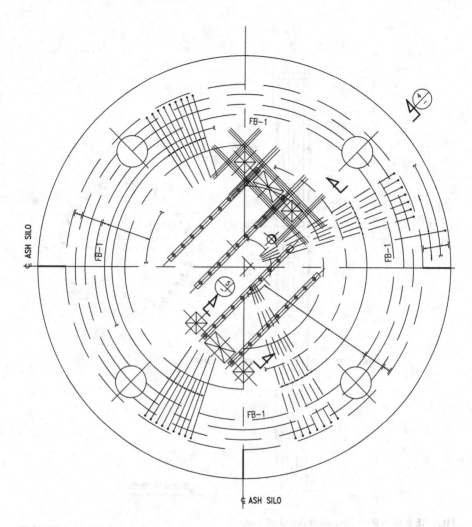

**FIGURE 9.15** Reinforcement arrangement at operating floor slab.

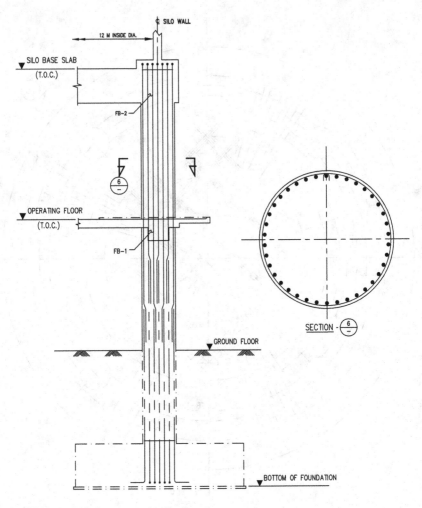

**FIGURE 9.16** Reinforcement detail of a column.

# Reinforced Concrete and Associated Work

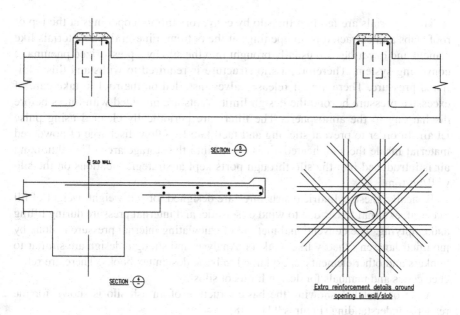

**FIGURE 9.17** Sectional views and details.

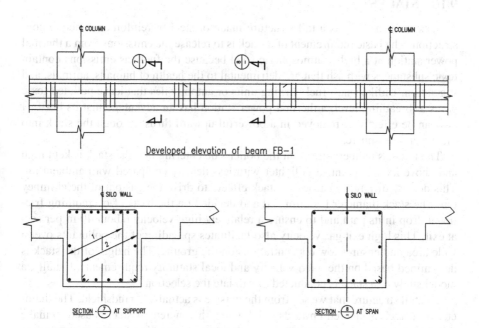

**FIGURE 9.18** Reinforcement arrangement in beam FB1.

The materials are fed into the silo by conveyors through openings at the top of roof slabs and extracted from openings at the bottom. Fine-grained materials like cement and dry ashes are usually brought into the silo by a pressurized pneumatic conveying system. Therefore, a silo structure is required to withstand this additional pressure. There are air release valves installed on the roof to take care of excessive pressure beyond the design limit. Vents are provided with filters before discharging to the atmosphere. The filters are periodically cleaned using pulse jet air. In order to prevent sticking and facilitate free flow discharge of powdered material inside the silo, heated air is blown into the storage area. This fluidizing air is introduced into the silo through ports kept at strategic locations on the silo wall/bottom.

Basically, these cylindrical structures are designed for self-weight, weight of fill material, external forces due to wind or seismic, and internal pressure during filling and emptying. The conventional method of calculating internal pressure is done by modified Janssen's theory like bunkers. Analysis and strength design are similar to bunkers and other elements as explained in basic design textbooks. There are reference codes and manuals for design limits of silos.

A set of sketches showing the basic structure of an ash silo is shown for the reader's understanding (Figures 9.11–9.18).

## 9.16 STACKS

A stack or a chimney is a tall structure made of steel or reinforced concrete construction. The basic requirement of a stack is to release the emissions from a thermal power station at a higher atmospheric level because the fuel gas emissions contain toxic substances with ash that are detrimental to the health of humans, animals, and vegetation. Unlike small fuel burning units or automobiles that leave polluted toxic gases at breathing level, a thermal power station has an advantage that its hot fuel gas can be effectively removed in a powerful upward thrust through the stack into the upper atmosphere.

The flue gas temperature from the boiler side entering into the stack is kept high and above its dew point. It is lighter with less density compared with ambient air. This density difference causes a "stack effect" to drive the gas out of the chimney top. The stack diameter is optimized and decided on the basis of overcoming frictional drop in its path and to ensure a relatively high velocity (about 40 m per sec) at exit. This high exit gas velocity also facilitates spreading of the pollutants over a wide area and ensures low concentrations on the ground. The height of the stack is determined based on the wind velocity and local statutory requirements. Usually, a model study of a stack is conducted to validate the selection.

The tall structure that we see from the outside is actually a windshield. The diameter of a stack is tapered upwards. High-strength concrete or steel is used to make this tall structure. Since concrete is weak to resist corrosion from aggressive fuel gases that condense into acid at lower temperatures at the top of the stack, it is protected by a heat and acid-resistant brick lining inside with a cast iron annular cap at the top. This brick liner is also shaped like the stack, but the height is broken into regular segments of approximately 10 m. Each segment rests on a projected corbel from the main concrete shell. There is a ventilation gap of 75–100 mm kept between the inner face of the concrete shell and the outer face of the brick liner. A series of small openings are kept in the bottom of the cell that attracts atmospheric air and is heated on mixing with the leaked gas inside the air gap and thus creates an upward flow for release through similar holes at the top. This is a typical description of a brick-lined single-flue stack.

A large thermal station with multiple generating units needs stacks of 275 m or higher for individual units. The civil cost of such tall stacks is considerably higher, which has a definite impact on the total project cost. So, the flue gas from individual generating units is conveyed through individual steel flues of cylindrical cross-section, say two or three, which climbs up and is housed in a single concrete windshield. This is called a multi-flue stack.

These flue cans are supported (by hanging) from steel platforms built inside the concrete windshield. One or two expansion compensators are usually provided to take care of the thermal expansion of the steel flues. Since the top portion comes in contact with rainwater, a stainless-steel can is provided at the stack exit to combat corrosion. Thermal insulation is provided on the outer surface of the steel flues to prevent the heat loss of flue gas and thereby falling temperatures.

The diameters of such multi-flue stacks are reduced with heights being tapered but must be adequate to house flue cans, stairs, elevators, and space for maintenance.

The outside surface of the stack is painted with acid-resistant paint at upper portions and normal cement painting below. Red and white bands of synthetic enamel paint at the outer face at the top to serve as an aviation warning. In addition, aviation warning lights and lightning protection is provided as per rules.

The construction of such a tall structure is a specialized job and done by a skilled and experienced construction company. The casting of a windshield is done in a continuous process using slip form units.

**272** Design of Industrial Structures

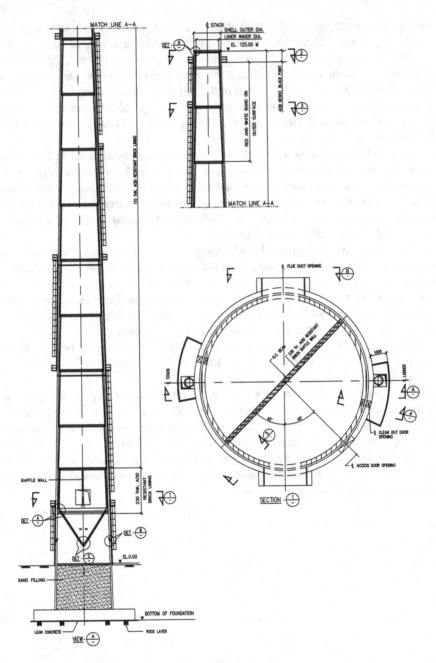

**FIGURE 9.19** RCC stack – plan and sectional elevation.

# Reinforced Concrete and Associated Work

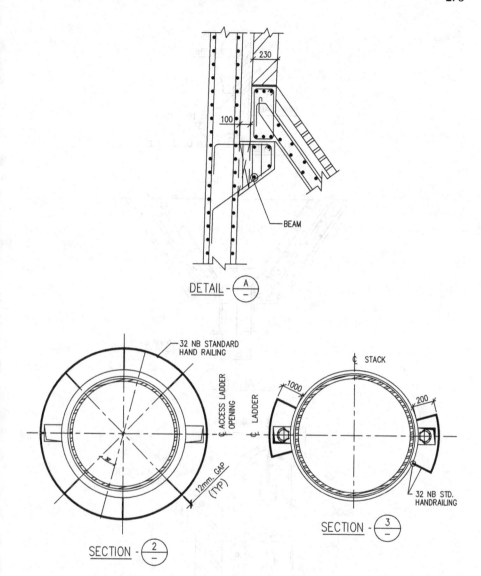

**FIGURE 9.20** RCC stack – plan and details.

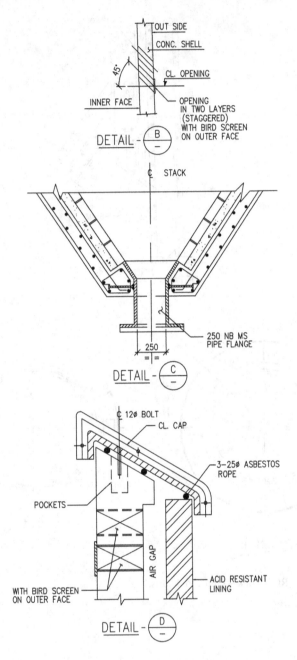

**FIGURE 9.21** RCC stack – details.

# Reinforced Concrete and Associated Work

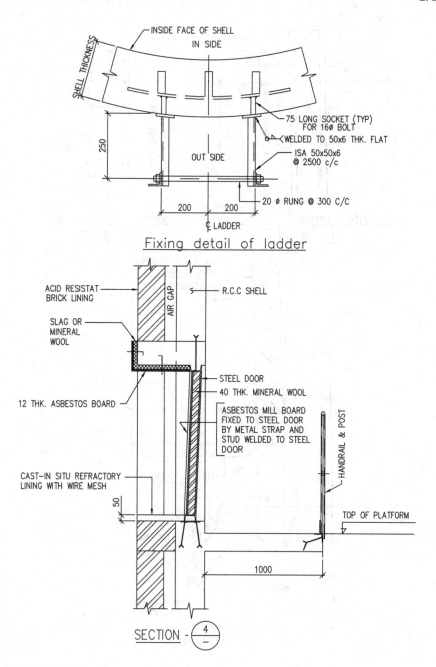

**FIGURE 9.22** RCC stack – misc. sections and details.

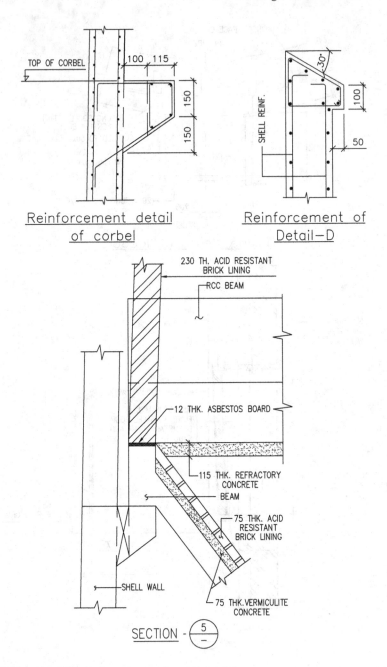

**FIGURE 9.23** RCC stack – misc. section and details.

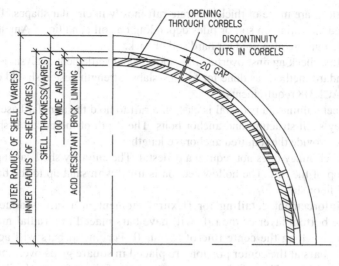

Detail showing opening through corbel

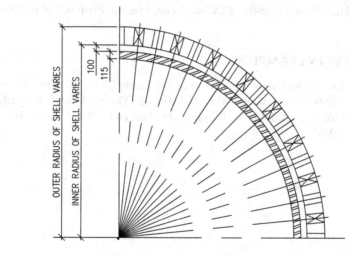

Detail showing opening through shell

**FIGURE 9.24** RCC stack – misc. details.

Foundations are made of thick concrete raft mostly in circular shapes. The mats are designed as resting on soil or piles depending on soil conditions. Annular ring-type foundations are also used for hard soil or rock.

A stability check against overturning, sliding, and settlement analysis is carried out by standard methods as done for footing slabs. Strength design should conform to IS: 456/ACI 318 requirements.

For a steel chimney, a pedestal is cast on a raft to hold the annular base plate of the chimney shell structure and anchor bolts. The depth of the pedestal should be adequate to provide the required anchorage length.

An RCC chimney does not require a pedestal. The chimney shell casting begins from the top of the raft. The hollow section is filled with sand up to the bottom of the ground floor slab.

The reinforcement detailing for flexural moment in a raft is done in two layers. The bottom layer of the raft will have bars placed in a radial manner at equal spacing along the centerline of the shell. Peripheral bars will act as the binder. The bars at the center portion are placed in square grids overlapping the ends of radial bars. The reinforcement at the top of the raft can also be placed in a square grid placed along an orthogonal direction. Raft thickness may be increased to meet the shear stress requirements and thus eliminate using shear reinforcement.

Typical details of single-flue RCC stacks are given in Figures 9.19 to 9.24.

## 9.17 DESIGN EXAMPLES

In the section, we will show some worked-out examples as guidelines for member strength designs in accordance with ACI 318–19. The method of design for concrete structure in accordance with Indian standard code IS 456 (2000) is available in Chapter 12.

# Reinforced Concrete and Associated Work

*Example 9.17.1 : Design of a Singly reinforced RCC Beam.*

Reference Code :   ACI 318 - 19   US customary units (stress in psi)

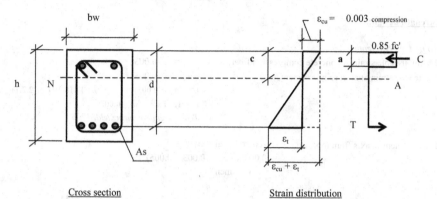

Cross section             Strain distribution

**Design parameters:**

| | | | | | | |
|---|---|---|---|---|---|---|
| fc' = | 3000 | psi | bw = | 10 | inch | |
| fy = | 60000 | psi | h = | 24 | inch | > span/16 |
| Es = | 29000000 | psi | d' = | 1.5 | inch | cover |
| | | | bar dia = | 0.75 | inch | |
| Span = | 19.68 | ft | d = | 22.5 | inch | |

**Design load:**        Factored load        | Load factor | 1 |

M span =    153 kip-ft    Mu span =    153 kip-ft
M support =    191 kip-ft    Mu support =    191 kip-ft
V =    46 kip    Vu =    46 kip

**Reinforcement :**                                                      | Astmin =    0.75 in$^2$ . |

**At Mid span**        bar #       dia        Ast
Top      2 nos      6    0.75 inch    0.88 in$^2$ .
Bot      6 nos      6    0.75 inch    2.64 in$^2$ .
Stirrup    2 legged    3    0.375 in dia @    10 in c/c(s)    Asv =    0.22 in$^2$/set.

**At support**        bar #       dia        Ast
Top (T)    2 nos      6    0.75 inch    0.88 in$^2$ plus ⎤    2.64 mm2 .
          4 nos      6    0.75 inch    1.76 in$^2$ . ⎦
Bot (C)    2 nos      6    0.75 inch    0.88 in$^2$ .

Stirrup  2 legged  3  0.375 in dia @  4 in c/c(s)   Asv =  0.22 in²/set.

## Strength design:

*To find out NA depth*
Maxm. useable strain at concrete compression fiber,   $\varepsilon_{cu}$ =  0.003
Net tensile strain at steel reinforcement,    $\varepsilon_t$ =  0.005
  [ >= $\varepsilon t y + 0.003$]    ACI 318-19 ( 21.2.2.1)
  $\varepsilon_{ty}$ = fy / Es =   0.002
  *Tension reinforcement yielded*

Depth of neutral axis from top,  c =  $\varepsilon cu \cdot d / (\varepsilon cu + \varepsilon t)$
  =  0.003 x 22.5 / (0.003 + 0.005)
  =  8.44 inch

  a =  $\beta \cdot c$   ACI 318-19 ( 21.2.2.4.1)
  $\beta$ =  0.85   ACI 318-19 (Table 22.2.2.4.3)
  a =  0.85 x 8.44  =  7.17 inch

*To find out Minimum reinforcement:*

In Flexure:  As min will be greater of  a) and  b)   ACI 318-19 ( 9.6.1.2)

a)  3 . (fc')^0.5 . bw. d / fy =  3 x 3000^0.5 x 10 x 22.5 / 60000 =   0.62 in².

b)  200 . bw. d / fy  =  200 x 10 x 22.5 / 60000 =   0.75 in².
   As min =  0.75 in².

Shear reinforcement :  Av min / s,  will be greater of  a) and  b)   ACI 318-19 ( Table 9.6.3.4)
s = spacing

a)  0.75 . (fc')^0.5 . bw  / fyt =  0.75 x 3000^0.5 x 10 / 60000 =   0.01  x spacing (in²).

b)  50 . bw  / fyt  =  50 x 10 / 60000 =   0.01  x spacing (in²).
   Av min =  0.01  x spacing (in²).

# Reinforced Concrete and Associated Work

## At Support
*Flexural reinforcement as singly reinforced beam*

| | | | |
|---|---|---|---|
| Strength reduction factor, | $\phi$ | 0.9 for moment | ACI 318-19 (Table 21.2.1) |
| | Mu | 191 kip-ft | tension controlled |
| Required nominal strength, $M_n$ = | Mu/$\phi$ | 212 kip-ft | |

$C = 0.85 fc' \cdot a \cdot bw = $ 0.85 x 3000 x 7.17 x 10 / 1000 = 183 kips

$Mn_1 = C(d - a/2) = $ 183 x (22.5 - 0.5 x 7.17) / 12 = 288.45 kip-ft

> Mn rqd; Singly reinforced.

$As1$ rqd = $Mn /fy \cdot (d- a/2) = $ 212x1000x12 / 60000x(22.5-0.5x7.17) = 2.24 in$^2$.

| | reqd | provided | |
|---|---|---|---|
| Total compression reinforcement, $As_2$ = | | 0.88 in$^2$. | Safe. |
| Total Tension reinforcement, As = | 2.24 in$^2$. | 2.64 in$^2$. | Safe. |

*for balanced section*

The maxm. area of reinforcing steel in a singly reinforced section = $As_1$ = $T/fs$

Here, $T = C = $ 183 kips

and $fs = fy = $ 60000 psi    $As1 = T/fy = $ 183 x 1000/60000 = 3.05 in$^2$.

## Shear reinforcement at support

| | | | |
|---|---|---|---|
| Strength reduction factor, | $\phi$ | 0.75 for shear | ACI 318-19 (Table 21.2.1) |
| | Vu | 46 kip-ft | |
| Required nominal strength, $V_{n\ rqd}$ = | Vn/$\phi$ | 61 kip-ft | |

Now, Vn = Vc + Vs, where Vc is the shear capacity of concrete beam section and Vs is capacity of of shear reinforcement in the form of stirrups or bent up bars.

$Vc = 2 \lambda fc'^{0.5} \cdot bw.d = $ 2 x 0.78 x (3000)^0.5 x 10 x 22.5 = **19** kips

ACI 318-19 (Table 22.5.5.1)

$\lambda = (2/(1+d/10))^{0.5} = $ 0.7845 <=1      ACI 318-19 (22.5.5.1.3)

Provide shear reinforcement.

Shear bars provided :   Stirrup     2 leggd #   3     0.375 in dia @     4 in c/c.

|  |  |  |  |
|---|---|---|---|
|  |  | Av = | 0.22 in² / set. |
|  | Av min = | 0.01 x 4 = | 0.04 in2 |

Now, Vs = Av. fyt. d / s =   0.22 x 60000 x 22.5 / 4=   74.25 kips   ACI 318-19 (22.5.8.5.3)
Vn = Vc + Vs =   19 + 74.25 =   94 kips   > Vn rqd; Safe,

Maxm. Allowable shear in this C/S of beam, Vu max =   74 kips.   ACI 318-19 (22.5.1.2)
Vu <= φ (Vc + 8. fc'^0.5. bw.d)
  Vu =   46 kips.   < Vu max; Ok.

## At Span
*Flexural reinforcement as singly reinforced beam*

Strength reduction factor,   φ   0.9  for moment   ACI 318-19 (Table 21.2.1)
  Mu   153 kip-ft   tension controlled
Required nominal strength, $M_{n\ rqd}$ =   Mu/φ   170 kip-ft
  from stress block above, a =   7.17 inch   c =   8.4 inch.   $\varepsilon_{cu}$ =   0.003

C = 0.85 fc' . a . bw =   0.85 x 3000 x 7.17 x 10 / 1000 =   183 kips
$Mn_1$ = C (d - a/2) =   183 x (22.5 - 0.5 x 7.17) / 12 =   288 kip-ft
  > Mn; Singly reinforced.

As1 rqd =   Mn /fy. (d- a/2) =   170 x 1000 x 12 / 60000 x (22.5 - 0.5 x 7.17) =   1.80 in².

|  | reqd | provided |  |
|---|---|---|---|
| Total compression reinforcement, $As_2$ = |  | 0.88 in². | Safe. |
| Total Tension reinforcement, As = | 1.80 in². | 2.64 in². | Safe. |

*for balanced section*
The maxm. area of reinforcing steel in a singly reinforced section = $As_1$   = T / fs
Here, T = C =   183 kips
and   fs = fy =   60000 psi   As1 = T / fy =   183 x 1000/60000 =   3.05 in².

End

# Reinforced Concrete and Associated Work

## Example 9.17.2 : Design of a Doubly reinforced RCC Beam.

Reference Code :  ACI 318 - 19    SI - metric stress in Mpa

*Cross section*  *Strain distribution*

### Design parameters:

| | | | | | |
|---|---|---|---|---|---|
| $f_c'$ = | 20 | Mpa | $b_w$ = | 250 | mm |
| $f_y$ = | 415 | Mpa | h = | 600 | mm > span/16 |
| $E_s$ = | 200000 | Mpa | d' = | 40 | mm cover |
| | | | bar dia = | 20 | mm |
| Span = | 6 | m | d = | 560 | mm |

### Design load:
M span = 207 kNm
M support = 334 kNm
V = 207 kN

### Factored load
Mu span = 207 kNm
Mu support = 334 kNm
Vu = 207 kN

Load factor  1

### Reinforcement :

$A_{st\,min}$ = 472 mm².

| At Mid span | | dia | | Ast | | |
|---|---|---|---|---|---|---|
| Top | 2 nos | 20 mm | | 628 mm². | | |
| Bot | 4 nos | 20 mm | | 1256 mm². | | |
| Stirrup | 2 legged | 8 mm φ @ | 300 mm c/c | $A_{sv}$ = | 100.56 mm²/set. | |
| **At support** | | | | Ast | | |
| Top (T) | 2 nos | 20 mm | | 628 mm² + | 2198 mm2 . | |
| | 5 nos | 20 mm | | 1570 mm². | | |
| Bot (C) | 2 nos | 20 mm | | 628 mm². | | |
| Stirrup | 2 legged | 8 mm φ @ | 100 mm c/c | $A_{sv}$ = | 100.56 mm²/set. | |

## Strength design:

### To find out NA depth

Maxm. useable strain at concrete compression fiber,  $\varepsilon_{cu}$ = 0.003
Net tensile strain at steel reinforcement,  $\varepsilon_t$ = 0.005
[ >= $\varepsilon_t$ y + 0.003]  ACI 318-19 ( 21.2.2.1)
$\varepsilon_{ty}$ = fy / Es = 0.002
*Tension reinforcement yielded*

Depth of neutral axis from top,  c = $\varepsilon_{cu}$ . d / ($\varepsilon_{cu}$ + $\varepsilon_t$)
= 0.003x560/(0.003+0.005)
= 210 mm
a = $\beta$ . c   ACI 318-19 ( 21.2.2.4.1)

$\beta$ = 0.85   ACI 318-19 (Table 22.2.2.4.3)
a = 0.85 x 210 = 179 mm

### To find out Minimum reinforcement:

In Flexure:  As min will be greater of  a) and b)   ACI 318-19 ( 9.6.1.2)

a)  0.25 . (fc')^0.5 . bw. d / fy =  0.25 x 20^0.5 x 250 x 560 / 415 =  377 mm$^2$.

b)  1.4 . bw. d / fy =  1.4 x 250 x 560 / 415 =  472 mm$^2$.
   As min = 472 mm$^2$.

Shear reinforcement :  Av min / s, will be greater of  a) and b)   ACI 318-19 ( Table 9.6.3.4)
s = spacing

a)  0.062 . (fc')^0.5 . bw  / fyt =  0.062 x 20^0.5 x 250 / 415 =  0.17  x spacing (mm$^{2)}$.

b)  0.35 . bw / fyt =  0.35 x 250 / 415 =  0.21  x spacing (mm$^{2)}$.
   As min = 0.21  x spacing (mm$^{2)}$.

# Reinforced Concrete and Associated Work

## At Support
*Flexural reinforcement :*

Strength reduction factor ,      $\phi$ =    0.9   for moment     ACI 318-19 (Table 21.2.1)
                                      Mu =    334 kNm                  tension controlled
Required nominal strength, $M_n$ =      Mu/$\phi$ =   371 kNm

      C = 0.85 fc' . a . bw =    0.85 x 20 x 179 x 250 / 1000 =         761 kN
      $Mn_1$ =    C (d - a/2) =    761 x (560 - 0.5 x 179) / 1000 =      358 kNm
                                                                                    *<Mn ; doubly reinforced.*

As1 rqd =      Mn /fy. (d- a/2) =     371 x 1000000 / 415 x (560 - 0.5 x 179) =         1901 mm$^2$.

*for balanced section*
*The maxm. area of reinforcing steel in a singly reinforced section* = $As_1$      = T / fs
Here, T = C =      761 kN
and fs = fy =      415 Mpa      $As1 = T / fy$ =    761 x 1000/415 =        1833 mm$^2$ .

*Provide flexural reinforcement as doubly reinforced beam*

| | | | Mu / $\phi$ | 371 | kip-ft |
|---|---|---|---|---|---|
| $Mn_2$ | = Mn rqd - $Mn_1$ =   371 - 358 =    13 kNm | | $Mn_1$ | 358 | kip-ft |

      $Mn_2$ = Cs . ( d - d' )     where    Cs = compression in steel at top
                                                 d' = top cover

      Cs = $Mn_2$ / (d - d') =    13 x 1000 / (560-40) =        25 kN

      $\varepsilon_s$ = $\varepsilon_{cu}$ . (c-d') / c =     where    $\varepsilon_s$ = strain in comp rebar at top
         =       0.0024          > ey , comp steel yielded.

Now, Stress in comp steel , fs' = fy =      415 Mpa

After deduction of equiv conc stress, Cs = ( fs'- 0.85fc') . $As_2$

      $As_2$ = Cs / (fs' - 0.85 fc') =     25 x 1000 / (415-0.85x 20) =         64 mm$^2$.

$As1 + As2 = 1901 + 64 = 1964$ mm$^2$.

| | reqd | provided | |
|---|---|---|---|
| Total compression reinforcement, $As_2 =$ | 64 mm$^2$. | 628 mm$^2$. | Safe. |
| Total Tension reinforcement, $As =$ | 1964 mm$^2$. | 2198 mm$^2$. | Safe. |

*Shear reinforcement at support*

| Strength reduction factor, | $\phi$ | 0.75 for shear | ACI 318-19 (Table 21.2.1) |
|---|---|---|---|
| | Vu | 207 kN | |
| Required nominal strength, $V_{n\,rqd} =$ | Vn/$\phi$ | 276 kN | |

Now, Vn = Vc + Vs, where Vc is the shear capacity of concrete beam section and Vs is capacity of of shear reinforcement in the form of stirrups or bent up bars.

$Vc = 0.17\,\lambda\,fc'^{\wedge}0.5.\,bw.d = 0.17 \times 0.79 \times (20)^{\wedge}0.5 \times 250 \times 560 =$ **84 kN**

ACI 318-19 (Table 22.5.5.1a)

$\lambda = (2/(1+0.004d))^{\wedge}0.5 = 0.7857 <=1$      ACI 318-19 (22.5.5.1.3)

Provide shear reinforcement.

| Shear bars provided : | Stirrup | 2 legged | 8 mm dia @ | 100 mm c/c. |
|---|---|---|---|---|
| | | | Av = | 100 mm$^2$ / set. |
| | | Av min = | 0.21 x 100 = | 21 in2 |
| Now, Vs = Av. fyt. d / s = | | 100 x 415 x 560 / 100= | 232 kN | ACI 318-19 (22.5.8.5.3) |
| Vn = Vc + Vs = | | 84 + 232 = | 316 kN. | Safe, |

Maxm. Allowable shear in this C/S of beam, Vu max =   310 kN      ACI 318-19 (22.5.1.2)
$Vu <= \phi\,(Vc + 0.66\,fc'^{\wedge}0.5.\,bw.d)$

Vu =    207 kN      Ok.

# Reinforced Concrete and Associated Work

## At Span
### Flexural reinforcement

Strength reduction factor,        $\phi =$    0.9   for moment     ACI 318-19 (Table 21.2.1)

                                  $M_u =$    207 kNm            tension controlled

Required nominal strength, $M_n =$    $M_u/\phi =$    230 kNm

*from stress block above,*    $a =$    *179 mm*     $c =$     *210 mm*     $\varepsilon_{cu} =$    *0.003*

       $C = 0.85 \, f'c \cdot a \cdot b_w =$    0.85 x 20 x 179 x 250 / 1000 =       761 kN

       $Mn_1 = C(d - a/2) =$    761 x (560 - 0.5 x 179) / 1000 =       358 kNm

                                                                                *> Mn; Singly reinforced.*

$As1$ rqd =    $M_n / f_y \cdot (d - a/2) =$    230x1000000 / 415x(560-0.5x179) =      1178 $mm^2$.

*for balanced section*

*The maxm. area of reinforcing steel in a singly reinforced section = $As_1$*         *= T / fs*

*Here, T = C =*      *761 kN*

*and   fs = fy =*      *415 Mpa*      *As1 = T / fy =*     *761 x 1000/415 =*        *1833 $mm^2$*.

                                                                  reqd       provided

             Total compression reinforcement, $As_2 =$                    628 $mm^2$.    Safe.

             Total Tension reinforcement, As =            1178 $mm^2$.    1256 $mm^2$.    Safe.

End

## Example 9.17.3 : Design of a RCC slab (US customary units).

Reference Code :   ACI 318 - 19    US customary units (stress in psi)

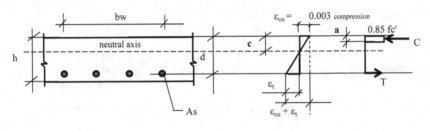

| Cross section | | Strain distribution |

**Design parameters:**

| | | | One way slab (both end fixed) | | | |
|---|---|---|---|---|---|---|
| fc' = | 3000 | psi | bw = | 12 | inch | |
| fy = | 60000 | psi | h = | 4.5 | inch   > span/28 | (**) |
| Es = | 29000000 | psi | d' = | 0.5 | inch   cover | |
| | | | bar dia = | 0.4 | inch | |
| Span L = | 10 | ft | d = | 4 | inch | |

| **Design load:** | | | **Factored load** | | | **Load factor** | 1 |
|---|---|---|---|---|---|---|---|
| M span = | 1.51 | kip-ft | Mu span = | 1.51 | kip-ft | | |
| M support = | 1.91 | kip-ft | Mu support = | 1.91 | kip-ft | | |
| V = | 0.6 | kip | Vu = | 0.6 | kip | | |

**Reinforcement :**

| **At Mid span** | bar # | dia | | spcg | | Ast |
|---|---|---|---|---|---|---|
| Bot | 3 | 0.375 inch @ | | 6 in c/c | | 0.22 in² / ft wide |

| **At support** | bar # | dia | | spcg | | Ast |
|---|---|---|---|---|---|---|
| Top | 3 | 0.375 inch @ | | 6 in c/c | | 0.22 in² / ft wide |

(**)  Table 7.3.1.1 Minm thickness of solid oneway slab

| Support conditon | Minimum h (for fy = 60000 psi) |
|---|---|
| Simply supported | L/20 |
| One end continuous | L/24 |

*[contd]*

# Reinforced Concrete and Associated Work

| Both end continuous | L/28 |
|---|---|
| Cantilever | L/10 |

**Strength design:**

*To find out NA depth*

Maxm. useable strain at concrete compression fiber,     $\varepsilon_{cu}$ =    0.003

Net tensile strain at steel reinforcement,     $\varepsilon_t$ =    0.005

     [ >= $\varepsilon t\, y$ + 0.003]      ACI 318-19 ( 21.2.2.1)

     $\varepsilon_{ty}$ = fy / Es =     0.002069

     Tension reinforcement yielded

Depth of neutral axis from top,   c =   $\varepsilon cu \cdot d / (\varepsilon cu + \varepsilon t)$

     =   0.003x4/(0.003+0.005)

     =   1.5 inch

     a =   $\beta \cdot c$      ACI 318-19 ( 21.2.2.4.1)

     $\beta$ =   0.85      ACI 318-19 (Table 22.2.2.4.3)

     a =   0.85 x 1.5    =    1.28 inch

*To find out Minimum reinforcement:*

In Flexure:    As min = 0.0018 Ag =    0.0018 x 12 x 4.5 =    0.10 $in^2$.    ACI 318-19 ( 7.6.1.1)

**At Support**

*Flexural reinforcement as singly reinforced slab*

Strength reduction factor,     $\phi$ =   0.9 for moment     ACI 318-19 (Table 21.2.1)

     Mu =   1.91 kip-ft      tension controlled

Required nominal strength, $M_n$ =   Mu/$\phi$ =   2 kip-ft

     C = 0.85 fc' . a . bw =   0.85 x 3000 x 1.275 x 12 / 1000 =    39.015 kips

     $Mn_1$ =   C (d - a/2) =   39.015 x (4 - 0.5 x 1.275) / 12 =    10.932 kip-ft

     > $Mn$; *Singly reinf.*

As rqd =    Mn /fy. (d- a/2) =    2x1000x12 / 60000x(4 - 0.5x1.28) =    0.13 $in^2$.

     Reinforcement Provided =   0.22 $in^2$.    Safe.

*For balanced section:*

The maxm. area of reinforcing steel in a singly reinforced section = $As_1$      = T / fs

   Here, T = C =    39.015 kips      $As1$ = T / fy =    39.015 x 1000/60000 =      0.65025 $in^2$.

   and   fs = fy =    60000 psi

### Shear reinforcement at support

Strength reduction factor, $\phi =$ 0.75 for shear    ACI 318-19 (Table 21.2.1)

$V_u =$ 0.6 kip

Required nominal strength, $V_n = V_n/\phi =$ 0.8 kip

$V_c = 2 \lambda f_c'^{0.5} \cdot b_w \cdot d = 2 \times 1 \times (3000)^{0.5} \times 12 \times 4 =$    5 kips    ACI 318-19 (Table 22.5.5.1)

$\lambda = (2/(1+d/10))^{0.5} =$    $1 <= 1$    ACI 318-19 (22.5.5.1.3)

Now, $V_n >= V_c$, where $V_c$ is the shear capacity of concrete slab section
<u>Shear reinf. not necessary.</u>

Maxm. Allowable shear in this C/S of slab, $V_u$ max =    15.8 kips.    ACI 318-19 (22.5.1.2)

$V_u <= \phi (V_c + 8 \cdot f_c'^{0.5} \cdot b_w \cdot d)$

$V_u =$    0.6 kips.    Ok.

## At Span

### Flexural reinforcement as singly reinforced slab

Strength reduction factor, $\phi =$ 0.9 for moment    ACI 318-19 (Table 21.2.1)

$M_u =$ 1.51 kip-ft    tension controlled

Required nominal strength, $M_n = M_u/\phi =$ 2 kip-ft

$C = 0.85 f_c' \cdot a \cdot b_w =$    $0.85 \times 3000 \times 1.275 \times 12 / 1000 =$    39.015 kips

$Mn_1 = C(d - a/2) =$    $39.015 \times (4 - 0.5 \times 1.275)/12 =$    10.932 kip-ft

> $M_n$; Singly reinforced.

As rqd =    $M_n/f_y \cdot (d - a/2) =$    $2 \times 1000 \times 12 / 60000 \times (4 - 0.5 \times 1.28) =$    0.10 in$^2$.

Reinforcement Provided =    0.22 in$^2$.    Safe.

<u>For balanced section:</u>
The maxm. area of reinforcing steel in a singly reinforced section $= As_1$     $= T/f_s$

Here, $T = C =$    39.015 kips

and $f_s = f_y =$    60000 psi    $As_1 = T/f_y =$    $39.015 \times 1000/60000 =$    0.65025 in$^2$.

End

# Reinforced Concrete and Associated Work

*Example 9.17.4 : Design of a RCC slab section (SI units).*

Reference Code :   ACI 318 - 19       SI - metric stress in Mpa

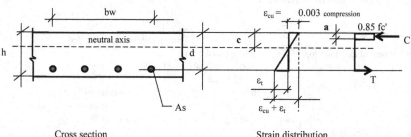

Cross section                          Strain distribution

**Design parameters:**                 One way slab (both end fixed)

| | | | | | | |
|---|---|---|---|---|---|---|
| fc' = | 20 | Mpa | bw = | 1000 | mm | |
| fy = | 415 | Mpa | h = | 115 | mm   > span/28 | (**) |
| Es = | 200000 | Mpa | d' = | 15 | mm   cover | |
| | | | bar dia = | 8 | mm | |
| Span L = | 3 | m | d = | 100 | mm | |

**Design load:**                Factored load                 | Load factor | 1 |

M span = 6.75 kNm        Mu span = 6.75 kNm
M support = 8.5 kNm      Mu support = 8.5 kNm
V = 9 kN                 Vu = 9 kN

**Reinforcement :**

**At Mid span**     dia        spcg       Ast
Bottom              8 mm @    150 mm c/c   335 mm$^2$ /m width.

**At support**                 Ast
Top                 8 mm       150 mm c/c   335 mm$^2$ /m width.

(**)   Table 7.3.1.1 Minm thickness of solid oneway slab

| Support conditon | Minimum h (for fy = 60000 psi) |
|---|---|
| Simply supported | L/20 |
| One end continuous | L/24 |
| Both end continuous | L/28 |
| Cantilever | L/10 |

## Strength design:

### To find out NA depth
Maxm. useable strain at concrete compression fiber,     $\varepsilon_{cu}$ =    0.003
Net tensile strain at steel reinforcement,     $\varepsilon_t$ =    0.005
               [ >= $\varepsilon t y$ + 0.003]        ACI 318-19 ( 21.2.2.1)
               $\varepsilon_{ty}$ = fy / Es =      0.002075
               Tension reinforcement yielded

Depth of neutral axis from top,   c =    $\varepsilon cu \cdot d / (\varepsilon cu + \varepsilon t)$
                             =    0.003 x 100 / (0.003 + 0.005)
                             =    37.5 mm
                           a =    $\beta \cdot c$                ACI 318-19 ( 21.2.2.4.1)
                           $\beta$ =    0.85            ACI 318-19 (Table 22.2.2.4.3)
                           a =    0.85 x 37.5    =    32 mm

### Minimum reinforcement:
In Flexure:      As min = 0.0018 Ag =    0.0018 x 1000 x 115 =     207 in$^2$.     ACI 318-19 ( 7.6.1.1)

## At Support
*Flexural reinforcement as singly reinforced slab*

Strength reduction factor ,          $\phi$     0.9   for moment         ACI 318-19 (Table 21.2.1)
                                  Mu        9 kNm               tension controlled
Required nominal strength, $M_n$ =    Mu/$\phi$     9 kNm

       C = 0.85 fc' . a . bw =    0.85 x 20 x 32 x 1000 / 1000 =         544 kN
       $Mn_1$ =    C (d - a/2) =    544 x (100 - 0.5 x 32) / 1000 =        46 kNm
                                                                           > Mn; *Singly reinforced.*
As rqd =     Mn /fy. (d- a/2) =    9 x 1000000 / 415 x (100 - 0.5 x 32) =     271 mm$^2$.
                                      Reinforcement Provided =    335 mm$^2$.    **Safe.**

*For balanced section:*
*The maxm. area of reinforcing steel in a singly reinforced section = $As_1$*         *= T / fs*
      Here, T = C =       544 kN
      and   fs = fy =       415 Mpa        As1 = T / fy =     544 x 1000/415 =          1311 mm$^2$.

### Shear reinforcement at support
Strength reduction factor ,          $\phi$     0.75   for shear         ACI 318-19 (Table 21.2.1)
                                  Vu        9 kN
Required nominal strength, $V_n$ =    Vn/$\phi$      12 kN

      Vc = 0.17 $\lambda$ fc'^0.5. bw.d =   0.17 x 1 x (20)^0.5 x 1000 x 100 =       76 kN
                                                                              ACI 318-19 (Table 22.5.5.1a)

# Reinforced Concrete and Associated Work

$\lambda = (2/(1+0.004d))^{0.5} = \quad 1 <=1$          ACI 318-19 (22.5.5.1.3)

Now, $Vn >= Vc$, where Vc is the shear capacity of concrete slab section

                                            ***Shear reinf. not necessary.***

Maxm. Allowable shear in this C/S of slab, Vu max =     221 kN         ACI 318-19 (22.5.1.2)
$Vu <= \phi (Vc + 0.66 fc'^{0.5}. bw.d)$

                                      Vu =       9 kN         Ok.

## At Span
*Flexural reinforcement as singly reinforced slab*

| | | | | | | |
|---|---|---|---|---|---|---|
| Strength reduction factor, | $\phi$ | 0.9 for moment | | ACI 318-19 (Table 21.2.1) | | |
| | Mu | 7 kNm | | tension controlled | | |
| Required nominal strength, $M_n$ = | Mu/$\phi$ | 8 kNm | | | | |
| *from stress block above*, a = | 32 mm | c = | 38 mm | $\varepsilon_{cu}$ = | 0.003 | |

        C = 0.85 fc' . a . bw =    0.85 x 20 x 32 x 1000 / 1000 =        544 kN
        $Mn_1$ =   C (d - a/2) =    544 x (100 - 0.5 x 32) / 1000 =       46 kNm
                                                                 > Mn; Singly reinforced.

As rqd =    Mn /fy. (d- a/2) =    8x1000000 / 415x(100 - 0.5x32) =      215 mm$^2$.
                                          Reinforcement Provided =    335 mm$^2$.    *Safe.*

*For balanced section:*
*The maxm. area of reinforcing steel in a singly reinforced section = $As_1$*          = T / fs
     *Here,* T = C =       544 kN
     *and* fs = fy =        415 Mpa      $As1 = T/fy$ =    544 x 1000/415 =       1311 mm$^2$.

                                                                                                End

## Example 9.17.5 : Design of a RCC Column.
Reference Code :     ACI 318 - 19          SI - metric stress in Mpa

**1.0 Column Mark/Grid :**     C1

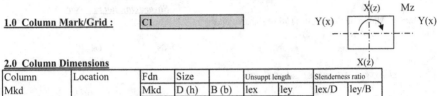

**2.0 Column Dimensions**

| Column Mkd | Location | Fdn Mkd | Size | | Unsuppt length | | Slenderness ratio | |
|---|---|---|---|---|---|---|---|---|
| | | | D (h) m | B (b) m | lex m | ley m | lex/D | ley/B |
| C1 | | | 0.6 | 0.4 | 5 | 2.5 | 8.33 | 6.25 |
| | | | | | | | Not Slender | Not Slender |

**3.0 Main Reinforcement**

| As1 | 3 | nos | 20 | dia | per sides |
|---|---|---|---|---|---|
| As2 | 2 | nos | 20 | dia | per sides |
| | | | | Safe | < 1; Safe. |

Ast = As1 + As2 =       3140 mm2     p =      1.3 %
Area of steel on shortfaces =     628 mm2 /per face
Area of steel on long faces =     942 mm2 /per face  Equal rebars on opposite sides.

Minimum Reinforcement = 0.01 Ag =     2400 mm2         ACI 318-19 ( 10.6.1.1)
Maxm. Reinforcement = 0.08 Ag =      19200 mm2

**4.0 Design parameters:**

Conc. grade = M25          fc' =     25 N/mm2       fy =     415 N/mm2
Clear Cover =   40 mm      Ag =     240000 mm2     Ast =    3140 mm2
         d' =   50 mm      Ec = 5700 (fc')^0.5 =   28500 N/mm2         γ =   0.8

**5.0 Design load:**

Pu     2200 kN         Mux    300 kNm         Muy    20 kNm

**6.0 Check for Slenderness effect :**                    ACI 318-19 ( 6.2.5)

     lex / D =     5000 / 600 =     8.33
     ley / B =     2500 / 6250 =    6.25

              < 22;  Short column. Slenderness effect is not to be considered.

# Reinforced Concrete and Associated Work

## *Slenderness effect:*

a. Slenderness effect (klu/r) is neglected if, lex/D or ley/B  <=     22     ACI 318-19 ( 6.2.5)

b. If, Slenderness ratio limit exceeds, moments value to be magnified as below:   ACI 318-19 (6.6.4)

Critical buckling load, $Pc = \pi^2 (EI)_{eff} /(klu)^2 =$     20232 kN     ACI 318-19 (6.6.4.4.2)

*about major axis:*
EI eff = 0.4. Ec. Ig / (1+ $\beta_{dns}$) = 0.25 Ec Ig    where,  (1+ $\beta_{dns}$) =   0.6    ACI 318-19 (6.6.4.4.4)
  = 0.25 x 28500 x (400x600^3 / 12)         ACI 318-19 (R6.6.4.4.4)
  = 5.13E+13    N-mm2
$(klu)^2 = 5000^2 =$     3E+07    mm2    for k = 1
   Pc =   20232    kN

*about minor axis:*
EI eff = 0.4. Ec. Ig / (1+ $\beta_{dns}$) = 0.25 Ec Ig    where,  (1+ $\beta_{dns}$) =   0.6    ACI 318-19 (6.6.4.4.4)
  = 0.25 x 28500 x (600x400^3 / 12)         ACI 318-19 (R6.6.4.4.4)
  = 2.28E+13    N-mm2
$(klu)^2 = 2500^2 =$     6E+06    mm2    for k = 1
   Pc =   35968    kN

Moment magnification factor,   $\lambda$ = Cm / ( 1 - Pu/0.75Pc)   >= 1     ACI 318-19 (6.6.4.5)
         Cm =   1           ACI 318-19 (6.6.4.3b)
         $\lambda$ =   1/(1 - 2200 / 0.75 x 20232) =    1.17

Amplified Moment for design , Mc = $\lambda$ $M_2$ , where $M_2$ is first order factored momnet

Note:  1  The column is Not Slender about major  axis.
       2  The column is Not Slender about minor  axis.

## 7.0  Capacity check by PCA load contour method:

| | | | | | |
|---|---|---|---|---|---|
| Pu  | 2200 | kN   | Pn = Pu/$\phi$ =  | 3385 | kN |
| Mux | 300  | kNm  | Mnx = Mux/$\phi$ = | 462  | kNm |
| Muy | 20   | kNm  | Mny = Muy/$\phi$ = | 31   | kNm |

$\phi$ = Strength reduction factor =    0.65       ACI 318-19 (Table 21.2.1&2)

$P_0$ =   Nominal axial strength at zero ecc          ACI 318-19 (Table 22.4.2.2)
   =   0.85 fc' (Ag - Ast) + fy Ast

= 0.85 x 25 x (240000-3140) + 415 x 3140) =     6336 kN
Pn max = 0.8 x P₀ =    0.8 x 6336 =    5069 kN         > Pn; Safe.

Mnox and Mnoy = Equivalent uniaxial bending moment strength
Assume    β =    0.65                                     PCA Note Chapter 7 ( Eqn 20)
          h =    600 mm
          b =    400 mm

Mnox = Mnx + Mny . h/b . ((1-β) / β)                      Mny/Mnx = 0.0667
  =    462 + 31 x (600 / 400) x (1-0.65) / 0.65
  =    486 kNm                                            Mnoy / Mnox = 0.40
Mnoy = Mny + Mnx . b/h . ((1-β) / β)
  =    31 + 462 x (400 / 600) x (1-0.65) / 0.65
  =    196 kNm
For, Mny/ Mnx = Mnoy /Mnox,     Equivalent uniaxial bending moment = Mnoy
     Mny/ Mnx <= Mnoy /Mnox,    Equivalent uniaxial bending moment = Mnox

Equiv BM, Mnox =    486.39 kNm

*To find out reinforcement ( ρ ) required as per interaction diagram as in SP-17 (14) Vol 3*

|  |  | SI unit |  | US customary Unit |  |
|---|---|---|---|---|---|
| Pn = | 3385 kN | γ = | 0.8 | γ = | 0.8 |
| Mnox = | 486 kNm  (=Pn.e) | fc'= | 25 N/mm2 | fc'= | 3625 psi |
| Ag = | 240000 mm2 | fy = | 415 N/mm2 | fy = | 60175 psi |
| h = | 600 mm. |  |  | Use diagram R3-60-8 | |

# Reinforced Concrete and Associated Work

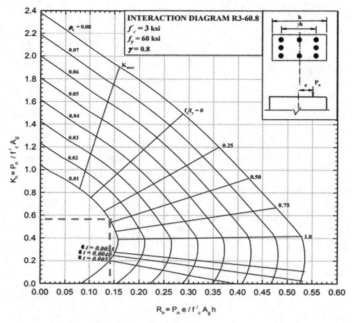

Kn = Pn / fc'. Ag =     3384615 / (25 x 240000)=     0.56
Rn = Pn.e / fc'.Ag.h =   486390533 / ( 25 x 240000 x 600) =    0.14

Reqd. % of reinforcement, ρ from chart =   **0.01**      ρ provided =   **0.013**     Safe

*Check for as per Load contour equation ,*

$$\left[\dfrac{Mnx}{Mnox}\right]^{\left[\dfrac{\log 0.5}{\log \beta}\right]} + \left[\dfrac{Mny}{Mnoy}\right]^{\left[\dfrac{\log 0.5}{\log \beta}\right]} \qquad \text{PCA Note Chapter 7 ( Eqn 14)}$$

$$= \left[\dfrac{462}{486}\right]^{\left[\dfrac{\log 0.5}{\log 0.65}\right]} + \left[\dfrac{31}{196}\right]^{\left[\dfrac{\log 0.5}{\log 0.65}\right]} \qquad \text{, where log 0.5/ log0.65 =} \quad 1.609$$

=    0.9191     +    0.0506    =    0.97    < 1; Safe.

End

# 10 Prestressed Precast Concrete

## 10.1 INTRODUCTION

Prestressed concrete structures are commonly used in bridges and prefabricated buildings. The PCI design handbook, *Precast and Pre-Stressed Concrete*, provides design procedures and standard details for building construction in accordance with ACI and ASCE codes.

Here, we will discuss how prestressed precast concrete is used in turbine buildings and similar industrial structures.

A turbine hall framework includes precast prestressed units as columns, main girders, floor beams, crane beams, and roof girders. The columns are built in a cast-in-situ foundation. The suspended floor slabs are made of precast prestressed planks with a cast-in concrete topping, including floor finish treatment. The span dimension is in multiples of 6 m in a transverse direction and the bay lengths in longitudinal directions are 6 m or 12 m.

Structural rigidity is achieved by fixity at the column base and hinged joints at member ends. The roof trusses are generally made of steel and have bolted or field welded connections at the top of precast columns with a steel embedment.

The concept of field joints and details are shown in Figures 10.1–10.3.

## 10.2 MEMBER DESIGN

A sample design of precast prestressed beam elements is given in the following section.

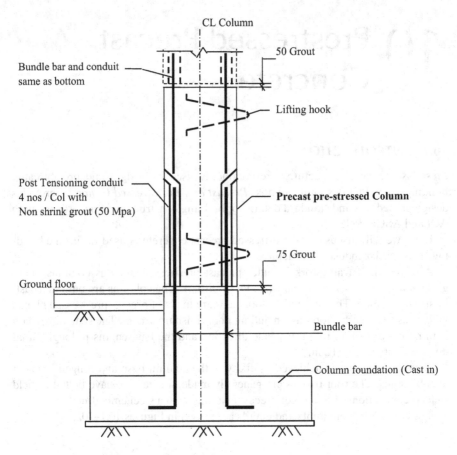

**FIGURE 10.1** Detail of a column.

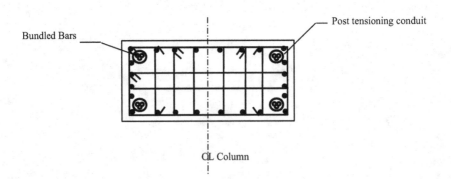

**FIGURE 10.2** Column reinforcement.

# Prestressed Precast Concrete

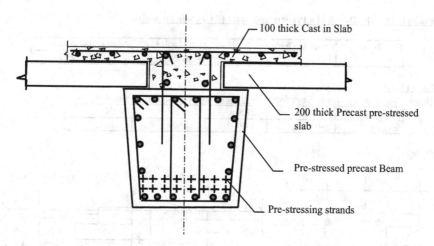

**FIGURE 10.3** Detail of a floor slab on a prestressed beam.

## Example 10.20.1  Design of a Pre-stressed Beam ( PCI Code Method)

| Post Tension | Concrete Grade : | fck | 40 | Mpa | Reinforcement bars: | | |
|---|---|---|---|---|---|---|---|
| | | Ec | 36050 | Mpa | fy | 390 | Mpa |

**Design load:**
*Member load from worst load cases [result from computer frame analysis] - Factor loads*

| Mx kNm | My kNm | Mz kNm | Fy kN | Fx kN |
|---|---|---|---|---|
| -1.58 | -0.98 | 1198 | 1053 | 189 |

**Sketch**

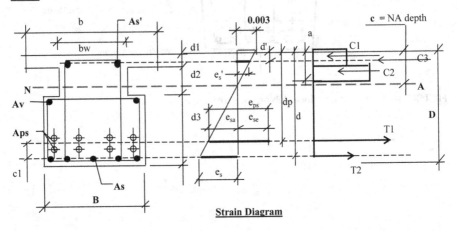

**Strain Diagram**

**Member Dimensions and Reinforcement Details :**          (MKS)

Concrete  
$f_c' =$   40 Mpa   (precast  $f_y =$  415 Mpa  
$f_c' =$   25          (topping)  
$e_{conc} =$  0.003

| Prestressing Strands, Aps | | | | Nos | Dia (mm) | Area (mm2) |
|---|---|---|---|---|---|---|
| Dia | 12.7 | mm | As' | 2 | 16 | 402 |
| Nos | 20 | | As | 8 | 25 | 3925 |
| Grade | 270 | ksi | Stirupps | | | |
| $A_{nom}$ | 98.7 | mm2 | leg | dia | spcg | area |
| Pull | 137.7 | kN | 2 | 12 | 140 | 226 |
| $f_{chrac}$ | 183.7 | kN [ f pu ] | Side face rebars | | | |
| $E_{ps}$ | 193053 | MPa | Av = | 2 | 16 | 402 |

| Beam Dimensions | | |
|---|---|---|
| B | 900 | mm |
| bw | 700 | mm |
| d1 | 100 | mm |
| d2 | 200 | mm |
| d3 | 493 | mm |
| D = | 793 | mm |

# Prestressed Precast Concrete

$E_s$    199948 MPa    Span, $l_n$ =   8.45 M.         Clear cover c =   40 mm
$f_{yt}$      415 Mpa     Spcg, $s_w$ =   5.30 M. clr dist betn webs     c1 =    85 mm

## Strength design of beam section

To find out flexural strength, $\phi M_n$:                   [ACI 318 14 - 6.3.1]

Effective width of flange    b =    2300 mm    [ min of (2×8 d1 + bw), 2×(span/8)+bw, sw/2 ],
                            d' =     48 mm
                           dp =    708 mm
                            d =     753 mm
                          fpu =    1861 Mpa    [ fpu = $f_{chrac}/A_{nom}$ ]

                          fse =    1117 Mpa    assuming 20% loss of prestress; Stand initially
                                                     tensioned to 75% of fpu,(fse = 0.8*0.75*fpu]

Maximum allowable strain :
     $e_{se} = f_{se} / E_{ps}$ = 0.0058     $e_y$ = fy / Es = 0.0021

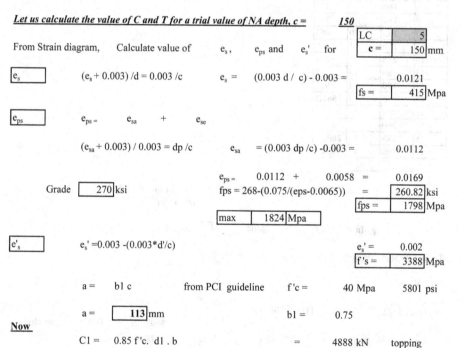

*Let us calculate the value of C and T for a trial value of NA depth, c =*    **150**

|  |  |  | LC | 5 |
|---|---|---|---|---|
| From Strain diagram, | Calculate value of | $e_s$, $e_{ps}$ and $e_s'$ for | c = | 150 mm |

$e_s$      $(e_s + 0.003)/d = 0.003/c$     $e_s$ = (0.003 d / c) - 0.003 =    0.0121
                                                                   fs =     415 Mpa

$e_{ps}$      $e_{ps}$ =    $e_{sa}$   +   $e_{se}$

        $(e_{sa} + 0.003)/0.003 = dp/c$     $e_{sa}$ = (0.003 dp /c) -0.003 =    0.0112

                                     $e_{ps}$ =   0.0112 +   0.0058 =   0.0169
     Grade    270 ksi       fps = 268-(0.075/(eps-0.0065))       =     260.82 ksi
                                                                             fps =    1798 Mpa
                                               max     1824 Mpa

$e_s'$      $e_s'$ = 0.003 - (0.003*d'/c)                                $e_s'$ =    0.002
                                                                          f's =    3388 Mpa

         a =    b1 c      from PCI guideline    f'c =    40 Mpa    5801 psi

         a =    113 mm                          b1 =     0.75

**Now**

         C1 =    0.85 f'c. d1 . b                       =     4888 kN     topping

# Design of Industrial Structures

|   |   |   |   |   |   |
|---|---|---|---|---|---|
| C2= | 0.85 f'c. (a - d1) . bw | | = | 298 kN |
| C3 = | As' . f' s | | = | 1362 kN |
| | | C = C1 + C2 + C3 | = | **6547 kN** |
| T1 = | Aps . fps = | 20x98.7x1798 N | = | 3550 kN |
| T2= | As . fs = | 3925x415 N | = | 1629 kN |
| | | T = T1 + T2 | = | **5179** |

so for c = $\boxed{150}$ we get, x = C-T = $\boxed{1368}$ kN

### Value of 'x' vs 'c'

| LC | c | LC | x |
|---|---|---|---|
|   |   | 5 | 1368 |
| 1 | 100 | 1 | 1600 |
| 2 | 113 | 2 | 1419.4 |
| 3 | 125 | 3 | 1277.2 |
| 4 | 145 | 4 | 1313.6 |
| 5 | 150 | 5 | 1368 |
| 6 | 155 | 6 | 1425 |
| 7 | **158** | 7 | **323** |
| 8 | 160 | 8 | 361 |
| 9 | 162 | 9 | 398 |
| 10 | 165 | 10 | 455 |
| 11 | 170 | 11 | 549 |
| 12 | 238 | 12 | 1832 |
| 13 | 180 | 13 | 737 |
| 14 | 190 | 14 | 926 |
| 15 | 275 | 15 | 2561.3 |
| 16 | 288 | 16 | 2807.7 |
| 17 | 300 | 17 | 3056.2 |
| 16 | 288 | 16 | 2807.7 |
| 17 | 300 | 17 | 3056.2 |
| | **158** | **7** | |

From the tabulated values of 'c' and 'x', we find the minimum value of x ( = C - T ), when NA depth is as follows:

c = $\boxed{158}$ mm

$$Mn = C1 (c-0.5d1) + C2. (c-d1-0.5(a-d1)) + C3(c-d') + T1 (dp - c) + T2 ( d - c )$$
$$= 3604 \text{ kNm}$$

$\phi$ Mn = 0.9 x 3604 = $\boxed{3243}$ kNm

### To find out shear strength, $\phi$ Vn:

Vn = Vc + Vs

a) as per CL 22.551    Vc = 2 l (fc')$^{0.5}$ bw × d    l = 1
                                                         fc' = 5802 psi    [fck = 40 Mpa ]
   for nonprestressed, Vc = 124 kips    (fc')$^{0.5}$ : 76 psi

# Prestressed Precast Concrete

|  |  |  |  |  |  |  |
|---|---|---|---|---|---|---|
|  | 554 kN | bw = | 27.56 in | [ bw = | 700 mm] |
|  |  | d = | 29.65 in | [ d = | 753 mm] |

b) as per 22.5.10.5.3  $V_s = A_v f_{yt} d / s$

|  |  |  |  |
|---|---|---|---|
|  |  | Av = | 226 mm2 |
|  |  | fyt = | 415 Mpa |
| Vs = | 505 kN | d = | 753 mm |
|  |  | s = | 140 mm |

$V_n$ =   554  +  505  =   **1058** kN

### Max shear capacity /cross-sectional dimension check

$V_u <= f(V_c + 8 (f_c')^{0.5} b_w d)$   <=   560 kips  =  2491 kN

|  |  |
|---|---|
| φ = | 0.9 |
| Vc = | 124 kips |
| $(f_c')^{0.5}$ = | 76 psi |
| bw = | 27.559 in |
| d = | 29.646 in |

### Check for Moment capacity:

Considering that flexural moment will be taken by the prestressing system and longitudinal reinforcement will be provided for lifting and handling stress due to self-weight, shear rebars will be provided for total shear.

### Design Moment, Mu

| Mx kNm | My kNm | Mz kNm | Fy kN | Fx kN |
|---|---|---|---|---|
| -1.58 | -0.98 | 1198 | 1053 | 189 |

Converting biaxial moments Mz and My into a single equivalent increased My moment acting about one axis only as per BS 8110 Part 1 (CL 3.8.4.5).

Mz = Moment about the major axis.   My = Moment about the minor axis.
Mx = Torsional Moment.

| | |
|---|---|
| h' = dt = | 753 mm |
| b' = B - c3 = | 860 mm |
| N = Fx = | 189 kN |
| fcu = fck = | 40 Mpa |
| N/b'h' fcu = | 0.0073 |

b = 1 - (7N/6b'h'$f_{cu}$) =   0.9915 [as per BS 8110 Table 3.22 ;   b = 1 ]

Mz/h' =   1591   Mz + b (Myh'/b') =   1199 kNm   Design Moment = 1199 kNm
My/b'     My + b (Mxb'/h') =   1 kNm
[ For Mz/h'>My/b", Mz' =Mz + b (My, h'/b'.  As per BS 8110 part 1 - 3.8.4.5 Biaxial bending]

Contribution of Torsional moment in the design of longitudinal reinforcement:

Mt = Tu ( 1 + D/b) / 1.7 =   2 kNm

$M_{design}$ = 1199 + 2 =   1201 kNm

Tu = Mx =   1.58 kNm   D =   793 mm
                       b =   900 mm     $M_{capacity}$ = 3243 kNm
                                                          Safe.

## Check for shear capacity

**Design Shear, Vu :**

LC   1         Mx =   -1.58 kNm
               Fy =   1053 kN

V =   1053 kN
1.6*(Tu/b) =   3.0 for torsional Moment Mx

$F_{design}$ =   1056 kN

$F_{capacity}$ = Vn =   1058 kN   Safe.

Transverse reinforcement check for Torsion

**Asv** = Tu sv / ( $b_1$ $d_1$ 0.87 ssv ) + Vu. sv / (2.5 $d_1$ 0.87 s sv) =   207 mm2   Safe

Tu = Mx =   1.58 kNm
Vu = Fy =   1053 kN
sv =   140 mm
b1 =   900 mm
d1 =   793 mm
$S_{sv}$ =   415 Mpa

End

# Prestressed Precast Concrete

## Example 10.2.2 Design of a Prestressed Beam (by IS Code Method)

| Post Tension | Concrete Grade : | fck | 30 Mpa | Reinforcement bars: | |
|---|---|---|---|---|---|
| | | Ec | 31220 Mpa | fy | 500 Mpa |

**Beam mkd       B1**                                                                 (IS 1343 : 2012)

### Design Parameters

**Prestressing strands:**
- Dia = 12.7 mm    (Aps)
- Nos = 20 nos
- Grade = 270
- Chrac strength = 184 kN    (fpu = fp)
- Nominal area = 99 mm2
- Pull * = 138 kN    (fpe)

(*) manufacturer's data.

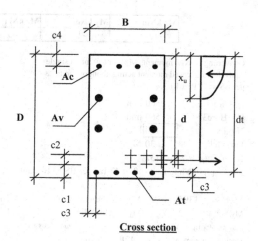

Cross section

**RCC Beam dimensions :**
- B = 600 mm
- D = 900 mm
- c1 = 100 mm
- c2 = 50 mm
- c3 = 40 mm
- c4 = 40 mm

**Beam rebars :**

|  | nos | dia | area | | |
|---|---|---|---|---|---|
| Ac | 4 | 20 | 1256 mm2 | | |
| At | 4 | 20 | 1256 mm2 | | |
| Stirrups : | 2 Leggd | 12 dia @ | 90 C/C. | | |
| Av = | 4 | 16 | 804 mm2 | | |

**Lifting arrangement:**

Length   6 M
N =   1.25 M

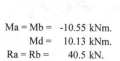

Ma = Mb =   -10.55 kNm.
Md =   10.13 kNm.
Ra = Rb =   40.5 kN.

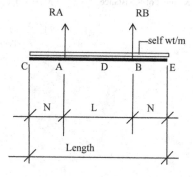

**Check for flexural strength:**

Considering that flexural moment will be taken by the prestressing system and longitudinal reinforcement will be provided for lifting and handling stress due to self-weight, shear rebars will be provided for total shear.

*Design Moment, $M_U$*

| Mx kNm | My kNm | Mz kNm | Fy kN | Fx kN |
|---|---|---|---|---|
| -1.6 | -1 | 1200 | 1055 | 190 |

Converting Biaxial moments Mz and My into a single equivalent increased moment acting about one axis as per BS 8110 Part 1 ( 3.8.4.5).

Mz = Moment about the major axis.
My = Moment about the minor axis.
Mx = Torsional moment.
Fy = Shear force
Fx = Axial load

$h'$ = dt =             860 mm
$b'$ = B - c3 =         560 mm
N = Fx =                190 kN
fcu = fck =             30 Mpa
N/b'h' fcu =            0.0132
b = 1 - (7N/6b'h'$f_{cu}$) =    0.9847  [as per BS 8110 Table 3.22    1.00 ]

Mz/h' =   1395        Mz' = Mz + b (Myh'/b') =       1202 kNm
My/b' =   2           My + b (Mxb'/h') =             1 kNm
[ for Mz/h' > My/b',  Mz' = Mz + b h'/b My ;   BS 8110 Part 1 - 3.8.4.5]
Design Moment, Mz' =    **1202**  kNm

Contribution of Torsional moment in design moment:
Mt = Tu ( 1 + D/b) / 1.7 =          2 kNm
                    $M_U$ = 1202 + 2 = 1204 kNm  > **Mn; Safe.**

Tu = Mx =      1.60 kNm    D =      900 mm
                           b =      600 mm

*To determine Moments of resistance ($M_n$) of a Rectangular Prestressed beam section.*

Moment of Resistance         $M_n$ =   $f_{pb} \cdot A_{ps} \cdot (d - 0.42 x_u)$         =    **2024.6** kNm

                             $M_n$ =   Moment of resistance of the section
                             $f_{pb}$ = Tensile stress in the tendon at failure          160 kN
                             $f_{pe}$ = effective prestress in tendon                    110 kN**

# Prestressed Precast Concrete 309

| | $A_{ps}$ = | Area of pretensioning in the tension zone | | 20 nos |
|---|---|---|---|---|
| | d = | Effective depth to the centroid of the steel area | | **775** mm |
| | | (900 - 100 - 0.5 x 50 =775) | | |
| | $x_u$ = | Neutral axis depth | 0.435 x 775 = | 337 mm |
| | $f_{pu}$ = | Characteristics tensile strength of tendon. | 12.7dia | 184 kN |

<u>From Table 11 in IS 1343 - 2012</u>             fpb =   0.87 x 183.7 =    160 kN
Aps. fpu / b.d. fck =    0.2268
fpb /0.87 fpu =           1
xu /d =                 0.435
[(**) as per IS code , fpe shall not be taken as greater than 0.6 fp;    0.6 fp =    110.2 Mpa ]

**Check for Shear Strength**

**Design Shear, Vu :**

| | | V = Fy = | 1055 kN | |
|---|---|---|---|---|
| From Torsional moment Mx, | 1.6*(Tu/b) = | | 4.3 | (1.6 x 1.6 / 0.6) |
| | | $F_{design}$ | **1059** kN | |

The Ultimate shear resistance of concrete alone, Vc should be considered at both conditions Cracked and Uncracked in flexure, the lesser value taken and if necessary, shear reinforcement provided.

<u>Section uncracked in flexure</u>

$$Vc = Vco = 0.67 \, b \, D \, [ \, (f_t^2 + 0.8 \, f_{cp} \, f_t) \, ]^{0.5} \quad = \quad \boxed{492} \, kN$$

| | b = B = | 600 mm |
|---|---|---|
| | D = | 900 mm |
| $f_t = 0.24 \times (f_{ck})^{0.5}$ = | 1.31 Mpa | (0.24 x 30^0.5) |
| | $f_{pe}$ = | 110 kN |
| | d= | 775 mm |
| fcp = [($f_{pe}$/d) * D/2 ]/B.D = | 0.119 Mpa | |

$f_{cp}$ = Compressive stress at the centroidal axis due to prestress taken as positive.

<u>Section cracked in flexure</u>

The ultimate shear resistance of a section uncracked in flexure = Vc

$$Vc = Vcr = [ \, 1 - 0.55 \, f_{pe} / f_p \, ] \, \zeta_c \, b \, d + M_0 \, (V/M)$$

     where,
         fpe = Effective prestress after all losses, which shall not be greater than 0.6 fp.
         fp = Characteristic strength of prestressing steel
         Ap = Area of prestressing tendon

$\zeta_c$ = Ultimate shear capacity of concrete obtained from Table 8 - IS 1343 2012

fpt = Stress due to prestress only at a depth of 'd' and distance 'y' from the centroid of the conc section

| | | | | | | |
|---|---|---|---|---|---|---|
| fpe = | 110 kN | | 100Ap / b.d = | 0.27 | | |
| fp = | 184 kN | | $\zeta_c$ = | **0.38** Mpa | | |
| Ap = | 1266 mm2 | (20 × 3.14 × 12.7^2 / 4) | fpt = | 1.39 Mpa | (138 × 1000/99) | |
| dia | 12.7 mm | | y = | 325 mm | (775 - 600/ 2) | |
| nos | 20 | | I = | 36450000000 mm4 | (600×900^3 / 12) | |
| b = B = | 600 mm | | | | | |
| d = | 775 mm | | $M_0$ = 0.8. fpt ( I / y ) = | 125 kNm | | |

Design shear  $F_{design}$ =    Vu = 1059 kN
Design moment  $M_{design}$ =    $M_U$ = 1204 kNm

As calculated   $V_{cr}$ =   228 kN
Vcr should be taken as not less than
0.1 b d fck^0.5 =         255 kN

Vcr =   **255** kN

So ,
Ultimate shear resistance , Vc =   **255** kN   By concrete   [ Min (255,492) ]
Vu - Vc =   940        By shear reinforcement
Vc =   **1195** kN
Design shear  =   Vu =   1059 kN   **< Vc; Safe.**

Stirrups provided :

**2 Legged 12 dia @ 90 c/c** [ Shear capacity = 940 kN.]
<u>Minimum Shear Reinforcement :</u>

Asv / b sv  = 0.4 / 0.87 fy
Asv = 0.4 × b × sv /0.87 fy    =    50 mm2

<u>Shear Reinforcement : when Vu exceeds Vc</u>

Vu- Vc = Shear Reinforcement capacity =    Asv . 0.87 fy dt  / sv    =    939.74 kN

Asv =  Total cross-sectional area of stirrups legs effective in shear =   **226** mm2   (2×3.14×12^2 /4)
b =   Breadth of the member.            =   600 mm
sv =   Stirrup spacing along the length of member.   =   90 mm
fy =   Characteristic strength of the stirrup reinforcement that   =   500 N/mm2

# Prestressed Precast Concrete

shall not be taken greater than 415 N/mm2

Maximum shear stress

  Concrete Grade = M30   Design shear = 1059 kN.

| | | | |
|---|---|---|---|
| B = | 600 mm | $S_{sh}$ = | 2.0 N/mm2 |
| D = | 900 mm | $S_{max\ allow}$ = 3.5 N/mm2` | > Ssh; OK |

To check transverse reinforcement Asv, for torsion

  **Asv** = Tu sv / ( $b_1$ $d_1$ 0.87 ssv ) + Vu. sv / (2.5 $d_1$ 0.87 $\sigma_{sv}$ ) =  **98** mm2

                          **> Asv Provided, OK.**

| | | |
|---|---|---|
| Tu = Mx = | 1.6 kNm | |
| Vu = Fy = | 1055 kN | |
| sv = | 90 mm | |
| b1 = | 600 mm | |
| d1 = | 900 mm | |
| $\sigma_{sv}$ = | 500 Mpa | |

Side face reinforcement, Av              ( IS 456 2000 - 26.5.1.3)

    $Av_{rqd}$ =  Total area shall of be less than 0.1% web area.
      =  0.001 × 600 × 900
      =  540 mm2

    Av provided =   4 nos   16 mm dia =    804 mm2  **> Av rqd;OK.**

**Check for Lifting Moment and Shear**

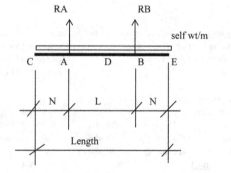

| | | | |
|---|---|---|---|
| Length= | 6 M | | |
| N = | 1.25 M | | |
| L = | 3.5 M | | |

Beam size
| | | | |
|---|---|---|---|
| B = | 600 mm | 0.6 m | |
| D = | 900 mm | 0.9 m | |

Self wt, w = 13.5 kN/m

Ma = Mb = -10.55 kNm.

Md = 10.1 kNm.

Ra = Rb = 40.5 kN.

n = N/L = 0.36

**Deflections:** ( Steel designer's manual ELBS - 4th ED)

$$d_C = d_E = (w L^3 N /24 E I) \times ( 1 - 6 n^2 - 3 n^3 ) = \quad 0.0026 \text{ mm}$$
$$d_D = (w L^4 / 384 E I) \times ( 5 - 24 n^2 ) = \quad 0.009 \text{ mm}$$

---

**Flexural design for lifting / handling**

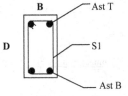

B = 0.6 m           fck = 30 MPa
D = 0.9 m.          fy = 500 Mpa
cover = 40 mm.
Span = 6 m.

**At span**                Load factor = 1.5 [ factored load ]
Mspan = 10.12 kNm          Mu span = 15.18 kNm
Msupport = -10.55 kNm      Mu supp = 15.82 kNm
Fsh = 40.5 kN              F supp = 60.75 kN.

**Reinforcement :**                                   Astmin =         1080 mm2
**At mid span**                         Ast
Top         4 nos        20 dia         1256 mm2 .
Bot         4 nos        20 dia         1256 mm2 .
Stirrup     12 dia @     90 mm c/c.     ( Asv =    226.28 mm2 /set)
**At support**                          Ast
Top         4 nos        20 dia         1256 mm2 .

Bottom      4 nos        20 dia         1256 mm2 .
Stirrup     2 legged     12 dia @       90 mm c/c.    ( Asv =    226.28 mm2 / set ).

**Design check:**

| | | | | | |
|---|---|---|---|---|---|
| $M_U$ Limit = | 1837 kNm. | **Singly reinf.** | | Mulim/bd^2= | 4.14 |

Span Moment = 15.188 kNm.
dprov = 860 mm
$M_u/bd^2$ = 0.03

| | | | | | | |
|---|---|---|---|---|---|---|
| pt = | 0.20 M30 | | | pc = | 0.200 M30 | |
| Ast reqd. = | 1032 mm2/m width | | Ast reqd. = | 1032 mm2/m width | |
| Ast provided = | 1256 mm2/m width | **Ok.** | Ast provided = | 1256 mm2/m width | **Ok.** |
| Ast min = | 1080 mm2/m width | | Ast min = | 1080 mm2/m width | |

pport Moment = 15.82 kNm.    $M_U$ Limit = 1837 kNm.    **Singly reinf.**
dprov = 860 mm
$M_u/bd^2$ = 0.04

| | | | | | | |
|---|---|---|---|---|---|---|
| pt = | 0.20 M30 | | | pc = | 0.20 M30 | |
| Ast reqd. = | 1032 mm2/m width | | Ast reqd. = | 1032 mm2/m width | |
| Ast provided = | 1256 mm2/m width | **Ok.** | Ast provided = | 1256 mm2/m width | **Ok.** |
| Ast min = | 1080 mm2/m width | | Ast min = | 1080 mm2/m width | |

End shear = 61 kN
Tc.b.d = 170 kN.
Vs = 0 kN
Vus = 941 kn    **Ok.**

End

# 11 Plant Area Roads and Drainage

## 11.1 ROADS AND PAVEMENTS

The following are types of roads and pavements that are used inside the plant boundary.

a) Flexible pavement
b) Rigid pavement
c) Interlocking concrete block pavement

### 11.1.1 FLEXIBLE PAVEMENT

Asphalt or bituminous concrete topping pavements are common types of flexible pavement. The construction of a flexible pavement is done in the following steps.

*Subgrade.* The stretch of ground selected for the road width is excavated (box cutting) to a shallow depth, say about 500 mm. The soil is replaced with good soil (preferably granular) and compacted in layers. The surface is rolled and compacted up to 90% of its Proctor dry density at optimum moisture content. A minimum of six passes of 8–10 T road rollers should be used to compact the subgrade.

*Subbase.* This layer is formed with coarse aggregates (40–90 mm in size) of broken stone or brick ballasts; the total thickness is about 300 mm and compacted in two layers.

*Base course.* This layer is water-bound macadam of a compacted thickness of 150 mm. It is done by using crushed stone (40 mm and below) compacted up to 95% of modified Proctor density.

*Top course.* The top course or wearing course is made with bituminous/asphaltic concrete. This layer is placed on a binder course (hot bitumen layer) laid over the base course. The thickness of this layer is generally 50 mm.

### 11.1.2 RIGID PAVEMENT

Rigid pavements are cement concrete pavements with reinforcing steel designed for an appropriate wheel load. This type of pavement is also used as slab-on-grade in an open store area.

This road width/pavement is constructed on compacted *subgrade* underlain by granular *subbase* of course aggregate. The top surface of the subbase is blinded with a thin layer of plain cement concrete. The thickness of this blinding concrete varies from 50 mm to 75 mm and more to provide camber and slope. RCC slabs of design thickness are cast on this blinding concrete layer. The thickness of the RCC slab should not be less than 250 mm and be reinforced in a single or double layer as per design. The strength design should be done for the applicable wheel load and environmental temperature stresses.

### 11.1.3  Interlocking Paver Blocks

These are concrete paver blocks laid on a sand base. This type of construction is generally used for walkways and footpaths. It remains dry in the rainy season because the surface runoff is quickly absorbed through joints and dissipates into the sand cushion below. These paver block walkways are very popular for their fast construction and ease of maintenance.

The width of plant area roads varies from 4 m, 6 m, 10 m, and 12 m depending on the number of lanes. The road along the plant boundary is 3 m wide. Plant roads are generally constructed between raised curbs and sloping towards gully pits for stormwater drainage. The top of the road is provided with a cross slope (about 1.5–2%) for drainage. Raised curbs with walkways run parallel on both sides with curbstones and drainage gulley pits at intervals. These gully pits are connected to underground pipelines of the plant drainage network. These road curbs are also provided at the same level of the road surface to allow surface runoff to pass over the paved sidewalks. The water is collected into open drains outside, running parallel to the road width.

The design and construction of roads and pavement shall be done using Indian Road Congress (IRC)/The American Association of State Highways and Transportation Officials (AASHTO) standards.

Pavements and hard standings in open area storage, fuel oil unloading, and equipment unloading areas are designed for appropriate loads (25 kN/m² and upward) with settlement limitations. For coal storage areas on soft soil, structural piles or CDM piles are used.

Typical details of plant roads are shown in Figures 11.1 through 11.3.

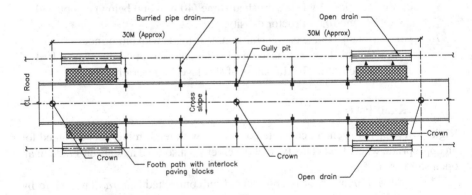

**FIGURE 11.1**  Part plan of a plant road.

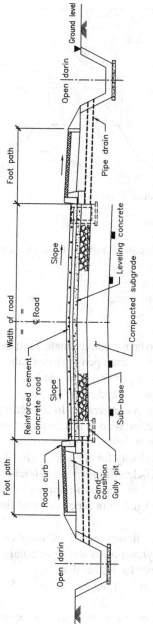

**FIGURE 11.2** A typical section of a reinforced concrete plant road.

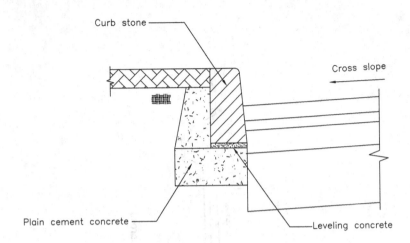

**FIGURE 11.3** Detail of a road curb.

## 11.2 DRAINAGE

Plant area drainage may be classified into the following categories:

a) Storm drains
b) Oily waste or plant waste drains
c) Sanitary sewerage

Storm drains are considered to be not contaminated and discharged into the terminal pit or pond or river. Storm drains are open surface drains – rectangular/trapezoidal shaped or buried underground pipes or a combination of both. The buried pipes are connected by manholes at intervals and corners.

A gravity flow system is always preferable. In case gravity flow is not possible, lift pumps with collection pits are installed at suitable intervals.

The storm drainage system is designed for maximum hourly rainfall.

The maximum allowable velocity for open drains is 1.8 m/sec, and for pipe drains it is 2.4 m/sec. Minimum self-cleansing velocity should not be less than 0.6 m/sec for open drains. For pipe drains, the minimum velocity should not be less than 0.7 m/sec at design flow and 1 m/sec in pipe full conditions as per British standard BS EN 752-4:1998.

The materials used for pipe drains are UPVC (unplasticized polyvinyl chloride)/ HDPE (high-density polyethylene) pipes for an internal diameter of up to 400 mm and cement concrete spun pipes for higher diameters.

The spacing of manholes should not exceed 50 m for buried underground pipe drains.

Plant waste drains considered contaminated are discharged into a monitoring basin and treatment plant. Oily drains are routed through an oil/water separator unit and then discharged into a storm drain. Ductile iron/UPVC pipes are used for plant waste. Oily drains are made of UPVC and concrete pipes.

# Plant Area Roads and Drainage

Sanitary waste drains are contaminated drains and are conveyed to a septic tank/cesspool/sewage treatment plant. These are buried underground. Concrete pipe/UPVC pipes are generally used. A minimum velocity should not be less than 0.9 m/sec for sewer lines. Sewage lift pumps with pits are used when gravity drains are not feasible.

## 11.3 DESIGN STEPS FOR STORMWATER DRAINS WITH BURIED PIPELINES

### 11.3.1 Reference Documents for Design Information

The reference documents include owner specifications and plant layout drawings. The design basis report is prepared from these reference documents. The information to be included is as follows: layout of the building, roads, and catchment areas to be considered for the calculation of surface runoff quantity; location and elevation of the terminal point or final discharge point; daily rainfall intensity; runoff coefficient for the roof, surface areas, and hard standings; type and materials to be used for drain pipelines; type and grade of concrete for manholes, materials, and covers for gulley pits and collection catch pits, and so on.

### 11.3.2 Preparation of Catchment Plan and the Network for Design

This drawing is prepared according to the plot plan drawing. The plot area will be divided into segments with individual area markings. Each plot will have a part network of branch drainage pipelines, which will be finally connected to the nearest main line. There will be an array of main drain lines usually selected along the center of the plant main road for uninterrupted passage. These mainlines are connected to the terminal pit after collecting all the discharges from branch drainages in various segments. The pipelines are connected to manholes at the start and end. The inverted elevations are matched with the flow direction to keep the hydraulic gradient as per the design with a minimum permissible slope. For guidelines, refer to BS EN 752 Parts 3 and 4 for drains and sewer systems outside buildings.

### 11.3.3 Pipe Size Calculation

According to the flow in different segments, the pipe size and gradient, i.e., invert elevation at the start point and end point, are selected by the designer, who maintains minimum earth cover as per the plant datum level. The velocity of flow is then calculated and checked with acceptable limits.

Minimum self-cleansing velocity = 0.7 m/sec at design flow and 1.0 m/sec in pipe full conditions as per BS EN 752-4:1998.

The diameter of the pipe and the gradient are adjusted to meet the acceptable limits. This is an interlinked process, and the elevation of upstream and downstream lines are linked together to get the best results.

The Colebrook–White equation may be used for the velocity calculation:

$$V = -2 \times (8gR_h J_E)^{1/2} \times \log_{10}\left[k/3.71/4/R_h + 2.51\nu/4/R_h/(8gR_h J_E)^{1/2}\right]$$

g = 9.80 m/sec
$R_h$ = hydraulic radius = A/P
$J_E$ = Pipe slope (gradient)
$\nu$ = kinematic viscosity of water = $[(1.304+1.004)/2] \times 10^{-6}$ = $1.154 \times 10^{-6}$ m²/sec
k = hydraulic pipeline roughness in meters = 0.0006

# 12 Design of Industrial Structures

In the previous chapters, we have had detailed discussions on design, construction, and project procedures followed in industrial practices. In this chapter, we provide complete design calculations of plant buildings to educate readers about the steps and methodology followed in professional practice.

## 12.1 DESIGN OF A RCC FOUR-STORY BUILDING FOR A PACKAGING PLANT

LIST OF CONTENTS

| Sl No | Item Descriptions |
|---|---|
| 12.1.1 | Introduction |
| 12.1.2 | References |
| 12.1.3 | Material |
| 12.1.4 | Load on the Building |
| 12.1.5 | Structural Analysis |
| 12.1.6 | Design of the Columns |
| 12.1.7 | Design of the Foundations |
| 12.1.8 | Design of the Grade Beams |
| 12.1.9 | Design of the Floor and Roof Beams |
| 12.1.10 | Design of the Slabs and Staircase |
| 12.1.11 | Computer Model |
| 12.1.12 | Drawings |

### 12.1.1 INTRODUCTION

This document contains analysis and design of a four-story reinforced concrete building to be used for the printing and packaging industry. The ground floor will be used for machines and offices. Floors above the ground floor will be used for lightweight machines and storage. The building is well ventilated and illuminated. There are walkways, access stairs, and elevators for handling goods and the movement of plant operating workers. The building framework has been designed to meet all the requirements of building codes and Indian standard practices. All foundations are resting on soil. Analysis of the framework has been done using STAAD Pro software and member strength design calculation in Excel conforming to Indian standard codes and practices.

### 12.1.2 REFERENCES

i) IS 456: 2000 Plain and Reinforced Cement Concrete
ii) IS 875: 1987 Design Loads for Building and Structures
iii) IS 1893: 2002 Criteria for Earthquake Resistant Design of Structure
iv) SP 16 Design Aids to IS 456
v) Equipment layout and general arrangement drawing
vi) Soil investigation report

### 12.1.3 MATERIAL

a. Grade of concrete: M25 (Characteristic strength = 25 MPa cube strength at 28 days) for foundation and substructure and M20 for superstructure.
b. Reinforcement bars: Fe 415 as per IS: 1786 (Yield stress = 415 MPa)

# Design of Industrial Structures

## 12.1.4 LOAD ON THE BUILDING

**Design parameters:**

|  |  |  |  | Elevations |  | Elevations |
|---|---|---|---|---|---|---|
| SBP = | 150 kN/m2 | Grade level EGL = | -0.6 M | | 2nd floor = | 9.3 M |
| $\gamma$soil = | 19 kN/m3 | Ground floor FGL = | 0 M | | 3rd floor = | 13.02 M |
| $\gamma$conc= | 24 kN/m3 | 1st floor = | 5.58 M | | Roof = | 16.74 M |
|  |  |  |  |  | Stair roof = | 19.74 M |

*[SBP = Safe Bearing Pressure; FGL = Finished Ground floor level ; EGL = Existing Grade level.]*

**Load intensity:**

| | | | **DL** | | **LL** |
|---|---|---|---|---|---|
| **Roof** | 120 mm | slab | 2.88 kN/m2 | | |
| | | Roof treatment | 0.36 kN/m2 | | |
| | | plaster | 0.24 kN/m2 | | |
| | | misc | 0.08 kN/m2 | | |
| | | | 3.56 kN/m2 | | 1.5 kN/m2 |
| | | Design load Intensity = | 3.56 kN/m2 (*1) | | |

| | | | **DL** | | **LL** |
|---|---|---|---|---|---|
| **Floors** | 140 mm | Slab including floor finishes | 3.36 kN/m2 (*2) | | |
| | 12 mm | plaster | 0.24 kN/m2 | | |
| | | Partition wall | 1 kN/m2 | | |
| | | | 4.6 kN/m2 | | 2.5 kN/m2 |
| | | Design load Intensity = | 4.6 kN/m2 (*1) | | |

| | | **DL** | | **LL** |
|---|---|---|---|---|
| **Stair roof** | | | | |
| Slab including roof treatment and finishes | | 3.56 kN/m2 | | 1.5 kN/m2 |

**Stairs**

| Waist slab | | | **DL** | | **LL** | | | 250 *Tread* |
|---|---|---|---|---|---|---|---|---|
| | 220 mm | slab with steps | 5.28 kN/m2 (in plan) | | | | 150 | |
| | 25 mm | Floor finish | 0.6 kN/m2 | | | | *Riser* | 292 |
| | 12 mm | plaster | 0.24 kN/m2 | | | | | *Steps* |
| | | | 6.12 kN/m2 | | 3 kN/m2 | | | |

| Landing slab | | | **DL** | | **LL** |
|---|---|---|---|---|---|
| | 125 mm | slab | 3 kN/m2 | | |
| | 25 mm | Floor finish | 0.6 kN/m2 | | |
| | 12 mm | plaster | 0.24 kN/m2 | | |
| | | | 3.84 kN/m2 | | 3 kN/m2 |

*Notes:*
*(\*1) Excluding Self weight of framing beams and columns, which are included as selfweight in frame analysis (STAAD)*
*(\*2) Using non-metalic hardner on top.*

Weight of tank including fill = 25 kN.
Equiv UDL (load/area) 1.4 kN/m2 [ 25 / 6 x 3 ]

Stair well dimension in plan :   6 m x   3 m.

**Brick wall Load:**

|  | 20% opng | | Ht | | Thk | | Unit wt | | |
|---|---|---|---|---|---|---|---|---|---|
| Parapet = | 1 | x | 1 | x | 0.25 | x | 20 | = | 5 kN/m |
| Ext. Wall above 3rd fl = | 0.8 | x | 3.12 | x | 0.25 | x | 20 | = | 12 kN/m |
| Ext. Wall above 2nd fl = | 0.8 | x | 3.12 | x | 0.25 | x | 20 | = | 12 kN/m |
| Ext. Wall above First fl = | 0.8 | x | 3.12 | x | 0.25 | x | 20 | = | 12 kN/m |
| Ext. Wall above GF = | 0.8 | x | 4.98 | x | 0.25 | x | 20 | = | 20 kN/m |

**Wind load**

Wind pressure =   1.21 kN/m2   Force coeff =   1.2
Building height above EGL =   17 m
Building length =   30 m   Wind force = 1.21 x17.34 x 30 x1.2
Building width =   28 m   =   **755** kN
   <Seismic Base shear = 926 kN

*Determination of design wind pressure as per IS 875 (Part3): 2015*

$V_b$ =   50 m/s
Wind Speed, $V_z$ = $V_b$ k1 k2 k3 k4 =   55.78 m/s
Wind pressure , $p_z$ = 0.6 $V_z^2$ =   1.87 kN/m2

Design wind pressure, $p_d$ = Kd Ka Kc $p_z$ =   **1.21** kN/m2

k1 =   1
k2 =   0.97   Kd =   0.9
k3 =   1   Ka =   0.8
k4 =   1.15   Kc =   0.9

**Wind force is less than Seismic, hence not considered for analysis; Seismic load governs.**

# Design of Industrial Structures

## Seismic load
*Determination of Seismic load by Seismic coefficient method as per IS 1893 part 1 : 2002*

Depth frame girder =    **0.6** m
Effective Height, h =    **16.44** m    [CG of girder above FFL]
$T_a = 0.075 \, h^{0.75} =$    0.6123

| $\alpha_h =$ | Z.I.Sa/(2 R g) | = | 0.04 |

Z = 0.16
I = 1
R = 5 SMRF
Sa/g = 2.5

### Roof / Floor dimension:
Len =  **30** m.    width=  **28** m.    Total area =    840 sqm.

## Seismic Weight calculation for frame design.

### Roof:

DL(kN)  LL(kN)  Sum(kN)

Roof :  Design load Intensity =  3.56 kN/m2  Area =  840 sqm    2990    0
    (3.56 x 840 =2990)
            nos    L(m)    W(m)    ht(m)    % opng
Parapet        1    116    0.25    1            580
    (1 x 116 x 0.25 x 1 x 20 =580)
Stair well        2    18    0.25    2.4    0.8        346
    (2 x 18 x 0.25 x 2.4 x 0.8 x 20 =346)
Water tank on Roof - 2 Nos                    50
    (2 x 25 = 50)
Columns        36    0.6    0.4    3.42            739
    (36 x 0.6 x 0.4 x 3.42 x 25 =739)
                            1715    0    1715

No of nodes =    **36**

### Floor:

DL(kN)  LL(kN)  Sum(kN)

Floor :  Design load Intensity =    4.6 kN/m2    Area =    840 sqm    3864    630
        LL intensity =    0.75 kN/m2    [**]
    DL    (4.6 x 840 = 3864)
    LL    (0.75 x 840 = 630)
        nos    L(m)    W(m)    ht(m)    % opng

Outer wall        1    116    0.25    3.72    0.8        1726
    (1 x 116 x 0.25 x 3.72 x 0.8 x 20 =1726)
Columns        36    0.6    0.45    3.72            904
    (36 x 0.6 x 0.45 x 3.72 x 25 = 904)
                            6494    630    7124

No of nodes =    **36**
** [30% reduction for Seismic case as per IS 1893 considering avg LL < 3kN/m2]

**Roof over stair:**

|  |  |  |  |  |  |  |  | DL(kN) | LL(kN) | Sum(kN) |
|---|---|---|---|---|---|---|---|---|---|---|
| Roof: | Design load Intensity = | 3.56 kN/m2 |  | Area = | 18 sqm |  |  | 64 |  |  |

|  | nos | L(m) | W(m) | ht(m) | % opng |  |
|---|---|---|---|---|---|---|
| Parapet | 1 | 18 | 0.25 | 0.5 |  | 45 |
|  | (1 x 18 x 0.25 x 0.5 x 20 =) |  |  |  |  |  |
| Columns | 4 | 0.3 | 0.3 | 1.5 |  | 14 |
|  | (4 x 0.3 x 0.3 x 1.5 x 25 =) |  |  |  |  | 59  0  58.5 |

No of nodes =   4  for each sets of stairs roof.


**Seismic Load distribution**

|  | DL | LL | $h_i$ | $W_i$ | $W_i h_i^2$ | $Q_i$ | Nodes | Load / Node |
|---|---|---|---|---|---|---|---|---|
| W = Roof | 1774 kN | 0 kN | 16.44 | 1774 | 479330 | 354 | 44 | 8 |
| 3rd floor = | 6494 | 630 | 12.72 | 7124 | 1152652 | 850 | 36 | 24 |
| 2nd floor = | 6494 | 630 | 9.00 | 7124 | 577044 | 426 | 36 | 12 |
| 1st floor = | 6494 | 630 | 5.28 | 7124 | 198606 | 147 | 36 | 4 |
| sum = | 21256 | + 1890 |  | $\Sigma W_i h_i^2 =$ | 1254980 |  |  |  |
|  |  | = 23146 kN |  |  |  |  |  |  |

$Q_i = V_b \cdot ( W_i h_i^2 / \Sigma W_i h_i^2 )$

| Seismic Base shear | $V_b = \alpha_h \times W =$ | = 0.04 x 23146 = | 926 kN |
|---|---|---|---|

| For frame analysis in STAAD : |  |  |
|---|---|---|
| Applied mass at Roof nodes = | 8 | kN |
| Applied mass at 3rd floor nodes = | 24 | kN |
| Applied mass at 2nd floor nodes = | 12 | kN |
| Applied mass at 1st floor nodes = | 4 | kN |

Design of Industrial Structures

### Load Combinations

1 DL
2 LL     on floor
3 LLR    on Roof
4 SLX    along transverse direction
5 SLZ    along longitudinal direction

6 DL + LL+ LLR
7 DL + SLZ
8 DL + SLX            ⎱ Working Stress (Unfactored load)
9 0.75 ( DL + 0.5LL+SLX )
10 0.75 ( DL + 0.5LL+SLZ )

11 1.5 DL + 1.5 LL
12 1.5 DL + 1.5 SLX
13 1.5 DL + 1.5 SLZ
14 1.5 DL -1.5 SLZ            ⎱ Limit State ( Factored )
15 1.2 ( DL + LL - SLX )
16 1.2 ( DL + LL - SLZ )

### 12.1.5 Structural Analysis

Structural analysis has been done using STAAD PRO software. Computer input for analysis have not been furnished in this example. Analysis result output data have been sorted out and worst load cases are considered for design of member sections.
    Sample design of members are presented below.

## 12.1.6 Design of Columns

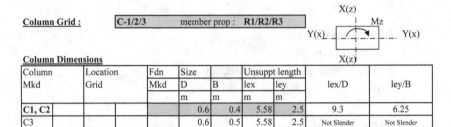

| Column Grid : | C-1/2/3 | member prop : R1/R2/R3 |

### Column Dimensions

| Column Mkd | Location Grid | Fdn Mkd | Size | | Unsuppt length | | lex/D | ley/B |
|---|---|---|---|---|---|---|---|---|
| | | | D m | B m | lex m | ley m | | |
| C1, C2 | | | 0.6 | 0.4 | 5.58 | 2.5 | 9.3 | 6.25 |
| C3 | | | 0.6 | 0.5 | 5.58 | 2.5 | Not Slender | Not Slender |

### Column Reinforcement

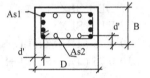

| As1 | 4 nos | 25 dia | per face |
|---|---|---|---|
| As2 | 3 nos | 25 dia | per face |
| Total | 14 nos | 25 dia | |

As = As1 + As2 =   6869 mm2   p =   2.9 %

Area of steel on short faces =   1472 mm2
Area of steel on long faces =   1963 mm2

Conc. grade = **M20**      fck=   20 N/mm2      fy =   415 N/mm2
Clear Cover =   40 mm   Ac =   233131 mm2      d' =   52.5 mm

Puz =   0.45 fck Ac + 0.75 fy As =      4236 kN

**Design of Columns**   [ As per SP-16 and IS: 456 ]

**Column Load :**                              Load factor =   1 [factored load]
[ Load from STAAD Analysis output file]

| SL no | | Beam | L/C | Node | Fx kN | Fy kN | Fz kN | Mx kNm | My kNm | MzkNm |
|---|---|---|---|---|---|---|---|---|---|---|
| 1 | Max Fx | 202 | 12 1.5DL+1.5SLX | 126 | 2206.4 | 74.43 | 17.582 | 0.965 | -11.82 | 239.5 |
| 2 | Min Fx | 881 | 15 1.2(DL+LL-SLX) | 60 | 265.16 | 61.923 | -3.617 | 0.162 | -7.516 | -132.4 |
| 3 | Max Fy | 908 | 11 1.5DL+1.5LL | 504 | 600.68 | 111.91 | -1.296 | 0.228 | -1.168 | 153.67 |
| 4 | Min Fy | 835 | 11 1.5DL+1.5LL | 457 | 530.37 | -104.7 | -9.134 | -0.71 | 9.95 | -144.7 |
| 5 | Max Fz | 202 | 14 1.5DL-1.5SLZ | 126 | 1947.4 | 12.598 | 90.922 | 0.941 | -127.5 | 11.158 |
| 6 | Min Fz | 65 | 13 1.5DL+1.5SLZ | 31 | 1811.9 | -13.07 | -87.73 | 0.929 | 126.2 | -11.98 |
| 7 | Max Mx | 209 | 14 1.5DL-1.5SLZ | 133 | 970.97 | -41.35 | 74.361 | 2.923 | -132.5 | -84.94 |
| 8 | Min Mx | 77 | 13 1.5DL+1.5SLZ | 43 | 1205 | -45.09 | -63.47 | -2.85 | 111.29 | -91.2 |
| 9 | Max My | 159 | 13 1.5DL+1.5SLZ | 97 | 1256.3 | -10.75 | -60.21 | 0.085 | 168.39 | -13.13 |
| 10 | Min My | 203 | 14 1.5DL-1.5SLZ | 127 | 1293.7 | -12.02 | 61.37 | -0.2 | -170.7 | -15.73 |
| 11 | Max Mz | 208 | 12 1.5DL+1.5SLX | 132 | 1905.3 | 106.02 | 2.478 | -0.622 | -8.828 | 283.17 |
| 12 | Min Mz | 208 | 12 1.5DL+1.5SLX | 138 | 1860.1 | 106.02 | 2.478 | -0.622 | 4.386 | -282.2 |

# Design of Industrial Structures

**Design Load for Columns**

| Sl no | Load Case nos and combinations [As per STAAD Load Comb Case] | Axial* Pu kN | Moment* Mz kNm | My kNm |
|---|---|---|---|---|
| 1 | 12 1.5DL+1.5SLX | 2206.4 | 239.5 | -11.82 |
| 2 | 15 1.2(DL+LL-SLX) | 265.16 | -132.4 | -7.516 |
| 3 | 11 1.5DL+1.5LL | 600.68 | 153.67 | -1.168 |
| 4 | 11 1.5DL+1.5LL | 530.37 | -144.7 | 9.95 |
| 5 | 14 1.5DL-1.5SLZ | 1947.4 | 11.158 | -127.5 |
| 6 | 13 1.5DL-1.5SLZ | 1811.9 | -11.98 | 126.2 |
| 7 | 14 1.5DL-1.5SLZ | 970.97 | -84.94 | -132.5 |
| 8 | 13 1.5DL+1.5SLZ | 1205 | -91.2 | 111.29 |
| 9 | 13 1.5DL+1.5SLZ | 1256.3 | -13.13 | 168.39 |
| 10 | 14 1.5DL-1.5SLZ | 1293.7 | -15.73 | -170.7 |
| 11 | 12 1.5DL+1.5SLX | 1905.3 | 283.17 | -8.828 |
| 12 | 12 1.5DL+1.5SLX | 1860.1 | -282.2 | 4.386 |

[ * Note : Pu = Fx; Mux = Mz; Muy = My ]

**Design as per SP16 Charts** [Compression member subject to Biaxial bending]

| L/C | Pu/Puz | $\alpha_n$ | d'/D | p/fck | Pu / fckBD | AsperChart 32 Mux / fckBD2 | Mu = Muz kNm | d'/B | AsperChart 32 Muy / fckDB2 | Muy = Muy kNm | Stress Interaction | Remarks |
|---|---|---|---|---|---|---|---|---|---|---|---|---|
| 1 | 0.52 | 1.535 | 0.1 | 0.14 | 0.46 | 0.17 | 489.6 | 0.1 | 0.17 | 326.4 | 0.34 | Safe |
| 2 | 0.06 | 0.771 | 0.1 | 0.14 | 0.06 | 0.22 | 633.6 | 0.1 | 0.22 | 422.4 | 0.34 | Safe |
| 3 | 0.14 | 0.903 | 0.1 | 0.14 | 0.13 | 0.24 | 691.2 | 0.1 | 0.24 | 460.8 | 0.26 | Safe |
| 4 | 0.13 | 0.875 | 0.1 | 0.14 | 0.11 | 0.24 | 691 | 0.1 | 0.24 | 460.8 | 0.29 | Safe |
| 5 | 0.46 | 1.433 | 0.1 | 0.14 | 0.41 | 0.19 | 547.2 | 0.1 | 0.19 | 364.8 | 0.23 | Safe |
| 6 | 0.43 | 1.380 | 0.1 | 0.14 | 0.38 | 0.2 | 576 | 0.1 | 0.2 | 384 | 0.22 | Safe |
| 7 | 0.23 | 1.049 | 0.1 | 0.14 | 0.20 | 0.24 | 691.2 | 0.1 | 0.24 | 460.8 | 0.38 | Safe |
| 8 | 0.28 | 1.141 | 0.1 | 0.14 | 0.25 | 0.23 | 662.4 | 0.1 | 0.23 | 441.6 | 0.31 | Safe |
| 9 | 0.30 | 1.161 | 0.1 | 0.14 | 0.26 | 0.23 | 662.4 | 0.1 | 0.23 | 441.6 | 0.34 | Safe |
| 10 | 0.31 | 1.176 | 0.1 | 0.14 | 0.27 | 0.23 | 662.4 | 0.1 | 0.23 | 441.6 | 0.34 | Safe |
| 11 | 0.45 | 1.416 | 0.1 | 0.14 | 0.40 | 0.19 | 547.2 | 0.1 | 0.19 | 364.8 | 0.40 | Safe |
| 12 | 0.44 | 1.399 | 0.1 | 0.14 | 0.39 | 0.2 | 576 | 0.1 | 0.2 | 384 | 0.37 | Safe |

Stress Interaction factor :

$(Mux/Mux1)^{\alpha n} + (Muy/Muy1)^{\alpha n} <= 1$   OR   $(Mz/Muz)^{\alpha n} + (My/Muy)^{\alpha n} <= 1$

**_Design of other columns are done following above procedure._**

## 12.1.7 Design of Foundations

| Foundation Mkd : | | F1, F1B & F1C - on all grids except E1,D1,B6 & C6 | | | | | |
|---|---|---|---|---|---|---|---|
| INPUT | | Load data as per STAAD analysis result | | | | | |
| Case | A | | Horizont | Vertical | Horizont | Moment G | | H |
| | Node | L/C no & Description | Fx kN | Fy kN | Fz kN | Mx kNm | My kNm | Mz kNm |
| 1 | 6 | 10 0.75(DL+.5LL+SLZ) | -16.44 | 471.84 | -6.609 | -86.14 | -0.488 | 9.134 |
| 2 | 36 | 8 DL+SLX | -39.11 | 1115.6 | -3.756 | -2.891 | -0.591 | 153.07 |
| 3 | 125 | 6 DL+LL+LLR | 0.286 | 1663 | 0.069 | -0.591 | -0.026 | 1.178 |
| 4 | 152 | 7 DL+SLZ | -3.092 | 1184.4 | -36.41 | -144.3 | 0.3 | 2.697 |

| OUTPUT | | Summary of design results | | | | | |
|---|---|---|---|---|---|---|---|
| Load Case | SBP max | min | Avg | Slab Th | Long bars | Transv bars | shear | Uplift |
| 1 | OK | OK | OK | OK. | OK. | OK. | OK. | |
| 2 | OK | OK | OK | OK. | OK. | OK. | OK. | |
| 3 | OK | OK | OK | OK. | OK. | OK. | OK. | |
| 4 | OK | OK | OK | OK. | OK. | OK. | OK. | |

*Sample calculation for Load case 3*

**SBP =**    150 kN/m2
**G sur =**    0 kn/m2
**FGL=**    0 m
**BOF=**    -2.5 m
**EGL =**    -0.6 m

Col dimension:
**l =**    0.6 m
**b =**    0.4 m
Footing dimension:
**L =**    3.6 m
**b =**    3.6 m
**Th =**    0.45 m avg
**cover=**    75 mm

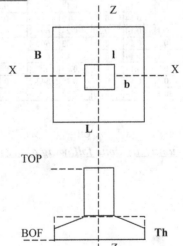

# Design of Industrial Structures

**Reinforcement:**
At bottom layer
Along X - Long Bars    16 dia @    100 mm c/c.    ( Ast =    2011 mm2 /m)
Along Z - Transv Bars    16 dia @    100 mm c/c.    ( Ast =    2011 mm2 /m)

Load at bottom of foundation:

| Case | Description | Fx kN | Fy kN | Fz kN | Mx kNm | My kNm | Mz kNm |
|---|---|---|---|---|---|---|---|
| 3 | 6 DL+LL+LLR | 0.286 | 1663 | 0.069 | -0.591 | -0.026 | 1.178 |

S/W of footing & backfill =    497
                        P =    2160 kN

**Soil pressure:**
Area =    12.96 m2    Zz =    7.776 m3    Zx =    7.776 m3
P/A =    167 kN/m2    Mz/Zz =    0.151        Mx/Zx =    0.076 kN/m2

Maxm GBP =    167.23 kN/m2    >Allow GBP;OK    (167 + 0.151 + 0.076)
Min GBP =    166.77 kN/m2    >Allow GBP;OK    (167 - 0.151 - 0.076)
Avg BP =    167 kN/m2    >Allow GBP;OK
Allowable GBP =    186 kN/m2  ( = 150 + 19 x 1.9 )
Net upward bearing pressure =    167 - 11 - 36 = 139 kN/m2.    (avg BP - slab - backfill)

**Footing Slab design:**                  Load factor =   1.5

**Long Bars:**    along X

    Projection of slab =    1.5 m.
    maxm proj =    1.5 m
    Moment at face of col =    156 kNm /m    (139 x 1.5^2 / 2)
    d reqd=    361 mm    dprov =    375 mm    **> drqd; OK.**
    $M_u/bd^2$ =    1.67
    pt =    0.52    M25
    Ast reqd. =    1931 mm2/m width
    Ast provided =    2011 mm2/m width    **> Ast reqd/Ast min; OK.**
    Ast min =    450 mm2/m width
    Tc.b.d =    345 kN.    OK.
    Vs =    156 kN

**Transverse Bars:** along Z

$$
\begin{aligned}
\text{Projection of slab} &= 1.6 \text{ m.} \\
\text{Moment at face of col} &= 151 \text{ kNm /m} \\
\text{d reqd} &= 355 \text{ mm} \quad \text{dprov} = 375 \text{ mm} \quad \text{> dprov; OK.} \\
M_u/bd^2 &= 1.61 \\
pt &= 0.50 \quad \text{M25} \\
\text{Ast reqd.} &= 1871 \text{ mm2/m width} \\
\text{Ast provided} &= 2011 \text{ mm2/m width} \quad \text{OK.} \\
\text{Ast min} &= 450 \text{ mm2/m width}
\end{aligned}
$$

*Design of other foundations are done by the above method of calculation.*

Design of Industrial Structures

## 12.1.8 Design of Grade beams

| Beam mkd | | Summary of Design loads are from STAAD analysis output file. | | | | | | | | |
|---|---|---|---|---|---|---|---|---|---|---|
| GB1,1A&3 | | member property : R5 ( 500 x 300 ) | | | | | | | | |
| LC | | Member | L/C no and description | Node | Fx kN | Fy kN | Fz kN | MxkNm | MykNm | MzkNm |
| 1 | Max Fx | 361 | 14 1.5DL-1.5SLZ | 127 | 46.0 | 141.7 | 0.1 | 2.3 | -0.3 | 202.4 |
| 2 | Min Fx | 389 | 12 1.5DL+1.5SLX | 184 | -30.8 | -58.6 | 0.0 | 4.4 | 0.1 | -26.8 |
| 3 | Max Fy | 366 | 14 1.5DL-1.5SLZ | 132 | 45.1 | 147.8 | 0.1 | -4.0 | 0.0 | 203.6 |
| 4 | Min Fy | 385 | 12 1.5DL+1.5SLX | 8 | 32.7 | -152.8 | -0.3 | 2.0 | -0.2 | 222.7 |
| 5 | Max Fz | 225 | 13 1.5DL+1.5SLZ | 161 | 22.7 | 122.6 | 1.7 | 19.1 | -2.7 | 129.8 |
| 6 | Min Fz | 395 | 14 1.5DL-1.5SLZ | 187 | 0.2 | -4.4 | -1.3 | -0.4 | 0.2 | -26.1 |
| 7 | Max Mx | 396 | 12 1.5DL+1.5SLX | 188 | 15.9 | -48.3 | 0.8 | 19.4 | -0.2 | -21.5 |
| 8 | Min Mx | 396 | 15 1.2(DL+LL-SLX) | 188 | 10.8 | -46.2 | -1.0 | -24.2 | 0.3 | -23.1 |
| 9 | Max My | 385 | 14 1.5DL-1.5SLZ | 8 | 22.9 | -122.3 | 1.6 | 18.1 | 2.7 | 129.9 |
| 10 | Min My | 225 | 13 1.5DL+1.5SLZ | 161 | 22.7 | 122.6 | 1.7 | 19.1 | -2.7 | 129.8 |
| 11 | Max Mz | 385 | 12 1.5DL+1.5SLX | 8 | 32.7 | -152.8 | -0.3 | 2.0 | -0.2 | 222.7 |
| 12 | Min Mz | 181 | 12 1.5DL+1.5SLX | 131 | -30.0 | -24.6 | 0.8 | -5.7 | -1.0 | -116.2 |

OUTPUT - Summary of design results of the beam at all load cases

| LC | MU limit kNm | | AstT(Sp) | AstB(Sp) | AstT(Sup) | AstB(Sup) | Stirrups |
|---|---|---|---|---|---|---|---|
| 1 | 219 | D Reinf. | Ok. | Ok. | Ok. | Ok. | |
| 2 | 219 | D Reinf. | Ok. | Ok. | Ok. | Ok. | |
| 3 | 219 | D Reinf. | Ok. | Ok. | Ok. | Ok. | |
| 4 | 219 | D Reinf. | Ok. | Ok. | Ok. | Ok. | |
| 5 | 219 | D Reinf. | Ok. | Ok. | Ok. | Ok. | |
| 6 | 219 | D Reinf. | Ok. | Ok. | Ok. | Ok. | |
| 7 | 219 | D Reinf. | Ok. | Ok. | Ok. | Ok. | |
| 8 | 219 | D Reinf. | Ok. | Ok. | Ok. | Ok. | |
| 9 | 219 | D Reinf. | Ok. | Ok. | Ok. | Ok. | |
| 10 | 219 | D Reinf. | Ok. | Ok. | Ok. | Ok. | |
| 11 | 219 | D Reinf. | Ok. | Ok. | Ok. | Ok. | |
| 12 | 219 | D Reinf. | Ok. | Ok. | Ok. | Ok. | |

## Sample Calculation
**GB1,1A&3**    R5    ( 500 x 300 )    On grids

| LC | 11 |

B = 0.3 m.    fck = 25 Mpa
D = 0.5 m.    fy = 415 Mpa
cover = 40 mm.
Span = 6 m.    Ht of wall, w = 5.1 m.

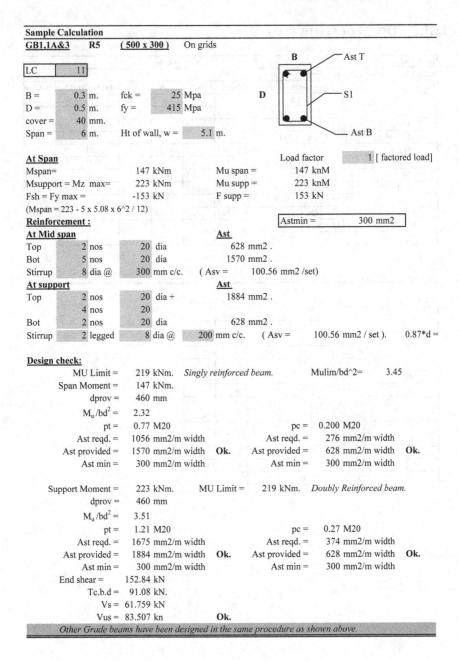

### At Span
Load factor    1 [ factored load]
Mspan= 147 kNm    Mu span = 147 knM
Msupport = Mz max= 223 kNm    Mu supp = 223 knM
Fsh = Fy max = -153 kN    F supp = 153 kN
(Mspan = 223 - 5 x 5.08 x 6^2 / 12)

### Reinforcement :
Astmin = 300 mm2

**At Mid span**                        Ast
Top      2 nos    20 dia         628 mm2 .
Bot      5 nos    20 dia        1570 mm2 .
Stirrup  8 dia @  300 mm c/c.  ( Asv =  100.56 mm2 /set)

**At support**                         Ast
Top      2 nos    20 dia +     1884 mm2 .
         4 nos    20
Bot      2 nos    20 dia         628 mm2 .
Stirrup  2 legged  8 dia @   200 mm c/c.  ( Asv = 100.56 mm2 / set ).    0.87*d =

### Design check:
MU Limit = 219 kNm.    *Singly reinforced beam.*    Mulim/bd^2= 3.45
Span Moment = 147 kNm.
dprov = 460 mm
$M_u /bd^2$ = 2.32
pt = 0.77 M20                              pc = 0.200 M20
Ast reqd. = 1056 mm2/m width              Ast reqd. = 276 mm2/m width
Ast provided = 1570 mm2/m width  **Ok.**   Ast provided = 628 mm2/m width  **Ok.**
Ast min = 300 mm2/m width                  Ast min = 300 mm2/m width

Support Moment = 223 kNm.    MU Limit = 219 kNm.    *Doubly Reinforced beam.*
dprov = 460 mm
$M_u /bd^2$ = 3.51
pt = 1.21 M20                              pc = 0.27 M20
Ast reqd. = 1675 mm2/m width              Ast reqd. = 374 mm2/m width
Ast provided = 1884 mm2/m width  **Ok.**   Ast provided = 628 mm2/m width  **Ok.**
Ast min = 300 mm2/m width                  Ast min = 300 mm2/m width
End shear = 152.84 kN
Tc.b.d = 91.08 kN.
Vs = 61.759 kN
Vus = 83.507 kn        **Ok.**

*Other Grade beams have been designed in the same procedure as shown above.*

## 12.1.9 DESIGN OF FLOOR & ROOF BEAMS

**FIRST FLOOR BEAMS**

| Beam mkd | | | Summary of Design loads are from STAAD analysis output file. | | | | | | |
|---|---|---|---|---|---|---|---|---|---|
| FB11,21 & RB31 | | | member property : | R7 (400 x 250) | | | | | |
| LC | | Member | L/C no and description | Node | Fx kN | Fy kN | Fz kN | Mx kNm | My kNm | MzkNm |
| 1 | Max Fx | 751 | 13 1.5DL+1.5SLZ | 386 | 29.371 | 84.8 | -2.263 | 0.61 | 6.988 | 90.271 |
| 2 | Min Fx | 751 | 14 1.5DL-1.5SLZ | 386 | -17.98 | 79.779 | 2.431 | 0.501 | -7.475 | 80.127 |
| 3 | Max Fy | 671 | 14 1.5DL-1.5SLZ | 346 | 1.933 | 97.569 | 1.594 | -0.828 | -4.688 | 105.11 |
| 4 | Min Fy | 747 | 14 1.5DL-1.5SLZ | 385 | -2.12 | -97.59 | 2.382 | 0.891 | 7.168 | 105.24 |
| 5 | Max Fz | 929 | 13 1.5DL+1.5SLZ | 511 | 0.271 | 39.28 | 5.148 | -2.336 | -8.286 | 1.944 |
| 6 | Min Fz | 929 | 14 1.5DL-1.5SLZ | 511 | -1.507 | 39.279 | -5.987 | 0.353 | 9.546 | 1.903 |
| 7 | Max Mx | 409 | 13 1.5DL+1.5SLZ | 194 | -0.146 | 15.339 | 1.309 | 3.744 | -2.124 | -3.85 |
| 8 | Min Mx | 419 | 14 1.5DL-1.5SLZ | 201 | -0.637 | 41.874 | -1.673 | -4.008 | 2.723 | 0.233 |
| 9 | Max My | 929 | 14 1.5DL-1.5SLZ | 511 | -1.507 | 39.279 | -5.987 | 0.353 | 9.546 | 1.903 |
| 10 | Min My | 749 | 14 1.5DL-1.5SLZ | 385 | -8.784 | 74.543 | 3.487 | -0.181 | -9.116 | 86.767 |
| 11 | Max Mz | 729 | 13 1.5DL+1.5SLZ | 376 | -17.68 | -91.31 | 2.213 | 0.41 | 6.47 | 115.14 |
| 12 | Min Mz | 407 | 12 1.5DL+1.5SLX | 193 | -0.791 | 16.536 | -0.23 | 0.039 | 0.412 | -16.85 |

## OUTPUT : Summary of design results at all load cases

| LC | MU limit kNm | | AstT (Span) | AstB (Span) | AstT (Sup) | AstB (Sup) | Stirrups |
|----|---|---|---|---|---|---|---|
| 1  | 89.424 | D Reinf. | Ok. | Ok. | Ok. | Ok. | Ok. |
| 2  | 89.424 | D Reinf. | Ok. | Ok. | Ok. | Ok. | Ok. |
| 3  | 89.424 | D Reinf. | Ok. | Ok. | Ok. | Ok. | Ok. |
| 4  | 89.424 | D Reinf. | Ok. | Ok. | Ok. | Ok. | Ok. |
| 5  | 89.424 | D Reinf. | Ok. | Ok. | Ok. | Ok. | Ok. |
| 6  | 89.424 | D Reinf. | Ok. | Ok. | Ok. | Ok. | Ok. |
| 7  | 89.424 | D Reinf. | Ok. | Ok. | Ok. | Ok. | Ok. |
| 8  | 89.424 | D Reinf. | Ok. | Ok. | Ok. | Ok. | Ok. |
| 9  | 89.424 | D Reinf. | Ok. | Ok. | Ok. | Ok. | Ok. |
| 10 | 89.424 | D Reinf. | Ok. | Ok. | Ok. | Ok. | Ok. |
| 11 | 89.424 | D Reinf. | Ok. | Ok. | Ok. | Ok. | Ok. |
| 12 | 89.424 | D Reinf. | Ok. | Ok. | Ok. | Ok. | Ok. |

### Sample Calculation
**FB11,21 & RB31**  (400 x 250)  **R7**

LC  11

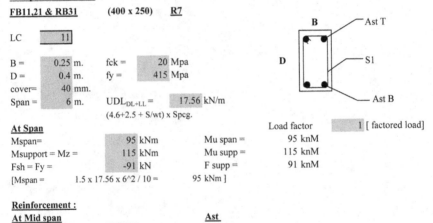

B = 0.25 m.   fck = 20 Mpa
D = 0.4 m.    fy = 415 Mpa
cover = 40 mm.
Span = 6 m.   $UDL_{DL+LL}$ = 17.56 kN/m
(4.6+2.5 + S/wt) x Spcg.

**At Span**                                    Load factor  1 [ factored load]
Mspan = 95 kNm          Mu span = 95 knM
Msupport = Mz = 115 kNm  Mu supp = 115 knM
Fsh = Fy = -91 kN        F supp = 91 knM
[Mspan =  1.5 x 17.56 x 6^2 / 10 = 95 kNm ]

**Reinforcement :**
**At Mid span**                    **Ast**
Top     2 nos   20 dia    628 mm2 .
Bot     3 nos   20 dia    942 mm2 .

Stirrup  8 dia @  300 mm c/c.  ( Asv = 100.56 mm2 /set)

**At support**                     **Ast**
Top     2 nos   20 dia +   942 mm2 .
        1 nos   20
Bot     2 nos   20 dia    628 mm2 .
Stirrup  2 legged  8 dia @ 200 mm c/c.  ( Asv = 100.56 mm2 / set ).

# Design of Industrial Structures

**Design check:**

| | | | | | |
|---|---|---|---|---|---|
| MU Limit = | 89 kNm. | **Doubly Reinf.** | | Mulim/bd^2= | 2.76 |
| Span Moment = | **95 kNm.** | | Astmin = | 200 mm2 | |
| dprov = | 360 mm | | (0.2 x 0.25 x 0.4 x 10^6 / 100) | | |
| $M_u/bd^2$ = | 2.93 | | | | |
| pt = | 0.84 M20 | | pc = | 0.200 M20 | |
| Ast reqd. = | 758 mm2/m width | | Ast reqd. = | 180 mm2/m width | |
| Ast provided = | 942 mm2/m width | **Ok.** | Ast provided = | 628 mm2/m width | **Ok.** |
| Ast min = | 200 mm2/m width | | Ast min = | 200 mm2/m width | |
| Support Moment = | **115 kNm.** | MU Limit = | 89.424 kNm. | **Doubly Reinf.** | |
| dprov = | 360 mm | | | | |
| $M_u/bd^2$ = | 3.55 | | | | |
| pt = | 1.00 M20 | | pc = | 0.26 M20 | |
| Ast reqd. = | 896 mm2/m width | | Ast reqd. = | 233 mm2/m width | |
| Ast provided = | 942 mm2/m width | **Ok.** | Ast provided = | 628 mm2/m width | **Ok.** |
| Ast min = | 200 mm2/m width | | Ast min = | 200 mm2/m width | |
| End shear = | 91 kN | | | | |
| Tc.b.d = | 54 kN. | Vs = 37.307 kN | Vus = 65.353 kn | | **Ok.** |

*Design of all the floor and roof beams are done by the same procedure as shown above.*

## 12.1.10 Design of Slabs & Stair Case

**Roof slab**

UDL1 =   **5.06** kN/m2   DL+LL  (3.56 + 1.5 )

Span =   **2** m.

Moment span =        2.02 kNm    (5.06 x 2^2 / 10)     R1 =   5.06 kN
Moment support =     1.69 kNm    (5.06 x 2^2 / 12)     R2 =   5.06 kN

Try with    **120** mm   slab    B =    **1** m.              fck =         **25** Mpa
                                 D =    **0.12** m.           fy =          **415** Mpa
                                 cover =   15 mm.             Load factor = **1.5**

Rebars   Span :    **8** dia @   150 mm c/c.   ( Ast =   335 mm2 /m)
         Support:  **8** dia @   150 mm c/c.   ( Ast =   335 mm2 /m)

### Design check:

| | | | | | |
|---|---|---|---|---|---|
| MU Limit = | 38.04 kNm. | **Singly reinforced.** | Mulim/bd^2= | 3.45 | |
| **Span Moment =** | 3.03 kNm. | (2.02 x 1.5) | | | |
| dprov = | 105 mm | | | | |
| $M_u/bd^2$ = | 0.27 | | | | |
| pt = | 0.12 | M20 | pc = | 0.200 | M20 |
| Ast reqd. = | 144 mm2/m width | | Ast reqd. = | 240 mm2/m width | |
| Ast provided = | 335 mm2/m width | **Ok.** | Ast provided = | 335 mm2/m width | **Ok.** |
| Ast min = | 144 mm2/m width | | Ast min = | 144 mm2/m width | |

| | | | | | |
|---|---|---|---|---|---|
| **Support Moment =** | 2.54 kNm. | (1.69 x 1.5) | | | |
| dprov = | 105 mm | MU Limit = | 38.036 kNm. | **Singly reinforced.** | |
| $M_u/bd^2$ = | 0.23 | | | | |
| pt = | 0.12 | M20 | pc = | 0.20 | M20 |
| Ast reqd. = | 144 mm2/m width | | Ast reqd. = | 240 mm2/m width | |
| Ast provided = | 335 mm2/m width | **Ok.** | Ast provided = | 335 mm2/m width | **Ok.** |
| Ast min = | 144 mm2/m width | | Ast min = | 144 mm2/m width | |
| End shear = | 5.06 kN | | | | |
| Tc.b.d = | 21.0 kN. | **Ok.** | | | |

### Floor slab

UDL1=   7.1 kN/m2  (DL+LL)   (4.6 + 2.5 )

SPAN=   2 M.

| | | | | | |
|---|---|---|---|---|---|
| Moment span = | 2.84 kNm | (7.1 x 2^2 / 10) | R1 = | 7.1 kN | |
| Moment support = | 2.37 kNm | (7.1 x 2^2 / 12) | R2 = | 7.1 kN | |

| | | | | | | |
|---|---|---|---|---|---|---|
| Try with | 120 mm | slab | B | 1 m. | fck | 25 Mpa |
| | | | D | 0.12 m. | fy | 415 Mpa |
| | | | cover | 15 mm. | Load factor | 1.5 |

| | | | | | | |
|---|---|---|---|---|---|---|
| Rebars | Span: | 8 dia @ | 150 mm c/c. | ( Ast = | 335 mm2 /m) | |
| Support: | 8 dia @ | 150 mm c/c. | ( Ast = | 335 mm2 /m) | | |

# Design of Industrial Structures

**Design check:**

| | | | | | |
|---|---|---|---|---|---|
| MU Limit = | 38.036 kNm. | **Singly reinforced.** | | Mulim/bd^2= | 3.45 |
| Span Moment = | 4.26 kNm. | (2.84 x 1.5) | | | |
| dprov = | 105 mm | | | | |
| $M_u/bd^2$ = | 0.39 | | | | |
| pt = | 0.09 | M20 | pc = | 0.20 | M20 |
| Ast reqd. = | 113 mm2/m width | | Ast reqd. = | 240 mm2/m width | |
| Ast provided = | 335 mm2/m width | **Ok.** | Ast provided = | 335 mm2/m width | **Ok.** |
| Ast min = | 144 mm2/m width | | Ast min = | 144 mm2/m width | |

| | | | | | |
|---|---|---|---|---|---|
| Support Moment = | 3.56 kNm. | (2.37 x 1.5) | | | |
| dprov = | 105 mm | MU Limit = | 38.036 kNm. | **Singly reinforced.** | |
| $M_u/bd^2$ = | 0.32 | | | | |
| pt = | 0.08 M20 | | pc = | 0.20 M20 | |
| Ast reqd. = | 98 mm2/m width | | Ast reqd. = | 240 mm2/m width | |
| Ast provided = | 335 mm2/m width | **Ok.** | Ast provided = | 335 mm2/m width | **Ok.** |
| Ast min = | 144 mm2/m width | | Ast min = | 144 mm2/m width | |
| End shear = | 7.1 kN | | | | |
| Tc.b.d = | 21.0 kN. | **Ok.** | | | |

## Stair slab

**Waist slab :**           TREAD= 250 mm    RISE= 167 mm
                           nos = 11         nos = 12

**Stairs**

waist slab                        **DL**              **LL**           250 *Tread*
    **125** mm   slab on slope =   3.61 kN/m2 (in plan)      167
    [(125 x 301 / 250) x 24 / 1000 ]                    *Riser*  301
    40 mm      Floor finish =   0.96                         *Steps*
    12 mm      plaster =       0.24
steps       (0.5 x 167 x 250 / 250) x 24 / 1000 =  2.00 kN/m2
                                          6.82 kN/m2              5 kN/m2

landing slab

| | | | DL | | LL |
|---|---|---|---|---|---|
| 125 mm | slab | | 3.00 kN/m2 | | |
| 40 mm | Floor finish | | 0.96 | | |
| 12 mm | plaster | | 0.24 | | |
| | | | 4.20 kN/m2 | | 5 kN/m2 |

**Flight 1**

| | DL | LL |
|---|---|---|
| UDL1 = | 6.82 | 5 kN/m2 |
| UDL2 = | 4.20 | 5 kN/m2 |

SPAN= 4.4 M.

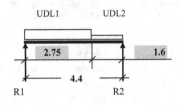

Moment span = 20.33 knm
Moment support = 10.05 knm
R1 = 28.47 kN    R2 = 23.23 kN

Try with  125 mm  slab   B = 1 m.         fck = 25 Mpa
                          D = 0.125 m.     fy = 415 Mpa
                          cover = 25 mm.   Load factor = 1.5

*Main reinforcements*

Span (at Bottom layer) :   12 dia @  100 mm c/c ( Ast = 1131 mm2 /m)   Ok.

Support (at top layer) :   12 dia @  150 mm c/c ( Ast = 754 mm2 /m)    Ok.

**Design check:**

MU Limit =     34.5 kNm.  **Singly reinforced.**    Mulim/bd^2=  3.45
Span Moment =  30 kNm.    (20.33 x 1.5)
dprov =        100 mm
$M_u/bd^2$ =   3.05
pt =           0.87      M20              pc = 0.200      M20
Ast reqd. =    1085 mm2/m width           Ast reqd. =  250 mm2/m width
Ast provided = 1131 mm2/m width  **Ok.**  Ast provided = 754 mm2/m width  **Ok.**
Ast min =      150 mm2/m width            Ast min = 150 mm2/m width

Support Moment = 15 kNm.  (10.05 x 1.5)
dprov =        100 mm           MU Limit = 34.500 kNm.  **Singly reinforced.**
$M_u/bd^2$ =   1.51
pt =           0.40      M20              pc = 0.20       M20
Ast reqd. =    495 mm2/m width            Ast reqd. =  250 mm2/m width
Ast provided = 754 mm2/m width  **Ok.**   Ast provided = 1131 mm2/m width  **Ok.**
Ast min =      150 mm2/m width            Ast min = 150 mm2/m width
End shear =    28.469 kN
Tc.b.d =       39.0 kN.   **Ok.**

# Design of Industrial Structures

## 12.1.11 Computer Model

Structural analysis has been done using STAAD PRO software. Computer model used for analysis are furnished here. Member and joints nos are not printed for clarity. Analysis results have been sorted out to select worst loading cases for design of member sections.

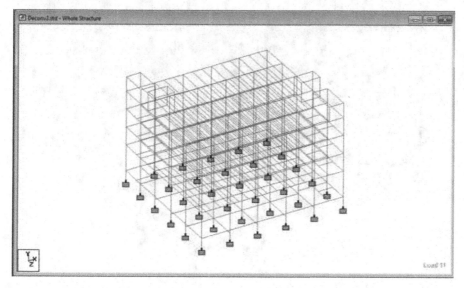

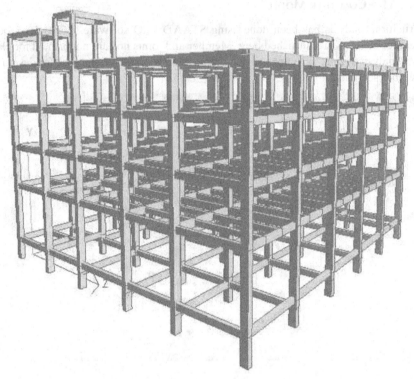

**3D VIEW OF MAIN FRAME WORK**

# Design of Industrial Structures

## 12.1.12 Drawings

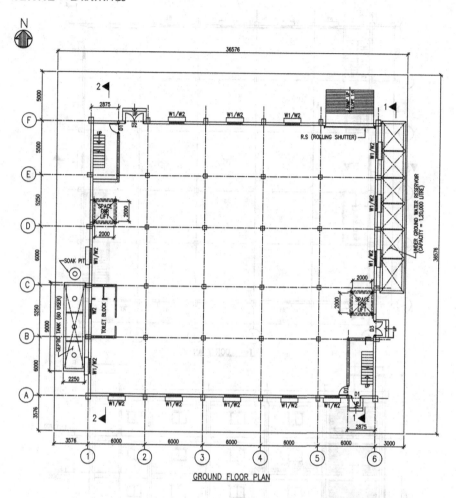

GROUND FLOOR PLAN

| DOOR & WINDOWS SCHEDULE | | | |
|---|---|---|---|
| MKD. | SIZE | MKD. | SIZE |
| D1 | 1000X2100 | W1 | 1500 X 1200 |
| D2 | 2000X2500 | W2 | 1500 X 600 |
| D3 | 1500X2100 | | |
| | | | |

NOTES :
1. ALL DIMENSIONS ARE IN MILLIMETRE AND ELEVATIONS ARE IN METER.
2. FOUNDATION BRICK WORK WILL BE FIRST CLASS BRICK WITH 1:6 CEMENT MORTAR.
3. 250TH. EXTERNAL WALLS WILL BE 1:6 CEMENT MORTR.
4. 25 TH. D.P.C. WILL BE 1:2:4 WITH PROPER WATER PROOFING COMPOUND.
5. ROOF TREATMENT OR LIME TERRACING WILL BE 100 TH. WITH THEIR PROPER MATERIALS AND MIXING.
6. CEILING AND ALL R.C. PLASTER WILL BE 12mm TH. WITH 1:4 CEMENT MORTAR.
7. ALL BUILDING MATERIALS WILL BE AS PER I.S. CODE & N.B.C. 1984.
8. THIS DRAWING SHOULD BE READ IN CONJUNCTION WITH RELEVANT ARCHITECTURAL & STRUCTURAL STRUCTURAL DRAWINGS.

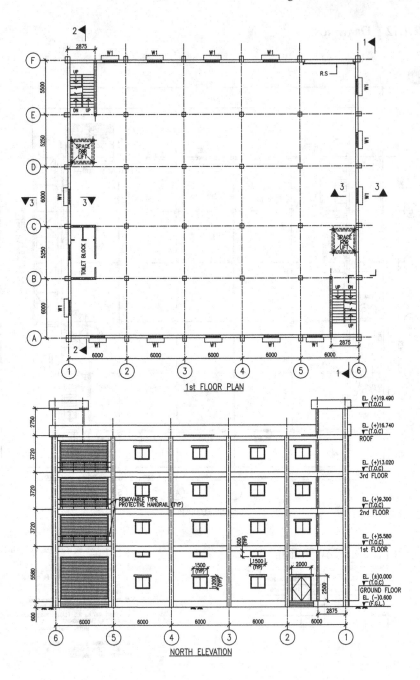

# Design of Industrial Structures

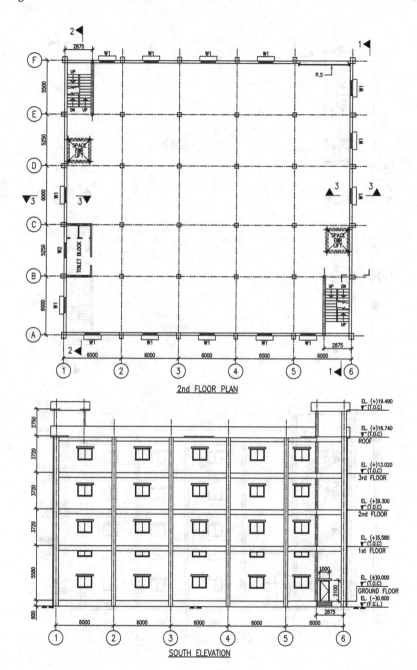

346  Design of Industrial Structures

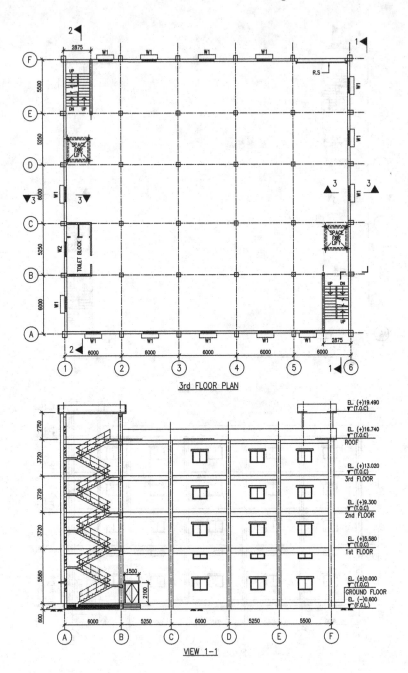

# Design of Industrial Structures

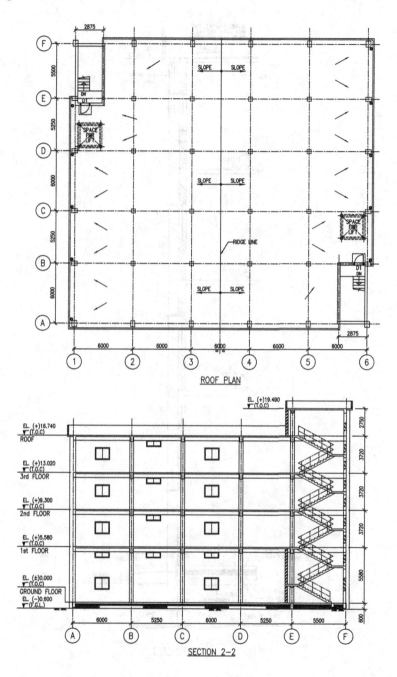

348  Design of Industrial Structures

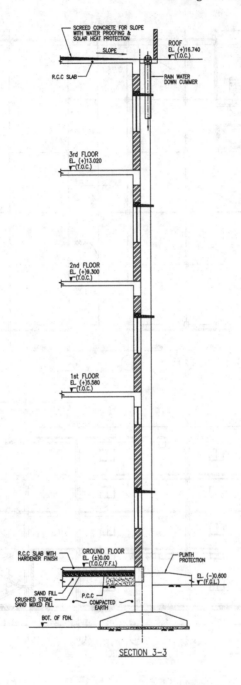

SECTION 3-3

# Design of Industrial Structures

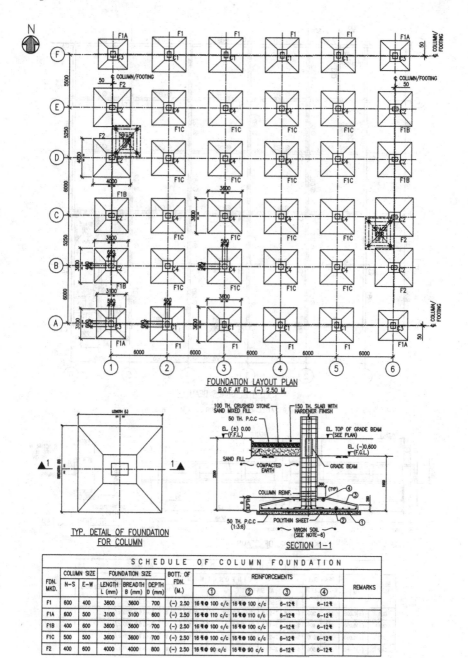

# 350 Design of Industrial Structures

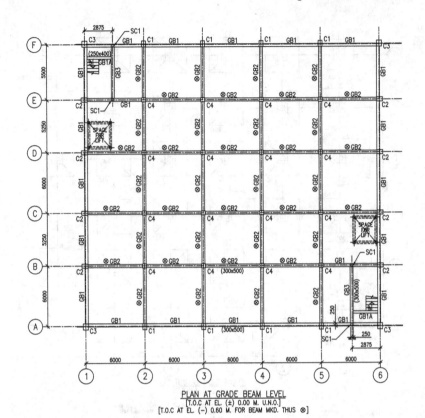

PLAN AT GRADE BEAM LEVEL
[T.O.C AT EL. (±) 0.00 M. U.N.O.]
[T.O.C AT EL. (−) 0.60 M. FOR BEAM MKD. THUS ⊗]

TYPICAL DETAIL OF BEAMS

TYP. SECTION OF BEAM

| BEAM MKD. | T.O.C (M.) | BEAM SIZE | | REINFORCEMENTS | | | | | | REMARKS |
|---|---|---|---|---|---|---|---|---|---|---|
| | | WIDTH (mm) | DEPTH (mm) | AT SUPPORT | | | AT MID SPAN | | | |
| | | | | TOP | BOT. | STIRRUP | TOP | BOT. | STIRRUP | |
| GB1 | (±)0.000 | 300 | 500 | 6−20⌀ | 2−20⌀ | 2L−8⌀ @ 200 c/c | 2−20⌀ | 5−20⌀ | 2L−8⌀ @ 200 c/c | |
| GB1A | (±)0.000 | 250 | 400 | 2−16⌀ | 2−16⌀ | 2L−8⌀ @ 200 c/c | 2−16⌀ | 2−16⌀ | 2L−8⌀ @ 200 c/c | |
| GB3 | (±)0.000 | 300 | 500 | 2−20⌀ | 2−20⌀ | 2L−8⌀ @ 200 c/c | 2−20⌀ | 5−20⌀ | 2L−8⌀ @ 200 c/c | |
| GB2 | (−)0.600 | 300 | 500 | 5−16⌀ | 2−16⌀ | 2L−8⌀ @ 250 c/c | 2−16⌀ | 4−16⌀ | 2L−8⌀ @ 250 c/c | |

SCHEDULE OF BEAMS

# Design of Industrial Structures

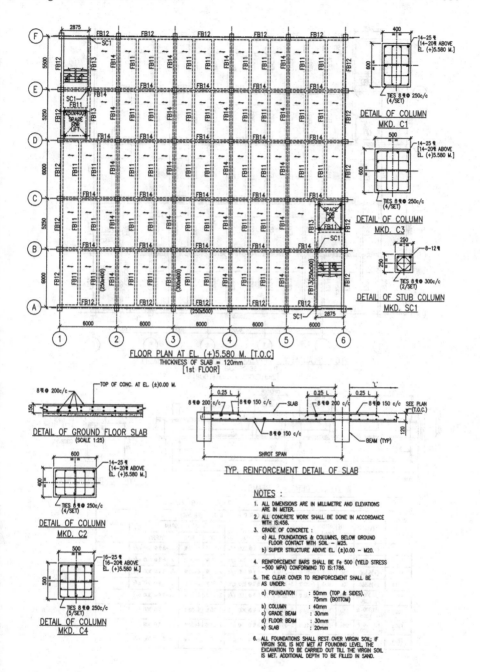

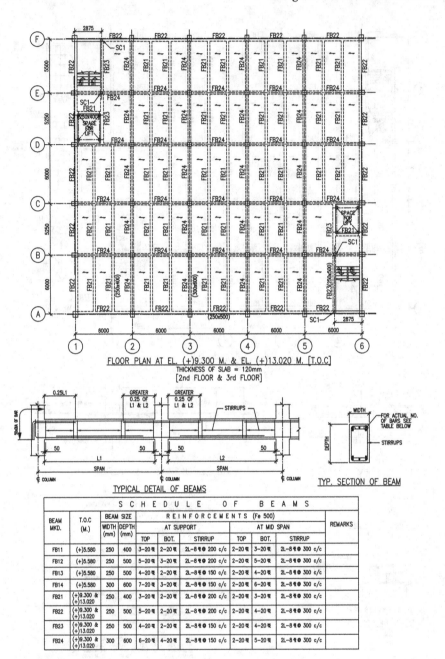

# Design of Industrial Structures

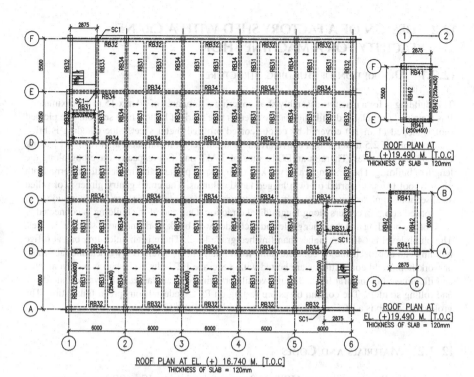

ROOF PLAN AT EL. (+) 16.740 M. [T.O.C]
THICKNESS OF SLAB = 120mm

ROOF PLAN AT EL. (+)19.490 M. [T.O.C]
THICKNESS OF SLAB = 120mm

### SCHEDULE OF BEAMS

| BEAM MKD. | T.O.C (M.) | BEAM SIZE | | REINFORCEMENTS (Fe 500) | | | | | REMARKS |
|---|---|---|---|---|---|---|---|---|---|
| | | WIDTH (mm) | DEPTH (mm) | AT SUPPORT | | | AT MID SPAN | | |
| | | | | TOP | BOT. | STIRRUP | TOP | BOT. | STIRRUP | |
| RB31 | (+)16.740 | 250 | 400 | 3–20⌀ | 2–20⌀ | 2L–8⌀@ 200 c/c | 2–20⌀ | 3–20⌀ | 2L–8⌀@ 300 c/c | |
| RB32 | (+)16.740 | 250 | 600 | 5–20⌀ | 2–20⌀ | 2L–8⌀@ 150 c/c | 2–20⌀ | 4–20⌀ | 2L–8⌀@ 300 c/c | |
| RB33 | (+)16.740 | 250 | 500 | 4–20⌀ | 2–20⌀ | 2L–8⌀@ 150 c/c | 2–20⌀ | 4–20⌀ | 2L–8⌀@ 300 c/c | |
| RB34 | (+)16.740 | 300 | 600 | 7–20⌀ | 4–20⌀ | 2L–8⌀@ 90 c/c | 2–20⌀ | 6–20⌀ | 2L–8⌀@ 300 c/c | |
| RB41 | (+)19.490 | 250 | 450 | 2–16⌀ | 2–16⌀ | 2L–8⌀@ 200 c/c | 2–16⌀ | 3–16⌀ | 2L–8⌀@ 200 c/c | |
| RB42 | (+)19.490 | 250 | 450 | 2–16⌀ | 2–16⌀ | 2L–8⌀@ 200 c/c | 2–16⌀ | 3–16⌀ | 2L–8⌀@ 300 c/c | |

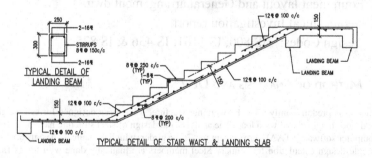

TYPICAL DETAIL OF LANDING BEAM

TYPICAL DETAIL OF STAIR WAIST & LANDING SLAB

End

## 12.2 DESIGN OF A FACTORY SHED WITH A CRANE FACILITY FOR A MACHINE SHOP

### 12.2.1 Description of Building Structure

This building is for a machine assembly shop covering a floor area of 50 m x 15 m approximately. It is a pitched roof cladded building made of steel frame work of hollow tube section. The picthed roof is sloped (1 in 10) covered by colour coated profiled metal sheet. Side cladding is brick manory work of 2.5 meter height above ground and 2m height metal sheet at eaves level leaving large opening at sides. There is an EOT crane of 3 MT on lifing capacity travelling along the length of building. The building frame work is a single span portal frame made with uniform depth pitched roof truss supported on twin legged laced columns. The crane gantry girder is of rolled steel section supported by steel brackets from columns. The portal frames are 14 meter wide span and spaced at 5.5 m centers along longitudinal direction. One end of the building is connected to RCC office building and the other gable end is cladded with brick work and metal sheet as in sides. The plinth level is 640 mm higher than finished grade level. There is an unloading bay at office building end approachable by concrete ramp from road. The structure is a welded and fabricated at workshop. Field connectiones are bolted with 8.8 grade HT bolts. The shed is well ventilated and illuminated by natural lights through ventilators,windows. and accessible by adequate doors and rolling shutters. The column bases are resting on RCC pedestals on spread foundation. The ground floor slab is a RCC slab with ready made mix hardener flooring.

### 12.2.2 Materials and Codes

1 Hollow sections - YST 310 ; conforming to IS: 4923
2 Structural steel - E 250 (Fe 410W) A, fy = 250MPa
3 Foundation concrete and base slab : M25 grade ; IS-456.
4 Reinforcement: High yield strength deformed bars - IS 1786

### 12.2.3 Reference Codes and Documents

1 Equipment layout and General arrangement dwg.
2 Geotechnical Investigation report.
3 Design Codes:   IS 800; IS 1161; IS 456 & IS 875.

### 12.2.4 Method of Ananlysis and Design

This structure is predominantly a wind governing structure. It has been designed for self weight, crane load, roof live load and wind load. The analysis and design of steel structure have been done using computer software STAAD-Pro, except for Crane girder, which is manually designed. However, sample design calculation for primary Steel members are manually done with forces from STAAD anlysis and furnished here for readers understanding. The design of foundation and anchor bolts are done by limit state method in accordance with IS 456.

# Design of Industrial Structures

## 12.2.5 Sketch

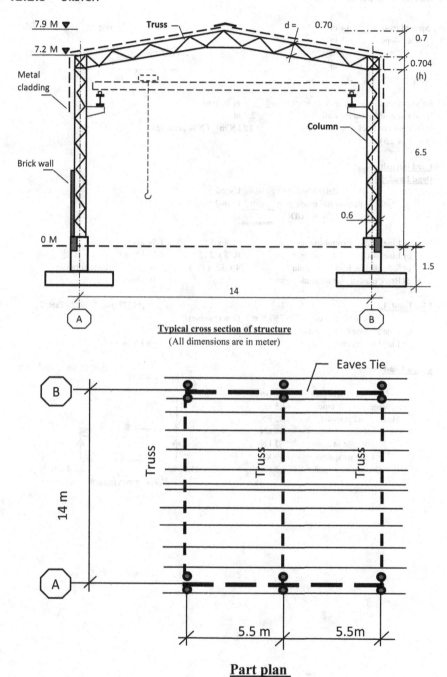

**Typical cross section of structure**
(All dimensions are in meter)

**Part plan**

## 12.2.6 LOADING

Span of truss = 14 m  
Roof slope = 1 in 10  
d = 0.7 m  
h = 0.704 m  
Spcg of truss = 5.5 m  

$\theta$ = 5.73 degree

Height of metal clading at sides = 4 m from roof  
Height of brick masonry wall = 2.5 m  
Unit weight of brick masonry wall = 10 kN/m   (20% opening)  
(5 x 0.8 x 2.5)

**Load intensity**
**Dead Load, DL:**

Metal Sheeting = 0.08 kN/m2  
Purlins and bracings = 0.2 kN/m2  
UDL = 0.28 kN/m2  

a) Line load on central trusses = 0.28 x 5.5 = 1.54 kN/m  
b) Line load on end trusses = 0.28 x 2.75 = 0.77 kN/m  
c) Side clading load on column = 4 x 5.5 x 0.15 = 6.16 kN   (concentrated)  
c) Brick masonry wt on grade beam = 10 kN/m  

**Live Load, LL:**    [IS 875 (part 2) 1987 - Table 2]

Live load on roof = 0.75 kN/m2  (inaccessible)  
a) Line load on central trusses = 0.75 x 5.5 = 4.13 kN/m  
b) Line load on end trusses = 0.75 x 2.75 = 2.06 kN/m  

**Crane Load**    [IS 875 (part 2) 1987 - 6.3]

No of wheels = 2 per side  
Wheel base distance = 2.5 m  
Maxm Load per wheel = 28.82 kN  
Weight of crane bridge = 50 kN  
Weight of crab = 3 kN  
Lifting capacity = 30 kN  
Crane girder span = 5.5 m  

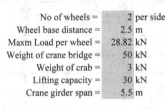
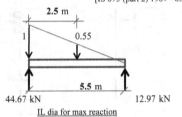

IL dia for max reaction

# Design of Industrial Structures

Maxm reaction on col = $28.82 \times (1 + 0.55) = \quad 44.67$ kN
Transverse surge wheel/side = $\quad 10\% \ (3 + 30) / 2 = \quad 1.65$ kN
Longitudinal surge on each track = $\quad 5\% \ (2 \times 28.82) = \quad 2.88$ kN

Column reaction for frame analysis:
*For crab position near row A :*

| Row | Vert | Tranv | Long |
|---|---|---|---|
| A | 44.67 | 1.65 | 2.88 |
| B | 19.38 | 1.65 | 1.94 |

**Wind load, WL :**

$V_z =$ 50 m/s
$p =$ 1.5 kN/m²

| ht | 7.2 |
|---|---|
| w | 14 |
| h/w | 0.51 |

### *a) Wind direction from Left to Right*

**On Roof**            [IS 875 (part 3) 1987 - Table 5]

| External pr + Internal pr. Co-eff | | | | External pr + Internal suction Coeff. | | |
|---|---|---|---|---|---|---|
| Roof | *Windward* | *Leeward* | | Roof | *Windward* | *Leeward* |
| extenal | -0.9 | -0.4 | | extenal | -0.9 | -0.4 |
| int pr | 0.2 | 0.2 | | int suc | -0.2 | -0.2 |
| Net pr | **-0.7** | **-0.2** | | Net pr | **-1.1** | **-0.6** |

$p =$ 1.5 kN/m²      frame spcg = 5.5 m

Line load on Truss rafter = $\quad -1.1 \times 1.5 \times 5.5 = \quad$ **-9.1** kN/m  *Upward on Windwardside.*
Line load on Truss rafter = $\quad -0.6 \times 1.5 \times 5.5 = \quad$ **-5.0**  *Upward on Leewardside.*

**On Side Cladding**            [IS 875 (part 3) 1987 - Table 4]

| External pr + Internal pr. Co-eff | | | | External pr + Internal suction Coeff. | | |
|---|---|---|---|---|---|---|
| wall | *Windward* | *Leeward* | | | *windward* | *Leeward* |
| ext | 0.7 | -0.25 | | | 0.7 | -0.25 |
| int pr | 0.2 | 0.2 | | int suc | -0.2 | -0.2 |
| | **0.9** | **-0.05** | | | **0.5** | **-0.45** |

$p =$ 1.5 kN/m²      frame spacing = 5.5 m

Line load on frame column = $\quad 0.9 \times 1.5 \times 5.5 = \quad$ **7.4** kN/m  *Pressure on Windward side.*
Line load on frame column = $\quad -0.05 \times 1.5 \times 5.5 = \quad$ **-0.4** kN/m  *Suction on Leeward side.*

**Gable wind**            [IS 875 (part 3) 1987 - Table 4]

| Pr coefficients | |
|---|---|
| **-0.6** | **-0.6** |

$p =$ 1.5 kN/m²      Gable col spcg = 3.43 mm

Line load on frame column = $\quad -0.6 \times 1.5 \times 3.43 = \quad$ **-3.1** kN/m  *Suction on Windward side.*
Line load on frame column = $\quad -0.6 \times 1.5 \times 3.43 = \quad$ **-3.1** kN/m  *Suction on Leeward side.*

## 12.2.7 ANALYSIS OF FRAME

This structure has been analysed and designed as 3D model using STAAD pro software.
The member sections, details of structure and foundation are given in drawing.
Sample calculation of strength design for primary members and foundation are given below.

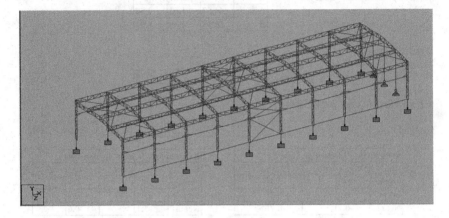

**Model used in STAAD.**

# Design of Industrial Structures

## 12.2.8 DESIGN OF PRIMARY MEMBERS

### 12.2.8.1 Purlin with Hollow Rectangular Section (TATA Section)

Member Mkd : purlin

| | | | |
|---|---|---|---|
| Span = | 5.5 m | | |
| spcg = | 1.45 m | | |
| Intensity = | 1.03 kN/m2 | DL+LL | |
| | -1.028 kN/m2 | 0.75(DL+WL) | |

| | | | |
|---|---|---|---|
| UDL = | 1.03 kN/m2 | max | |
| w = | 1.49 kN/m | | |
| M = | 5.65 kN-m | [$WL^2/8$ end span] | |
| 0.75(0.28-1.1x 1.5) | | | |

Try  Hollow  122 61 4.5    wt/m  9.82 kg/m

| Bending Stress | | |
|---|---|---|
| *as per IS: 800 2007- 8.2.2* | | |
| Section | 122 61 4.5 | |
| $r_y$ | 25.39 | mm |
| $Z_c$ | 49737 | mm3 |
| $L_{LT}$ | 1000 | mm |
| $\alpha Lt$ | 0.49 | |
| $g_{mo}$ | 1.1 | |
| $f_y$ | 310 | Mpa |
| E | 200000 | Mpa |
| $f_y/g_{mo}$ | 281.82 | |
| $I_y$ | 983582 | mm4 |
| $I_x$ | 2978007 | mm4 |
| $h_f$ | 91.2 | mm |
| $t_f$ | 4.8 | mm |
| $M_{cr}$ | 97480908 | Nmm |
| $l_T =$ | 0.44 | |
| $f_{LT} =$ | 0.65 | |
| $c_{LT} =$ | 0.88 | |
| fbd = | 248 | Mpa |
| fbt = | 205 | Mpa |
| *as per IS : 800 2007 - 11.4.1* | | |
| fabc = | 149 | Mpa |
| fabt = | 205 | Mpa |

| Working stress method (IS 800 2007 - 11.4.1) | |
|---|---|
| Design stress:  fbca | 114 Mpa |
| Allow. bending comp stress, fabc | 149 Mpa |
| Allow. bending tensile stress, fabt | 205 Mpa |
| | OK. |
| d max = | 29.88 mm |
| d allowable (L/150)= | 36.67 mm |
| | OK. |

## 12.2.8.2 Truss: Top chord with Hollow Rectangular Section (TATA Section)

The table below indicates the summary of load to be carried by the top chord of truss as per frame analysis computer output data.

| R10 | Beam | L/C | Node | Fx kN | Fy kN | Fz kN | Mx kNm | My kNm | Mz kNm |
|---|---|---|---|---|---|---|---|---|---|
| Max Fx | 739 | 6 DL+LLR | 391 | 102.44 | 0.737 | 0.036 | 0.018 | -0.029 | -0.121 |
| Min Fx | 693 | 8 DL + WL R TO L | 350 | -103.41 | 2.844 | -0.015 | -0.009 | -0.016 | 0.312 |
| Max Fy | 790 | 7 DL+ WL LTO R | 416 | 49.324 | 10.679 | 0.121 | 0.007 | 0.067 | -2.946 |
| Min Fy | 575 | 6 DL+LLR | 307 | -39.264 | -6.758 | -0.105 | -0.034 | 0.057 | 2.232 |
| Max Fz | 695 | 9 DL+0.5 WL R TO L+ CRLR | 362 | -13.9 | -1.142 | 0.676 | -0.077 | -0.174 | -0.409 |
| Min Fz | 682 | 9 DL + 0.5 WL R TO L+ CRLR | 359 | -5.422 | -0.509 | -0.995 | 0.115 | 0.219 | -0.228 |
| Max Mx | 939 | 6 DL+LLR | 495 | 1.717 | -0.283 | 0.09 | 0.164 | -0.052 | -0.116 |
| Min Mx | 959 | 6 DL+LLR | 506 | 1.783 | 0.448 | -0.087 | -0.164 | 0.052 | 0.312 |
| Max My | 777 | 8 DL + WL R TO L | 410 | 8.469 | -0.808 | -0.514 | -0.043 | 0.564 | 0.135 |
| Min My | 737 | 8 DL + WL R TO L | 390 | -83.535 | 4.568 | -0.312 | -0.03 | -0.48 | -0.286 |
| Max Mz | 693 | 6 DL+LLR | 362 | 80.659 | -3.021 | -0.024 | 0.003 | -0.036 | 2.341 |
| Min Mz | 790 | 7 DL+ WL LTO R | 416 | 49.324 | 10.679 | 0.121 | 0.007 | 0.067 | -2.946 |

Member Mkd : R10    Rafter

Span of Truss = 14 m        $L_x$ = 1.15 m      $L_y$ = 3.363 m
spcg = 5.5 m                 $K_x$ = 0.7 m       $K_y$ = 0.7 m
                             $KL_x$ = 0.805 m    $KL_y$ = 2.354 m

*Load*
Axial comp, P = **103.4** kN
Axial Tension, T = **102.4** kN
Moment, Mz = **2.9** kNm

Try   Hollow   100 100 6
               Tube YST310

| wt/m | 17.23 | kg/m | buckling class | b |
|---|---|---|---|---|
| rx | 39.10 | mm | α | 0.34 |
| ry | 39.10 | mm | γmo | 1.1 |
| A | 2200 | mm2 | Zx | 68781 mm3 |

*Slenderness ratio*

$KL_x/r_x$ = 805 /39.1 =   21
$KL_y/r_y$ = 2354.1 /39.1 =   60        Maxm KL/r =   60  < 250   ( IS 800 2007 - Table 3)

# Design of Industrial Structures

| *Allowable Comp Strength* | |
|---|---|
| IS 800 2007 - 7.1.2.1 | |
| Section | 100 100 6 |
| r min | 39.10 mm |
| KL/r | 60 |
| Ly | 3363 mm |
| α | 0.34 |
| $\gamma_{mo}$ | 1.1 |
| fy | 310 Mpa |
| E | 200000 Mpa |
| fy/$\gamma_{mo}$ | 281.82 |
| λ = | 0.7549 |
| φ = | 0.8793 |
| fcd = | 212 Mpa |
| IS 800 2007 - 11.3.1; 11.2.1 | |
| fac = | 127 Mpa |
| fat = | 186 Mpa |

| *Allowable Bending strength* | | |
|---|---|---|
| IS 800 2007 - 8.2.2 | | |
| Section | 100 100 6 | |
| ry | 39.10 | mm |
| Zx | 68781 | mm3 |
| $L_{LT}$ | 3363 | mm |
| αLt | 0.34 | |
| $\gamma_{mo}$ | 1.10 | |
| fy | 310 | Mpa |
| E | 200000 | Mpa |
| fy/$g_{mo}$ | 281.82 | |
| $I_y$ | 3335872 | mm4 |
| $h_f$ | 94 | mm |
| $t_f$ | 6 | mm |
| $M_{cr}$ | 43283372 | Nmm |
| $\lambda_T =$ | 0.7689 | |
| $\phi_{LT} =$ | 0.8923 | |
| $\chi_{LT} =$ | 0.7435 | |
| fbd = | 210 | Mpa |
| IS 800 2007 - 11.4.1 | | |
| fabc = fabt = | 169 | Mpa |

*Summary of design stresses:*

|  |  | Actual stress |  | permissible stress |  |
|---|---|---|---|---|---|
| Axial Comp: | fc = | 47 Mpa | f ac = | 127 Mpa | OK. |
| Axial Tension: | ft = | 47 MPa | f at = | 186 MPa | OK. |
| Bending : | fbc = fbt = | 43 MPa | fabc = fabt = | 169 MPa | OK. |

*Combination of Bending and Axial compression :*

fbc / f abc + fc / fac =   43 / 169 + 47 / 127 =   0.25 +   0.37 =   0.62
                                                                  < 1; OK.

*Combination of Bending and Axial Tension :*

fbt / f abt + ft / fat =   43 / 169 + 47 / 186 =   0.25 +   0.25 =   0.51
                                                                  < 1; OK.

## 12.2.8.3 Truss: Bottom Chord with Hollow Rectangular Section (TATA Section)

**Member load from computer output:**

| R11 | Beam | L/C | Node | Fx kN | Fy kN | Fz kN | Mx kNm | My kNm | M kNm |
|---|---|---|---|---|---|---|---|---|---|
| Max Fx | 36 | 8 DL + WL R TO L | 25 | 97.19 | 0.422 | 0.041 | -0.034 | -0.017 | 0.453 |
| Min Fx | 592 | 8 DL + WL R TO L | 297 | -113.703 | 3.279 | 0.135 | -0.029 | 0.088 | 0.332 |
| Max Fy | 948 | 7 DL+ WL L TO R | 502 | 7.059 | 7.006 | -0.127 | -0.067 | -0.042 | 1.767 |
| Min Fy | 975 | 7 DL+ WL L TO R | 553 | 13.454 | -7.044 | 0.015 | 0.023 | -0.005 | 6.056 |
| Max Fz | 493 | 9 DL + 0.5 WL R TO L+ CRLR | 256 | -14.307 | -0.948 | 0.803 | -0.098 | -0.157 | -0.358 |
| Min Fz | 480 | 9 DL + 0.5 WL R TO L+ CRLR | 253 | -8.23 | -0.537 | -1.103 | 0.133 | 0.252 | -0.245 |
| Max Mx | 75 | 9 DL + 0.5 WL R TO L+ CRLR | 41 | -14.58 | -2.591 | -1.059 | 0.165 | 0.358 | -0.906 |
| Min Mx | 49 | 7 DL+ WL L TO R | 32 | -16.505 | 0.065 | 0.134 | -0.147 | -0.097 | -0.098 |
| Max My | 74 | 9 DL + 0.5 WL R TO L+ CRLR | 41 | -27.633 | 1.17 | 0.387 | -0.069 | 0.58 | -0.958 |
| Min My | 794 | 8 DL + WL R TO L | 415 | -110.222 | 3.075 | -0.388 | 0.023 | -0.602 | -2.639 |
| Max Mz | 965 | 7 DL+ WL L TO R | 511 | 5.478 | 6.855 | -0.155 | 0.055 | 0.253 | 6.294 |
| Min Mz | 948 | 7 DL+ WL L TO R | 501 | 7.078 | 6.812 | -0.127 | -0.067 | -0.19 | -6.333 |

# Design of Industrial Structures

**Member Mkd :** R11    **Bottom chord**

| | | | | | | | |
|---|---|---|---|---|---|---|---|
| Span of Truss = | 14 m | Lx = | 1.15 m | Ly = | 3.36 m | | |
| spcg = | 5.5 m | Kx = | 0.7 m | Ky = | 0.7 m | | |
| | | KLx = | 0.81 m | Kly = | 2.35 m | | |

*Load*

Axial comp, P = **113.7** kN
Axial Tension, T = **97.2** kN
Moment, Mz = **6.3** kNm

Try    Hollow    **100 100 6**
                 Tube YST310

| wt/m | 17.23 | kg/m | buckling class | b |
|---|---|---|---|---|
| rx | 39.10 | mm | α | 0.34 |
| ry | 39.10 | mm | γmo | 1.1 |
| A | 2200 | mm2 | Zx | 68781 mm3 |

*Slenderness ratio*

KLx/rx = 805 /39.1 =    21
KLy/ry = 2354.1 /39.1 =    60    Maxm KL/r =    60  < 250   ( IS 800 2007 - Table 3)

| *Allowable Comp Strength* | |
|---|---|
| *IS 800 2007 - 7.1.2.1* | |
| Section | 100 100 6 |
| r min | 39.10 mm |
| KL/r | 60 |
| Ly | 3363 mm |
| α | 0.34 |
| $\gamma_{mo}$ | 1.1 |
| fy | **310** Mpa |
| E | 200000 Mpa |
| $fy/\gamma_{mo}$ | 281.82 |
| λ = | 0.7549 |
| φ = | 0.8793 |
| fcd = | 212 Mpa |
| *IS 800 2007 - 11.3.1; 11.2.1* | |
| fac = | **127** Mpa |
| fat = | **186** Mpa |

| *Allowable Bending strength* | | |
|---|---|---|
| *IS 800 2007 - 8.2.2* | | |
| Section | 100 100 6 | |
| ry | 39.10 | mm |
| Zx | 68781 | mm3 |
| $L_{LT}$ | 3363 | mm |
| αLt | 0.34 | |
| $\gamma_{mo}$ | 1.10 | |
| fy | **310** | Mpa |
| E | 200000 | Mpa |
| $fy/g_{mo}$ | 281.82 | |
| $I_y$ | 3335872 | mm4 |
| $h_f$ | 94 | mm |
| $t_f$ | 6 | mm |
| $M_{cr}$ | 43283372 | Nmm |
| $\lambda_T$ = | 0.7689 | |
| $\phi_{LT}$ = | 0.8923 | |
| $\chi_{LT}$ = | 0.7435 | |
| fbd = | 210 | Mpa |
| *IS 800 2007 - 11.4.1* | | |
| fabc = fabt = | **169** | Mpa |

*Summary of design stresses:*

|  | Actual stress |  | permissible stress |  |
|---|---|---|---|---|
| Axial Comp: $f_c$ = | 52 Mpa | $f_{ac}$ = | 127 Mpa | OK. |
| Axial Tension: $f_t$ = | 44 MPa | $f_{at}$ = | 186 MPa | OK. |
| Bending: $f_{bc} = f_{bt}$ = | 92 MPa | $f_{abc} = f_{abt}$ = | 169 MPa | OK. |

*Combination of Bending and Axial compression :*

$f_{bc} / f_{abc} + f_c / f_{ac}$ =   92 / 169 + 52 / 127 =   0.54 +   0.41 =   0.95
                                                                                  < 1; OK.

*Combination of Bending and Axial Tension :*

$f_{bt} / f_{abt} + f_t / f_{at}$ =   92 / 169 + 44 / 186 =   0.54 +   0.24 =   0.78
                                                                                  < 1; OK.

Design of Industrial Structures 365

## 12.2.8.4 Truss: Diagonals with Hollow Rectangular Section (TATA Section)

**Member load from computer output:**

| R2 | Beam | L/C | Node | Fx kN | Fy kN | Fz kN | Mx kNm | My kNm | MzkNm |
|---|---|---|---|---|---|---|---|---|---|
| Max Fx | 548 | 6 DL+LLR | 287 | 63.73 | -0.105 | -0.011 | -0.002 | 0.012 | -0.093 |
| Min Fx | 570 | 7 DL+ WL LTO R | 295 | -62.23 | -0.077 | 0.02 | -0.009 | -0.019 | -0.058 |
| Max Fy | 570 | 6 DL+LLR | 295 | 61.48 | 0.171 | -0.007 | 0.004 | 0.008 | 0.092 |
| Min Fy | 952 | 7 DL+ WL LTO R | 495 | 16.51 | -0.196 | 0.019 | -0.016 | 0.046 | 0.157 |
| Max Fz | 88 | 9 DL + 0.5 WL R TO L+ CRLR | 32 | 17.62 | 0.003 | 0.05 | -0.012 | -0.065 | -0.065 |
| Min Fz | 76 | 9 DL + 0.5 WL R TO L+ CRLR | 22 | 21.65 | 0.013 | -0.066 | 0.018 | 0.082 | -0.058 |
| Max Mx | 952 | 6 DL+LLR | 499 | 1.57 | 0.044 | -0.017 | 0.021 | -0.037 | -0.004 |
| Min Mx | 784 | 8 DL + WL R TO L | 393 | 40.54 | 0.018 | -0.041 | -0.025 | -0.003 | -0.068 |
| Max My | 76 | 9 DL + 0.5 WL R TO L+ CRLR | 22 | 21.65 | 0.013 | -0.066 | 0.018 | 0.082 | -0.058 |
| Min My | 986 | 6 DL+LLR | 499 | -5.53 | 0.026 | 0.047 | 0.004 | -0.069 | 0.01 |
| Max Mz | 582 | 6 DL+LLR | 287 | -52.56 | 0.088 | -0.012 | -0.007 | 0.004 | 0.187 |
| Min Mz | 795 | 7 DL+ WL LTO R | 403 | 53.91 | -0.056 | 0.017 | -0.002 | -0.033 | -0.205 |

Member Mkd : **R2**   End diagonals

Span of Truss = **14** m   $L_x$ = **1.35** m   $L_y$ = **1.35** m
spcg = **5.5** m   $K_x$ = **0.7** m   $K_y$ = **0.7** m
   $KL_x$ = **0.94** m   $KL_y$ = **0.94** m

*Load*
Axial comp, P = **62.2** kN
Axial Tension, T = **63.7** kN
Moment, Mz = **0.2** kNm

Try   Hollow   **72 72 3.2**
         Tube YST310

| wt/m | 6.79 | kg/m | buckling class | b |
|---|---|---|---|---|
| rx | 28.50 | mm | α | 0.34 |
| ry | 28.50 | mm | γmo | 1.1 |
| A | 900 | mm2 | Zx | 19780 mm3 |

*Slenderness ratio*

$KL_x/r_x$ = 940 /28.5 =   33
$KL_y/r_y$ = 940 /28.5 =   33   Maxm KL/r =   33 < 250   ( IS 800 2007 - Table 3)

*Allowable Comp Strength*

| IS 800 2007 - 7.1.2.1 | | |
|---|---|---|
| Section | 72 72 3.2 | |
| r min | 28.50 mm | |
| KL/r | 33 | |
| Ly | 1346 mm | |
| α | 0.34 | |
| $\gamma_{mo}$ | 1.1 | |
| fy | 310 | Mpa |
| E | 200000 | Mpa |
| fy/$\gamma_{mo}$ | 282 | |
| λ = | 0.41 | |
| φ = | 0.62 | |
| fcd = | 259 | Mpa |
| IS 800 2007 - 11.3.1; 11.2.1 | | |
| fac = | 156 | Mpa |
| fat = | 186 | Mpa |

*Allowable Bending strength*

| IS 800 2007 - 8.2.2 | | |
|---|---|---|
| Section | 72 72 3.2 | |
| ry | 28.50 | mm |
| Zx | 19780 | mm3 |
| $L_{LT}$ | 1346 | mm |
| αLt | 0.34 | |
| $\gamma_{mo}$ | 1.10 | |
| fy | 310 | Mpa |
| E | 200000 | Mpa |
| fy/$g_{mo}$ | 282 | |
| $I_y$ | 696246 | mm4 |
| $h_f$ | 69 | mm |
| $t_f$ | 3 | mm |
| $M_{cr}$ | 29032384 | Nmm |
| $\lambda_T$ = | 0.50 | |
| $\phi_{LT}$ = | 0.68 | |
| $\chi_{LT}$ = | 0.88 | |
| fbd = | 249 | Mpa |
| IS 800 2007 - 11.4.1 | | |
| fabc = fabt = | 169 | Mpa |

*Summary of design stresses:*

|  |  | Actual stress |  | permissible stress |  |
|---|---|---|---|---|---|
| Axial Comp: | fc = | 69 Mpa | f ac = | 156 Mpa | OK. |
| Axial Tension: | ft = | 71 MPa | f at = | 186 MPa | OK. |
| Bending : | fbc = fbt = | 10 MPa | fabc = fabt = | 169 MPa | OK. |

*Combination of Bending and Axial compression :*

fbc / f abc  +  fc / fac  =    10 / 169 + 69 / 156  =    0.06  +    0.44  =    0.50
                                                                                                                                                < 1; OK.

*Combination of Bending and Axial Tension :*

fbt / f abt  +  ft / fat  =    10 / 169 + 71 / 186  =    0.06  +    0.38  =    0.44
                                                                                                                                                < 1; OK.

## 12.2.8.5 Truss: Vericals with Hollow Rectangular Section (TATA Section)

**Member load from computer output:**

| R1 | Beam | L/C | Node | Fx kN | Fy kN | Fz kN | Mx kNm | My kNm | MzkNm |
|---|---|---|---|---|---|---|---|---|---|
| Max Fx | 1145 | 6 DL+LLR | 553 | 10.749 | -0.215 | 0 | 0 | 0 | 0.115 |
| Min Fx | 1145 | 8 DL + WL R TO L | 493 | -24.12 | 0.207 | 0.007 | -0.001 | -0.018 | 0.074 |
| Max Fy | 966 | 7 DL+ WL LTO R | 506 | 2.123 | 2.142 | -0.136 | 0.011 | 0.035 | 0.513 |
| Min Fy | 947 | 7 DL+ WL LTO R | 501 | -2.823 | -2.089 | 0.104 | 0.014 | 0.049 | 0.957 |
| Max Fz | 1145 | 7 DL+ WL LTO R | 493 | -2.461 | 0.211 | 1.441 | 0.009 | -0.401 | 0.061 |
| Min Fz | 969 | 7 DL+ WL LTO R | 504 | 0.095 | -0.381 | -0.147 | -0.011 | 0.037 | -0.093 |
| Max Mx | 947 | 7 DL+ WL LTO R | 496 | -2.863 | -2.085 | 0.104 | 0.014 | -0.023 | -0.496 |
| Min Mx | 945 | 6 DL+LLR | 495 | 0.128 | 0.068 | -0.054 | -0.014 | 0.01 | 0.024 |
| Max My | 1145 | 7 DL+ WL LTO R | 553 | -2.415 | 0.211 | 1.441 | 0.009 | 0.766 | -0.11 |
| Min My | 1145 | 7 DL+ WL LTO R | 493 | -2.461 | 0.211 | 1.441 | 0.009 | -0.401 | 0.061 |
| Max Mz | 947 | 7 DL+ WL LTO R | 501 | -2.823 | -2.089 | 0.104 | 0.014 | 0.049 | 0.957 |
| Min Mz | 966 | 7 DL+ WL LTO R | 511 | 2.163 | 2.138 | -0.136 | 0.011 | -0.059 | -0.978 |

Member Mkd : R1    Verticals

| | | | |
|---|---|---|---|
| Span of Truss = | 14 m | Lx 0.7 m | Ly 0.7 m |
| spcg = | 5.5 m | Kx 0.7 m | Ky 0.7 m |
| | | KLx 0.49 m | KLy 0.49 m |

*Load*

Axial comp, P = **24.1** kN
Axial Tension, T = **10.7** kN
Moment, Mz = **1.0** kNm

Try Hollow **49 49 4.5** Tube YST310

| wt/m | 6.00 | kg/m | buckling class | b | |
|---|---|---|---|---|---|
| rx | 18.70 | mm | $\alpha$ | | 0.34 |
| ry | 18.70 | mm | $\gamma_{mo}$ | | 1.1 |
| A | 761 | mm2 | Zx | 11425 | mm3 |

*Slenderness ratio*

KLx/rx = 490 /18.7 = 26

KLy/ry = 490 /18.7 = 26    Maxm KL/r = 26 < 250  ( IS 800 2007 - Table 3)

*Allowable Comp Strength*

| IS 800 2007 - 7.1.2.1 | |
|---|---|
| Section | 49 49 4.5 |
| r min | 18.70 mm |
| KL/r | 26 |
| Ly | 700 mm |
| α | 0.34 |
| $\gamma_{mo}$ | 1.1 |
| fy | 310 Mpa |
| E | 200000 Mpa |
| fy/$\gamma_{mo}$ | 282 |
| λ = | 0.33 |
| φ = | 0.58 |
| fcd = | 269 Mpa |
| IS 800 2007 - 11.3.1; 11.2.1 | |
| fac = | 161 Mpa |
| fat = | 186 Mpa |

*Allowable Bending strength*

| IS 800 2007 - 8.2.2 | | |
|---|---|---|
| Section | 49 49 4.5 | |
| ry | 18.70 | mm |
| Zx | 11425 | mm3 |
| $L_{LT}$ | 700 | mm |
| αLt | 0.34 | |
| $\gamma_{mo}$ | 1.10 | |
| fy | 310 | Mpa |
| E | 200000 | Mpa |
| fy/$g_{mo}$ | 282 | |
| $I_y$ | 267067 | mm4 |
| $h_f$ | 45 | mm |
| $t_f$ | 5 | mm |
| $M_{cr}$ | 31329902 | Nmm |
| $\lambda_T$ = | 0.37 | |
| $\phi_{LT}$ = | 0.60 | |
| $\chi_{LT}$ = | 0.94 | |
| fbd = | 264 | Mpa |
| IS 800 2007 - 11.4.1 | | |
| fabc = fabt = | 169 | Mpa |

*Summary of design stresses:*

|  | | Actual stress | | permissible stress | |
|---|---|---|---|---|---|
| Axial Comp: | fc = | 32 Mpa | f ac = | 161 Mpa | OK. |
| Axial Tension: | ft = | 14 MPa | f at = | 186 MPa | OK. |
| Bending : | fbc = fbt = | 86 MPa | fabc = fabt = | 169 MPa | OK. |

*Combination of Bending and Axial compression :*

fbc / f abc  +  fc / fac =     86 / 169 + 32 / 161 =     0.51  +     0.20  =     0.71
                                                                              < 1; OK.

*Combination of Bending and Axial Tension :*

fbt / f abt  +  ft / fat =     86 / 169 + 14 / 186 =     0.51  +     0.08  =     0.58
                                                                              < 1; OK.

Design of Industrial Structures

## 12.2.8.6 Column: Twin Legs with Hollow Rectangular Section (TATA Section)

**Member load from computer output:**

| R12 | Beam | L/C | Node | Fx kN | Fy kN | Fz kN | Mx kNm | My kNm | Mz kNm |
|---|---|---|---|---|---|---|---|---|---|
| Max Fx | 977 | 7 DL+ WL LTO R | 514 | 117.08 | 0.023 | -2.84 | 0.018 | 3.489 | 0.023 |
| Min Fx | 889 | 7 DL+ WL LTO R | 1006 | -156.45 | 0.002 | -5.305 | 0.031 | -0.638 | 0.009 |
| Max Fy | 498 | 9 DL+0.5 WL R TO L+ CRLR | 263 | 52.191 | 1.489 | -1.382 | 0.222 | -0.047 | 0.433 |
| Min Fy | 95 | 9 DL+0.5 WL R TO L+ CRLR | 52 | -53.651 | -1.492 | -0.929 | -0.398 | 1.465 | -1.091 |
| Max Fz | 789 | 7 DL+ WL LTO R | 419 | 54.585 | -0.192 | 9.596 | 0.038 | 1.935 | 0.004 |
| Min Fz | 170 | 8 DL + WL R TO L | 91 | 23.112 | 0.094 | -6.637 | -0.04 | 2.287 | 0.022 |
| Max Mx | 819 | 8 DL + WL R TO L | 431 | -99.126 | -0.146 | -6.245 | 0.296 | 1.767 | 0.011 |
| Min Mx | 95 | 9 DL+0.5 WL R TO L+ CRLR | 52 | -53.651 | -1.492 | -0.929 | -0.398 | 1.465 | -1.091 |
| Max My | 889 | 7 DL+ WL LTO R | 470 | -156.2 | 0.002 | -5.305 | 0.031 | 5.993 | 0.011 |
| Min My | 617 | 8 DL + WL R TO L | 309 | -99.232 | 0.162 | -6.603 | -0.105 | -4.661 | -0.121 |
| Max Mz | 603 | 9 DL+0.5 WL R TO L+ CRLR | 319 | -11.803 | 0.559 | -2.233 | 0.289 | 0.205 | 0.516 |
| Min Mz | 95 | 9 DL+0.5 WL R TO L+ CRLR | 52 | -53.651 | -1.492 | -0.929 | -0.398 | 1.465 | -1.091 |

**Member Mkd :**   R12     Column leg (each)

Height of col = 5.86 m    Lx = 0.875 m    Ly = 3 m
spcg = 5.5 m    Kx = 0.7 m    Ky = 0.7 m
Laced at 0.875 m    KLx = 0.6125 m    Kly = 2.1 m
Longitunal Tie at 3 m height

*Load*

Axial comp, P = **156** kN
Axial Tension, T = **117** kN
Moment, Mz = **6** kNm

Try   Hollow   130 130 5.4
Tube YST310

| wt/m | 20.77 | kg/m | buckling class | b |
|---|---|---|---|---|
| rx | 51.50 | mm | α | 0.34 |
| ry | 51.50 | mm | γmo | 1.1 |
| A | 2633 | mm2 | Zx | 109616 mm3 |

*Slenderness ratio*

KLx/rx = 612.5 /51.5 =     12
KLy/ry = 2100 /51.5 =      41       Maxm KL/r =    41  < 250   ( IS 800 2007 - Table 3)

*Allowable Comp Strength*

| IS 800 2007 - 7.1.2.1 | |
|---|---|
| Section | 130 130 5.4 |
| r min | 51.50 mm |
| KL/r | 41 |
| Ly | 3000 mm |
| α | 0.34 |
| $\gamma_{mo}$ | 1.1 |
| fy | 310 Mpa |
| E | 200000 Mpa |
| fy/$\gamma_{mo}$ | 282 |
| λ = | 0.51 |
| φ = | 0.68 |
| fcd = | 248 Mpa |
| *IS 800 2007 - 11.3.1; 11.2.1* | |
| fac = | **149** Mpa |
| fat = | **186** Mpa |

*Allowable Bending strength*

| IS 800 2007 - 8.2.2 | | |
|---|---|---|
| Section | 130 130 5.4 | |
| ry | 51.50 | mm |
| Zx | 109616 | mm3 |
| $L_{LT}$ | 3000 | mm |
| αLt | 0.34 | |
| $\gamma_{mo}$ | 1.10 | |
| fy | 310 | Mpa |
| E | 200000 | Mpa |
| fy/$g_{mo}$ | 282 | |
| $I_y$ | 6977046 | mm4 |
| $h_f$ | 125 | mm |
| $t_f$ | 5 | mm |
| $M_{cr}$ | 109364273 | Nmm |
| $\lambda_T$ = | 0.61 | |
| $\phi_{LT}$ = | 0.76 | |
| $\chi_{LT}$ = | 0.83 | |
| fbd = | 234 | Mpa |
| *IS 800 2007 - 11.4.1* | | |
| fabc = fabt = | **169** | Mpa |

*Summary of design stresses:*

|  | Actual stress | | permissible stress | |
|---|---|---|---|---|
| Axial Comp: | fc = | 59 Mpa | f ac = 149 Mpa | OK. |
| Axial Tension: | ft = | 44 MPa | f at = 186 MPa | OK. |
| Bending : | fbc = fbt = | 55 MPa | fabc = fabt = 169 MPa | OK. |

*Combination of Bending and Axial compression :*

fbc / f abc + fc / fac =   55 / 169 + 59 / 149 =   0.33 + 0.40 =   0.72
                                                                 < 1; OK.

*Combination of Bending and Axial Tension :*

fbt / f abt + ft / fat =   55 / 169 + 44 / 186 =   0.33 + 0.24 =   0.56
                                                                 < 1; OK.

Design of Industrial Structures

## 12.2.8.7 Column Bracing: Single Leg Hollow Rectangular Section (TATA Section)

**Member load from computer output:**

| R16 | Beam | L/C | Node | Fx kN | Fy kN | Fz kN | Mx kNm | My kNm | MzkNm |
|---|---|---|---|---|---|---|---|---|---|
| Max Fx | 1056 | 9 DL + 0.5 WL R TO L+ CRLR | 302 | 23.772 | -0.46 | 0.069 | 0.024 | 0.213 | 0.508 |
| Min Fx | 1055 | 9 DL + 0.5 WL R TO L+ CRLR | 319 | -14.711 | 0.463 | 0.001 | 0.029 | -0.023 | 0.538 |
| Max Fy | 2073 | 6 DL+LLR | 250 | -3.505 | 0.479 | 0.005 | -0.1 | -0.089 | 0.492 |
| Min Fy | 2094 | 6 DL+LLR | 312 | -3.002 | -0.48 | 0.004 | -0.1 | 0.033 | 0.501 |
| Max Fz | 1078 | 7 DL+ WL LTO R | 274 | -1.743 | 0.464 | 0.108 | 0.031 | -0.314 | 0.537 |
| Min Fz | 1077 | 7 DL+ WL LTO R | 327 | -11.157 | 0.467 | -0.079 | -0.025 | 0.197 | 0.549 |
| Max Mx | 2094 | 7 DL+ WL LTO R | 274 | 2.505 | 0.452 | 0.006 | 0.148 | -0.027 | 0.415 |
| Min Mx | 2093 | 7 DL+ WL LTO R | 327 | 8.986 | 0.459 | 0.048 | -0.147 | -0.139 | 0.437 |
| Max My | 1078 | 7 DL+ WL LTO R | 311 | -0.997 | -0.448 | 0.108 | 0.031 | 0.452 | 0.479 |
| Min My | 1077 | 7 DL+ WL LTO R | 258 | -10.41 | -0.446 | -0.079 | -0.025 | -0.367 | 0.473 |
| Max Mz | 1078 | 8 DL + WL R TO L | 274 | -12.855 | 0.469 | 0.073 | 0.094 | -0.145 | 0.555 |
| Min Mz | 2072 | 9 DL + 0.5 WL R TO L+ CRLR | 266 | -8.576 | 0.433 | 0.072 | -0.006 | -0.206 | 0.346 |

**Member Mkd :** R16

Length = 4.07 m

Lx = 2.03 m  Ly = 4.07 m
Kx = 0.7 m  Ky = 0.7 m
KLx = 1.42 m  Kly = 2.85 m

*Load*
Axial comp, P = 15 kN
Axial Tension, T = 24 kN
Moment, Mz = 1 kNm

Try   Hollow  100 100 6
              Tube YST310

| wt/m | 17.23 | kg/m | buckling class | b |
|---|---|---|---|---|
| rx | 39.10 | mm | α | 0.34 |
| ry | 39.10 | mm | γmo | 1.1 |
| A | 2184 | mm2 | Zx | 68781 mm3 |

*Slenderness ratio*

KLx/rx = 1420 /39.1 = 36
KLy/ry = 2850 /39.1 = 73   Maxm KL/r = 73 < 250  ( IS 800 2007 - Table 3)

| Allowable Comp Strength | |
|---|---|
| *IS 800 2007 - 7.1.2.1* | |
| Section | 100 100 6 |
| r min | 39.10 mm |
| KL/r | 73 |
| Ly | 4070 mm |
| α | 0.34 |
| $\gamma_{mo}$ | 1.1 |
| fy | 310 Mpa |
| E | 200000 Mpa |
| fy/$\gamma_{mo}$ | 282 |
| λ = | 0.91 |
| φ = | 1.04 |
| fcd = | 184 Mpa |
| *IS 800 2007 - 11.3.1; 11.2.1* | |
| fac = | **110** Mpa |
| fat = | **186** Mpa |

| Allowable Bending strength | | |
|---|---|---|
| *IS 800 2007 - 8.2.2* | | |
| Section | 100 100 6 | |
| ry | 39.10 | mm |
| Zx | 68781 | mm3 |
| $L_{LT}$ | 4070 | mm |
| αLt | 0.34 | |
| $\gamma_{mo}$ | 1.10 | |
| fy | 310 | Mpa |
| E | 200000 | Mpa |
| fy/$g_{mo}$ | 282 | |
| $I_y$ | 3335872 | mm4 |
| $h_f$ | 94 | mm |
| $t_f$ | 6 | mm |
| $M_{cr}$ | 33428400 | Nmm |
| $\lambda_T =$ | 0.87 | |
| $\phi_{LT} =$ | 1.00 | |
| $\chi_{LT} =$ | 0.68 | |
| fbd = | 191 | Mpa |
| *IS 800 2007 - 11.4.1* | | |
| fabc = fabt = | **169** | Mpa |

*Summary of design stresses:*

|  |  | Actual stress |  | permissible stress |  |
|---|---|---|---|---|---|
| Axial Comp: | fc = | 7 Mpa | f ac = | 110 Mpa | OK. |
| Axial Tension: | ft = | 11 MPa | f at = | 186 MPa | OK. |
| Bending : | fbc = fbt = | 8 MPa | fabc = fabt = | 169 MPa | OK. |

*Combination of Bending and Axial compression :*

fbc / f abc  +  fc / fac  =     8 / 169 + 7 / 110  =     0.05  +     0.06  =     0.11
                                                                              **< 1; OK.**

*Combination of Bending and Axial Tension :*

fbt / f abt  +  ft / fat  =     8 / 169 + 11 / 186  =    0.05  +     0.06  =     0.11
                                                                              **< 1; OK.**

# Design of Industrial Structures

## 12.2.8.8 Rafter Bracing: Single Leg Hollow Rectangular Section (TATA Section)

**Member load from computer output:**

| R16 | R17 | Beam | L/C | Node | Fx kN | Fy kN | Fz kN | Mx kNm | My kNm | MzkNm |
|---|---|---|---|---|---|---|---|---|---|---|
| Max Fx | Max Fx | 631 | 6 DL+LLR | 330 | 134.42 | 2.695 | 0.012 | -0.001 | -0.008 | 2.328 |
| Min Fx | Min Fx | 1125 | 8 DL + WL R TO L | 546 | -35.51 | 0.306 | -0.016 | -0.163 | 0.038 | 0.306 |
| Max Fy | Max Fy | 631 | 6 DL+LLR | 330 | 134.42 | 2.695 | 0.012 | -0.001 | -0.008 | 2.328 |
| Min Fy | Min Fy | 631 | 7 DL+ WL LTO R | 330 | 39.04 | -5.633 | 0.009 | 0 | -0.006 | -5.262 |
| Max Fz | Max Fz | 1138 | 9 DL + 0.5 WL R TO L+ CRLR | 550 | 3.06 | 0.338 | 0.052 | 0.011 | -0.074 | 0.118 |
| Min Fz | Min Fz | 1312 | 9 DL + 0.5 WL R TO L+ CRLR | 48 | -1.94 | 0.4 | -0.043 | 0.054 | 0.132 | 0.287 |
| Max Mx | Max Mx | 1124 | 8 DL + WL R TO L | 19 | -23.60 | 0.456 | 0.052 | 0.155 | -0.135 | 0.76 |
| Min Mx | Min Mx | 1125 | 8 DL + WL R TO L | 546 | -35.51 | 0.306 | -0.016 | -0.163 | 0.038 | 0.306 |
| Max My | Max My | 1138 | 9 DL + 0.5 WL R TO L+ CRLR | 516 | 3.12 | -0.44 | 0.052 | 0.011 | 0.169 | 0.356 |
| Min My | Min My | 1126 | 9 DL + 0.5 WL R TO L+ CRLR | 39 | 1.64 | 0.405 | 0.05 | -0.05 | -0.163 | 0.314 |
| Max Mz | Max Mz | 631 | 7 DL+ WL LTO R | 331 | 38.79 | -5.633 | 0.009 | 0 | 0.008 | 3.188 |
| Min Mz | Min Mz | 631 | 7 DL+ WL LTO R | 330 | 39.04 | -5.633 | 0.009 | 0 | -0.006 | -5.262 |

Member Mkd : R17

Length = 4.48 m

| | | | |
|---|---|---|---|
| Lx | 2.24 m | Ly | 4.48 m |
| Kx | 0.7 m | Ky | 0.7 m |
| KLx | 1.57 m | KLy | 3.14 m |

### Load
Axial comp, P = **36** kN  
Axial Tension, T = **134** kN  
Moment, Mz = **5** kNm

Try Hollow **100 100 6** Tube YST310

| wt/m | 17.23 | kg/m | buckling class | b | |
|---|---|---|---|---|---|
| rx | 39.10 | mm | α | | 0.34 |
| ry | 39.10 | mm | γmo | | 1.1 |
| A | 2184 | mm2 | Zx | 68781 | mm3 |

### Slenderness ratio

KLx/rx = 1570 /39.1 = 40  
KLy/ry = 3140 /39.1 = 80    Maxm KL/r = 80 < 250  ( IS 800 2007 - Table 3)

| Allowable Comp Strength | |
|---|---|
| IS 800 2007 - 7.1.2.1 | |
| Section | 100 100 6 |
| r min | 39.10 mm |
| KL/r | 80 |
| Ly | 4480 mm |
| α | 0.34 |
| $\gamma_{mo}$ | 1.1 |
| fy | 310 Mpa |
| E | 200000 Mpa |
| fy/$\gamma_{mo}$ | 282 |
| λ = | 1.01 |
| φ = | 1.14 |
| fcd = | 167 Mpa |
| IS 800 2007 - 11.3.1; 11.2.1 | |
| fac = | 100 Mpa |
| fat = | 186 Mpa |

| Allowable Bending strength | | |
|---|---|---|
| IS 800 2007 - 8.2.2 | | |
| Section | 100 100 6 | |
| ry | 39.10 | mm |
| Zx | 68781 | mm3 |
| $L_{LT}$ | 4480 | mm |
| αLt | 0.34 | |
| $\gamma_{mo}$ | 1.10 | |
| fy | 310 | Mpa |
| E | 200000 | Mpa |
| fy/$g_{mo}$ | 282 | |
| $I_y$ | 3335872 | mm4 |
| $h_f$ | 94 | mm |
| $t_f$ | 6 | mm |
| $M_{cr}$ | 29525494 | Nmm |
| $\lambda_T$ = | 0.93 | |
| $\phi_{LT}$ = | 1.06 | |
| $\chi_{LT}$ = | 0.64 | |
| fbd = | 181 | Mpa |
| IS 800 2007 - 11.4.1 | | |
| fabc = fabt = | 169 | Mpa |

*Summary of design stresses:*

|  | | Actual stress | | permissible stress | |
|---|---|---|---|---|---|
| Axial Comp: | fc = | 16 Mpa | f ac = | 100 Mpa | OK. |
| Axial Tension: | ft = | 62 MPa | f at = | 186 MPa | OK. |
| Bending : | fbc = fbt = | 77 MPa | fabc = fabt = | 169 MPa | OK. |

*Combination of Bending and Axial compression :*

fbc / f abc + fc / fac =   77 / 169 + 16 / 100 =   0.46 +   0.16 =   0.62
                                                                  < 1; OK.

*Combination of Bending and Axial Tension :*

fbt / f abt + ft / fat =   77 / 169 + 62 / 186 =   0.46 +   0.33 =   0.79
                                                                  < 1; OK.

# Design of Industrial Structures

## 12.2.8.9 Design of RCC Grade Beam

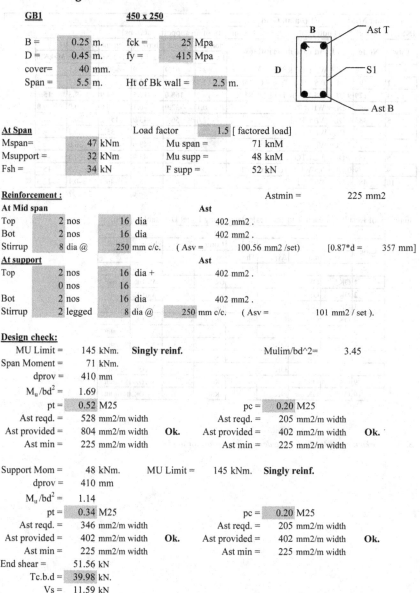

| GB1 | | 450 x 250 | | |
|---|---|---|---|---|
| B = | 0.25 m. | fck = | 25 | Mpa |
| D = | 0.45 m. | fy = | 415 | Mpa |
| cover= | 40 mm. | | | |
| Span = | 5.5 m. | Ht of Bk wall = | 2.5 m. | |

**At Span**        Load factor    1.5 [ factored load]
Mspan=    47 kNm      Mu span =    71 knM
Msupport =   32 kNm      Mu supp =    48 knM
Fsh =        34 kN        F supp =       52 kN

**Reinforcement :**                          Astmin =    225 mm2
**At Mid span**                 Ast
Top      2 nos    16 dia       402 mm2 .
Bot      2 nos    16 dia       402 mm2 .
Stirrup    8 dia @   250 mm c/c.  ( Asv =   100.56 mm2 /set)    [0.87*d =   357 mm]
**At support**                 Ast
Top      2 nos    16 dia +     402 mm2 .
       0 nos    16
Bot      2 nos    16 dia       402 mm2 .
Stirrup    2 legged   8 dia @   250 mm c/c.  ( Asv =   101 mm2 / set ).

**Design check:**
    MU Limit =    145 kNm.   **Singly reinf.**      Mulim/bd^2=    3.45
Span Moment =    71 kNm.
      dprov =    410 mm
    $M_u /bd^2$ =   1.69
        pt =    0.52 M25                   pc =    0.20 M25
   Ast reqd. =    528 mm2/m width        Ast reqd. =    205 mm2/m width
Ast provided =    804 mm2/m width   **Ok.**    Ast provided =    402 mm2/m width   **Ok.**
   Ast min =      225 mm2/m width          Ast min =    225 mm2/m width

Support Mom =    48 kNm.    MU Limit =    145 kNm.   **Singly reinf.**
     dprov =    410 mm
   $M_u /bd^2$ =   1.14
      pt =    0.34 M25                   pc =    0.20 M25
  Ast reqd. =    346 mm2/m width        Ast reqd. =    205 mm2/m width
Ast provided =    402 mm2/m width   **Ok.**    Ast provided =    402 mm2/m width   **Ok.**
   Ast min =      225 mm2/m width          Ast min =    225 mm2/m width
End shear =    51.56 kN
   Tc.b.d =    39.98 kN.
      Vs =    11.59 kN
    Vus =    59.54 kN     > 51.56 **Ok.**

## 12.2.8.10  Design of RCC Footing

Foundation Mkd :    F1 (on all Grid )
INPUT from STAAD output analysis result (summary of all support reactions)

| Case | Node | Load case descriptions | Horizont Fx kN | Vertical Fy kN | Horizont Fz kN | Moment Mx kNm | G My kNm | H MzkNm |
|---|---|---|---|---|---|---|---|---|
| 1 | 7 | 7 DL+ WL LTO R | 12.5 | 61.9 | -5.9 | -39.9 | -0.1 | 11.6 |
| 2 | 491 | 9 DL + 0.5 WL R TO L+ CRLR | -12.7 | 102.9 | -3.9 | -13.8 | 0.1 | -11.2 |
| 3 | 279 | **9 DL + 0.5 WL R TO L+ CRLR** | **-10.7** | **168.0** | **-12.5** | **-54.9** | **-7.3** | **15.2** |
| 4 | 551 | 8 DL + WL R TO L | 0.1 | -32.9 | -0.1 | 0.0 | 0.0 | 0.0 |
| 5 | 383 | 7 DL+ WL LTO R | 2.6 | 56.4 | -43.9 | -155.2 | -20.3 | -4.0 |
| 6 | 383 | 7 DL+ WL LTO R | 2.6 | 56.4 | -43.9 | -155.2 | -20.3 | -4.0 |
| 7 | 277 | 7 DL+ WL LTO R | -0.8 | 67.4 | -43.0 | -149.2 | 20.0 | 1.3 |
| 8 | 383 | 7 DL+ WL LTO R | 2.6 | 56.4 | -43.9 | -155.2 | -20.3 | -4.0 |
| 9 | 279 | 9 DL + 0.5 WL R TO L+ CRLR | -10.7 | 168.0 | -12.5 | -54.9 | -7.3 | 15.2 |
| 10 | 489 | 8 DL + WL R TO L | -12.0 | 46.3 | -8.9 | -46.4 | -0.1 | -12.3 |

*1        2        3        4        5        6        7        8        9        10        11        12*

Summary of Results for each load case (calculated using Excel Data table option)

| OUTPUT Case | SBP max | SBP min | SBP Avg | Slab Th | Long bars | Transv bars | shear | Uplift |
|---|---|---|---|---|---|---|---|---|
| 1 | OK | OK | OK | OK. | OK. | OK. | > Vs; OK. | No uplift |
| 2 | OK | OK | OK | OK. | OK. | OK. | > Vs; OK. | No uplift |
| 3 | OK | OK | OK | OK. | OK. | OK. | > Vs; OK. | No uplift |
| 4 | OK | OK | OK | OK. | OK. | OK. | > Vs; OK. | No uplift |
| 5 | OK | OK | OK | OK. | OK. | OK. | > Vs; OK. | No uplift |
| 6 | OK | OK | OK | OK. | OK. | OK. | > Vs; OK. | No uplift |
| 7 | OK | OK | OK | OK. | OK. | OK. | > Vs; OK. | No uplift |
| 8 | OK | OK | OK | OK. | OK. | OK. | > Vs; OK. | No uplift |
| 9 | OK | OK | OK | OK. | OK. | OK. | > Vs; OK. | No uplift |
| 10 | OK | OK | OK | OK. | OK. | OK. | > Vs; OK. | No uplift |

*Sample workout example of a footing design given below for load case 3*

# Design of Industrial Structures

### Design parameters

| | | |
|---|---|---|
| **SBP** = | 120 | kN/m2 |
| γconc = | 24 | kn/m2 |
| gsoil = | 18 | kn/m2 |
| γsur = | 5 | kn/m2 |
| **FGL** = | 0 | m |
| **BOF** = | -2.5 | m |
| **EGL** = | -0.6 | m |

Col dimension:
l = 1.2 m
b = 0.45 m

Footing dimension:
L = 3.5 m
B = 2.2 m
Th = 0.33 m avg
cover = 75 mm

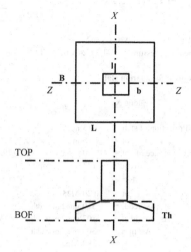

### Reinforcement :
At bottom layer
Along X - Long Bars    12 dia @    160 mm c/c.    ( Ast    707 mm2 /m)
Along Z - Transv Bars    12 dia @    160 mm c/c.    ( Ast    707 mm2 /m)

### Column load at bot of foundation:

| Case | Descrip | Fx kN | Fy kN | Fz kN | Mx kNm | My kNm | Mz kNm | h (m) | d Staad |
|---|---|---|---|---|---|---|---|---|---|
| 3 | 9 DL + 0.5 WL R TO L+ CRLR | -10.7 | 168 | -12.5 | -55 | -7.3 | 15.2 | 1 | -1.5 |

S/W of footing + Earth =    381 kN
Total vertical load at base of foundation, P =    549 kN

d Staad =    Bot of foundation assumed in analysis model is at EL    -1.5 m
h = d Staad - BOF =    1 m
Ecc. Moment from brick wall =    2.5 m ht x5 x 5.5m long x 0.575 m ecc =    39.53

### Soil bearing pressure at bottom of foundation:

Area = 3.5 x 2.2 =    7.7 m2
Zz = 3.5 x 2.2^2 =    2.82 m3
Zx = 2.2 x 3.5^2 =    4.49 m3

P/A =    549 /7.7 =                           71.30 kN/m2
Mz/Zz =    (15.2 (+) -10.7 x 1 + 39.53) / 2.82 =    15.61 kN/m2
Mx/Zx =    (-54.893 (+) -12.456 x 1) / 4.49 =    -15.00 kN/m2

Maxm GBP =   71.3 + 15.61 (+) -15 =           71.91 kN/m2   < SBP; OK
Min GBP =    71.3 - 15.61( -) -15 =           70.69 kN/m2   < SBP;OK    No uplift
                              Avg BP =        71.30 kN/m2   < SBP; OK

Allowable GBP =          120 kN/m2
Net upward bearing pressure =    71.3 - 24 x 0.33 - 1.9 x 18 =    **29.18** kN/m2

### Strength design of foundation slab

**Long Bars:**            along X                              Load factor=  1.5

    Projection of slab =   1.15 m.
    maxm proj =            1.15 m
    Moment at face of col = 19.30 kNm /m         (29.18 x 1.15^2 / 2)
    d reqd=                127 mm     dprov =   255 mm      **> dreqd; OK.**
    $M_u/bd^2$ =           0.45                 (1.5x19.3x10^6) / (1000X127^2)
    pt =                   0.20 **M25**
    Ast reqd. =            510 mm2/m width
    Ast provided =         707 mm2/m width   **> Ast rqd, Ast min; OK.**
    Ast min =              306 mm2/m width
    Tc =                   0.92 Mpa
    Tc.b.d =               235 kN.    **> Vs; OK.**   (0.92 x 1000 x 255 / 1000)
    Vs =                   26 kN


**Transverse Bars:**       along Z

    Projection of slab =   0.875 m.
    Moment at face of col = 9.49 kNm /m          (29.18 x 0.875^2 / 2)
    d reqd=                88.95 mm   dprov =   255 mm      **> dreqd; OK.**
    $M_u/bd^2$ =           0.22
    pt =                   0.20 **M25**
    Ast reqd. =            510 mm2/m width
    Ast provided =         707 mm2/m width   **> Ast rqd, Ast min; OK.**
    Ast min =              306 mm2/m width

# Design of Industrial Structures

## 12.2.8.11 Column Base and Anchor Bolt

**Max force per leg**

| | | |
|---|---|---|
| P = | 156 | kN |
| T = | 117 | kN |
| Col size = | 130 130 5.4 | |
| wide | 130 | mm |
| depth | 130 | mm |
| bolt crs, x = | 200 | mm |
| Base Pl size, B = | 300 | mm |
| D = | 300 | mm |
| Base Pl thick = | 20 | mm |

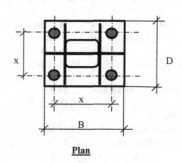

Plan

**Anchor bolt**

Base Pr = 156448 / (300 x 300) =   1.74 N/mm2        OK.< 6 kN/m2 (M20 Grout comp strength)

Provide Bolt     4 nos     24 dia     M24

Design strength of bolt, $T_{dn}$ = 0.9*An fu /$\gamma$m1           ( IS 800 2007 - 6.3.2)
 = 0.9 x 3.14x (0.85x24) ^2 x 0.25 x 410 / 1.25
 = 96.4 kN

Permissible tensile strength, $T_{nb}$ = 0.6 Tnb =   57.86 kN        ( IS 800 2007 - 11.6.2.3)

Total Bolt capacity  =   4 x 57.86 =   231 kN   **> T; safe.**

*Concrete breakout strength, Vb in lb :*        *(ACI 318-18)*

| | | |
|---|---|---|
| Embedded Length, lo = | 24 | in |
| Bolt OD , do = | 1 | in |
| fc' = | 3000 | psi   M20 |
| c1=c2= | 4 | in |

Vb =   8x(24/1)^2x1^0.5x4^1.5=   6619 Lb
                                                          3.00 MT
                                                          **30** kN

Actual T / Bolt = 117 / 4 =     **29.25** kN

                 **< Vb; OK.**                Pedestal

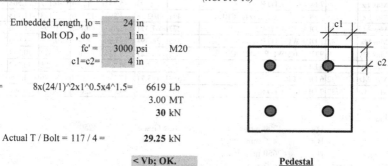

**Base Plate :**

Thickness, $t_s = (3w (a^2 - 0.3b^2) / f_{bs})^{0.5}$　　　　( IS 800 2007 - 11.3.3)

| | | |
|---|---|---|
| Base pressure, w = | 1.74 N/mm2 | |
| Projection, a = | 85 mm | |
| b = | 85 mm | |
| Perm. stress, fbs = 0.75fy = | 187.5 N/mm2 | |
| ts = | 11.86 mm | |
| Base Plate thickness = | 20 mm | |
| | > ts; OK. | |

| | | | |
|---|---|---|---|
| Actual tension / bolt = | 29.25 kN | (117 / 4) | |
| dist from col face = | 27 mm | [0.5 x (200 -130) - 8] | |
| effective width = | 54 mm | | |
| Ms = | 790 kNmm | (29.25 x 27) | |
| Ze = | 3600 mm3 | (54 x 20^2 /6) | |
| fbc = Ms/Ze = | 219 Mpa | | |
| Allowable stress, fbs = | 234 Mpa | (25% increased for Wind) | |
| | > fbs; OK. | | |

## 12.2.8.12　Design of RCC Pedestal

**Member load from computer output:**

| R6 | Beam | L/C | Node | Fx kN | Fy kN | Fz kN | Mx kNm | My kNm | Mz kNm |
|---|---|---|---|---|---|---|---|---|---|
| Max Fx | 531 | 9 DL + 0.5 WL R TO L+ CRLR | 279 | 168.03 | -12.456 | -10.699 | -7.309 | 15.209 | -54.893 |
| Min Fx | 3 | 8 DL + WL R TO L | 3 | -32.867 | -20.716 | 0.033 | -0.071 | -0.011 | -55.769 |
| Max Fy | 732 | 6 DL+LLR | 383 | 153.71 | 17.115 | 2.377 | 8.436 | -4.299 | 51.957 |
| Min Fy | 732 | 7 DL+ WL LTO R | 383 | 56.394 | -43.86 | 2.645 | -20.342 | -3.996 | -155.21 |
| Max Fz | 9 | 7 DL+ WL LTO R | 7 | 61.93 | -5.882 | 12.46 | -0.147 | 11.568 | -39.864 |
| Min Fz | 935 | 9 DL + 0.5 WL R TO L+ CRLR | 491 | 102.88 | -3.893 | -12.716 | 0.138 | -11.158 | -13.758 |
| Max Mx | 530 | 7 DL+ WL LTO R | 277 | 67.371 | -42.983 | -0.796 | 19.99 | 1.286 | -149.19 |
| Min Mx | 732 | 7 DL+ WL LTO R | 383 | 56.394 | -43.86 | 2.645 | -20.342 | -3.996 | -155.21 |
| Max My | 9 | 9 DL + 0.5 WL R TO L+ CRLR | 9 | 73.21 | 0.135 | 11.67 | 0.414 | 30.338 | -1.387 |
| Min My | 935 | 6 DL+LLR | 492 | 56.736 | -0.298 | -12.244 | -0.25 | -30.313 | -0.875 |
| Max Mz | 732 | 6 DL+LLR | 383 | 153.71 | 17.115 | 2.377 | 8.436 | -4.299 | 51.957 |
| Min Mz | 732 | 7 DL+ WL LTO R | 383 | 56.394 | -43.86 | 2.645 | -20.342 | -3.996 | -155.21 |

Pedestal size
　　　　D = 　1200 mm　　　Concrete Grade　　　M25
　　　　B = 　400 mm
　　　　H = 　2650 mm
Reinforcement　Asc =　3617 mm2　　18 nos　　16 dia

*Slenderness ratios*　　　　　　　　　　　　　　　　(IS 456 2000 - 25.1.2)
　　　　lex/D = 2 x 2650 / 1200 = 　　4.4
　　　　ley/B = 2650 / 400 = 　　　　6.6　　< 12; short column.

*Design forces*
| LC | P | M |
|---|---|---|
| 1 | 168 kN | 55 kNm |
| 2 | 56 kN | 155 kNm |

# Design of Industrial Structures

*Actual stresses in concrete*

| LC | P/A | +/- | M/Z | Compression (σbc) | Tension (σbt) |
|----|-----|-----|-----|-------------------|---------------|
| 1  | 0.35 | +/- | 0.57 | 0.92 N/mm2 | -0.22 N/mm2 |
| 2  | 0.12 | +/- | 1.62 | 1.73 N/mm2 | -1.50 N/mm2 |

*Permissible stresses in concrete*  (IS 456 2000 - Table 21)

Bending compression, σ abc =   8.5 N/mm2   Concrete grade =   **M25**
Tension, σ at =   3.2 N/mm2

*Observation*

Actual comp stress < Permissible comp stress; Safe
Actual tesile stress < Permissible tensile stress; Safe
Provide nominal reinforcement.   (IS 456 2000 - 26.5.3.1 h)

Minimum reinforcement in pedestal =   0.15 x 400 x 1200 /100 =   720 mm2
Reinforcement provided, Asc =   3617 mm2   **OK.**

## 12.2.8.13  Design of Steeel Crane Girder

**Design Parameter:**

Crane girder Span      L =    **5.5** m.

**Crane data:**
Lift (Main hook) =         **30** kN.
Weight of crab =           **3** kN.
Weight of Crane bridge =   **50** kN.
Crane Span =               **12.5** M.
Minm. Hook approach =      **1.2** M.
No of Wheel per side =     **2** nos.
Wheel base distance =      **2.5** m
Maxm Load per wheel =      **28** kN

**Elevation** (Span L, loads P, P)

Impact factor :                                        [IS 875 (part 2) 1987 - 6.3]
a. Vertical wheel load              **25** % of static wheel load
b. Horizontal surge per track       **10** % of weight of crab plus lifting load
c. Longitudinal surge per track     **5**  % of static wheel load.

Vertical load per wheel =          28 x 1.25 =        35 kN
Transverse surge /wheel/side =     10% (30 + 3) =     1.65 kN
Longitudinal surge on each track = 5% (28 x 2) =      2.80 kN

## Design Moment and shear

Wheel positions are set in an order that center line of girder is bisected by their resultant and a wheel, to get the maxm span moment.

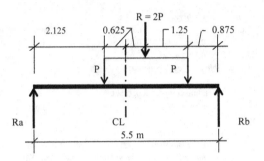

Ra = (2 x 35) x (1.25 + 0.875) /5.5 =    27.05 kN
Rb = 2 x 35 - 27.05 =    **42.95 kN**

Mv = 27.05 x 2.125 =    **57.48 kNm**
Mh= 57.48 x 1.65 / 35 =    **2.71 kNm**

### Strength design of member section      ( IS 800 1984 - 6.2.4)

Try with built up section :    **ISMB 250**    Web and    **ISMC 300**    Top flange

*To calculate Critical bending stress:*

Section size :
Top flange    **ISMC 300**    +    Web    **ISMB 250**

| | | | | | | |
|---|---|---|---|---|---|---|
| D = | 300 mm | | D = | 250 mm | | N.A |
| B = | 90 mm | | B = | 125 mm | | |
| tflng = | 13.6 mm | | tflng = | 12.5 mm | | |
| tweb = | 7.6 mm | | tweb = | 6.9 mm | | |
| d1 = | 273 mm | | d1 = | 225 mm | | |
| ry = | 2.61 cm | | ry = | 2.65 cm | | |
| Ixx = | 6363 cm4 | | Iyy = | 335 cm4 | | |
| A = | 45.64 cm2 | | A = | 47.55 cm2 | | **Section** |
| wt = | 35.8 kg/m | | Ixx = | 5132 cm4 | | |
| | | | wt = | 37.3 kg/m | | wt/m =    73.1 kg/m |

fy = 250 Mpa

T = k1 x T          T = Mean thickness of comp flng. =    18.81 mm
  = 19 mm                                                                            φ    | 1
L = 5500 mm                                           k1 =    | 1
ry = 10 x ((6363+335)/(45.64+47.55))^0.5 =    84.78 mm
D = 250 + 7.6 =    257.6 mm

Iyy of bot flange =    12.5x125^3/12=    2034505    mm4      ω =    0.97
Iyy of top flange =    63630000 + 2034505 =    65664505    mm4     k2 =    0.45

# Design of Industrial Structures

*To find out NA depth:*
$c_1 =$ (45.64 x (250 x 0.5 x 0.1 + 7.6 x 0.1) / (45.64 + 47.55)) x 10 =     68 mm
$c_2 =$ 250 + 7.6 - 68 =                190 mm

$I_x =$ (5132+47.55 x (0.5x250+0.76-6.8)^2+45.64x(6.8-3.45)^2)x10000 =     75270976    mm4
$I_y =$ (6363 + 335) x 10000 =      66980000    cm4

$Z_x$ top =      75270976 / 68 =         1106926     mm3
$Z_x$ bot =      75270976 / 190 =        396163      mm3
$Z_y$ =          66980000 / 150 =        446533      mm3

$Y = 26.5 \times 10^5 / (L/r_y)^2$   =   26.5 x 10^5 /(5500 / 84.78)^2 = =   630   Mpa
$X = Y[1 + (1/20).(LT/r_y D)^2]^{0.5}$ =  630 x (1+(1/20)x(5500x19 / 84.78/257.6)^2)^0.5 =
                                                                        =   923   Mpa
$f_{cb} = k_1(X + k_2 Y)(c_2/c_1)$   =   1 x (923 + 0.45 x 630) x (190/ 68) =   3371  Mpa

$n =$        1.4

*Permissible bending stress*
$f_{abc}$ = 0.66 (fcb .fy) / [ fcb ^n + fy ^n]^ 1/n
     = 0.66x(3371x250)/(3371^1.4+250^1.4)^0.71
     = **162** Mpa                        Max permissible stress =    165 Mpa

*Check design stress :*
**Bending**
   $f_{bc}$   $M_v / Z_x$ top =   57.48 x 10^6 / 1106926 =      52 Mpa
         $M_h / Z_y$ =        2.71x10^6 / 446533 =              6 Mpa
                                                               58 MPa      < fabc; Safe.

   $f_{bt}$ =      $M_v / Z_x$ bot =           145 Mpa              < fabc; Safe.

**Shear**
         $\tau_b = V_s / A_v =$ 42950 / (250 x 6.9) =    25 Mpa      < fabc; Safe.
         $\tau_{ab} = 0.40 f_y =$     100 Mpa

**Deflexion :**

         $P_1 =$      28 kN      $a =$    2125 mm    wheel 1    $L =$    5500 mm
         $P_2 =$      28 kN      $a =$    4625 mm    wheel 2    $I =$    75270976    mm4

         d center = $\Sigma (PL^3/48EI)[3a/L - 4(a/L)^3]$

         $d =$     5.985   +    0.929  =    **6.91** mm

         Allowable limit = L/750 =         **7.33** mm      Safe.

## 12.2.9 Design Drawings

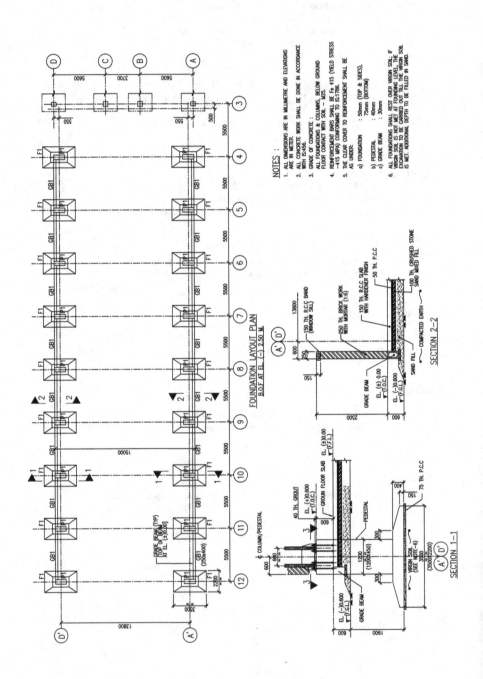

# Design of Industrial Structures

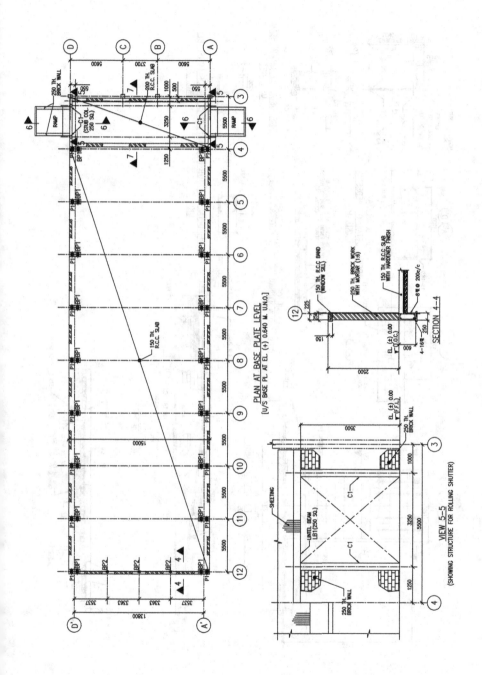

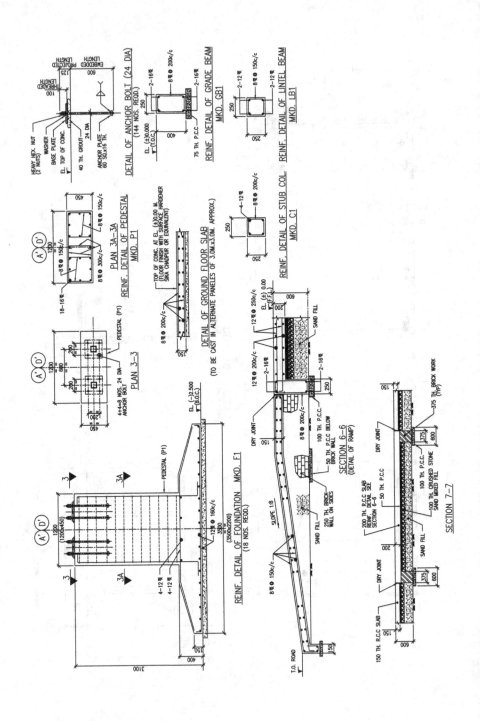

# Design of Industrial Structures

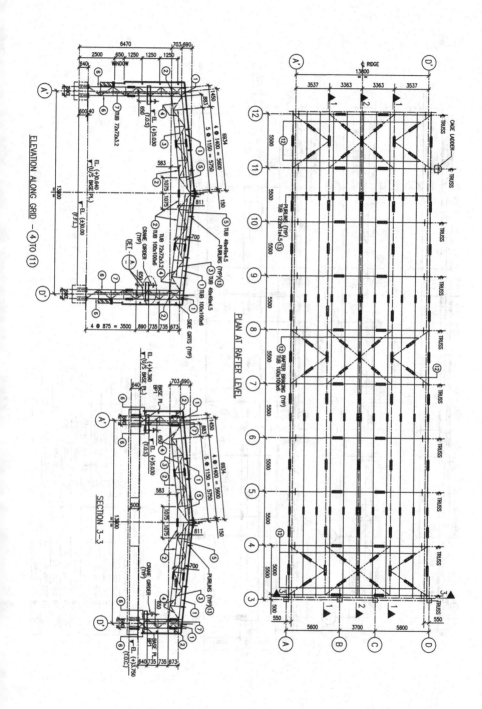

# 388  Design of Industrial Structures

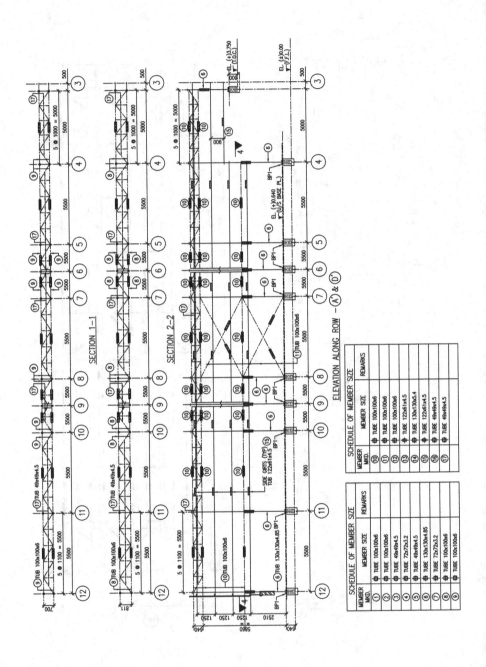

# Design of Industrial Structures

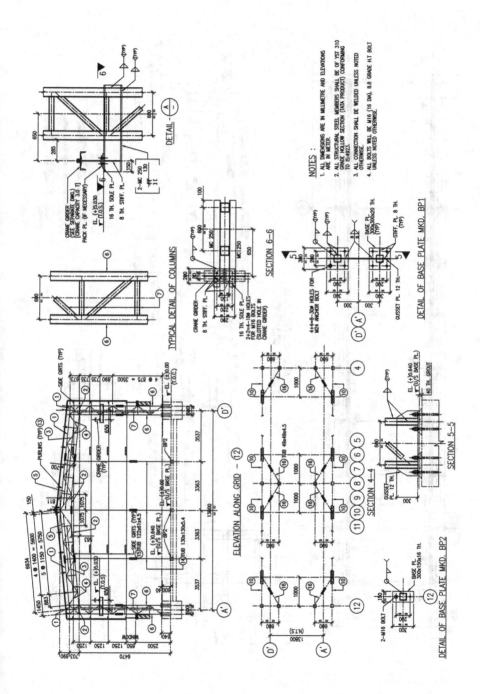

390  Design of Industrial Structures

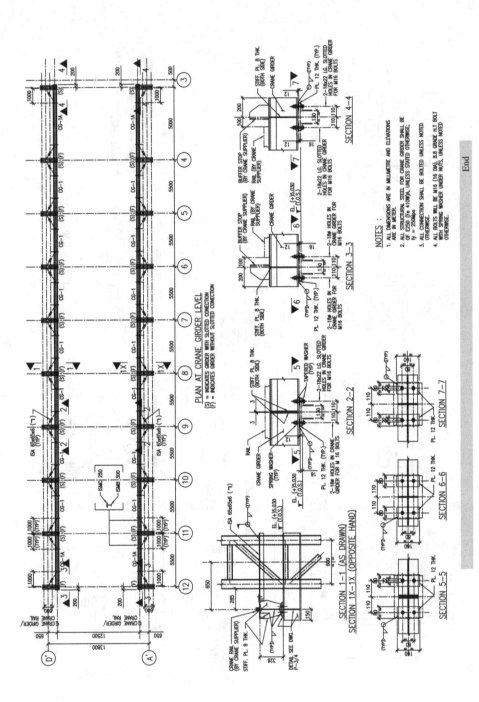

Design of Industrial Structures

## 12.3 DESIGN OF PIPE AND CABLE RACKS IN A POWER PLANT

### 12.3.1 SINGLE-TIER PIPE RACK FOR A RAW WATER PIPELINE

#### 12.3.1.1 Description

The designed pipe rack is a single tier structure spanning over columns carrying pipes and cables. The structure is a five span module, which can be repeated along length depending on actual requirement. The structure is designed for pipe & cable weight, wind load and self weight of structure. The maximum length of a single module is kept within 60 m consider allowable limit of temperature expansion gap. The columns are designed for lateral forces along both the axes, so column bracings are not necessary. It is a simple form of pipe rack structure and member design are done by working stress method according to old version of IS: 800 (1984) and the latest edition IS: 800 - 2007. The reader can observe the change in allowable stresses.

#### 12.3.1.2 Reference

    1.0   IS 800 : 1962; 1984; 2007
    2.0   IS 875 : 1987
    3.0   Steel designer's Manual 4ed - ELBS

#### 12.3.1.3 Material

| IS 2062 | | | | | | |
|---|---|---|---|---|---|---|
| Grade | E 250 (Fe 410 W) A | | | | | |
| Yield Stress | $f_y$ | 250 | 240 | 230 | MPa | (IS 800:2007 - Table 1) |
| d or t | | <20 | 20-40 | >40 | | |
| Tensile Stress | $f_u$ | 410 | 410 | 410 | MPa | |
| Elasticity | E | | | 200000 | MPa | (IS 800:2007 - 2.2.4.1) |
| Mat. safety fac | $\gamma_{m0}$ | | | 1.10 | | (IS 800:2007 - Table 5) |

#### 12.3.1.4 Dimensioning

    Span =     10 m
    Height =     4.5 m
    Col leg (c/c) =     0.9 m

## 12.3.1.5 Load

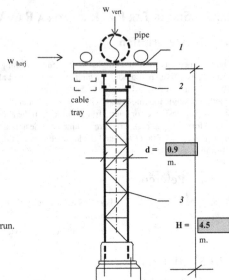

**Vertical:**

| | Nos | NB | Weight kN/m |
|---|---|---|---|
| Pipe | 1 | 250 | 0.92 |
| (IS 1161) | 1 | 80 | 0.15 |
| | 1 | 100 | 0.23 |
| Cable tray | 2 | | 2.20 |
| Self weight | | | 1.65 |
| | | $w_{vert}$ | 5.15 kN/m run. |

d = 0.9 m.

H = 4.5 m.

**Horizontal (lateral):**

*a) Wind :*
Intensity :   pwind =   1.5 kN/m2   (IS 875 Part 3)

| | Nos | proj mm | Shape factor | Force kN/m | |
|---|---|---|---|---|---|
| Pipe | 1 | 250 | 1.6 | 0.6 | (1x0.25x1.6x1.5) |
| | 1 | 80 | 1.6 | 0.19 | |
| | 1 | 100 | 1.6 | 0.24 | |
| Cable Tray | 2 | 100 | 1.6 | 0.48 | |
| Mem 2 | 2 | 300 | 2 | 1.8 | |

*b) Pipe friction :*
10% of vert load ($w_{vert}$)    0.52
[ Hot pipes]          $w_{horj}$    3.83 kN/m run.

**Section**

**Part plan at top**

# Design of Industrial Structures

## 12.3.1.6 Analysis

Structural system: *The pipe rack structure is a five span module. The members will be continuous over supporting column with hinged joints as shown.*

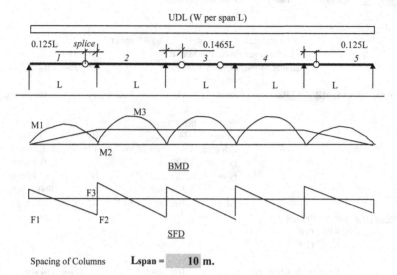

Spacing of Columns    **Lspan = 10 m.**

**Vertical moment and shear forces :**

W =   $w_{vert} \times L_{span}$ =    51.50 kN.
$M_1$ vert = 0.0957WL =    49.29 kNm.      F1 vert = 0.4357W =    22.44 kN.
$M_2$ vert = (- )WL/16 = (    32.188 kNm.      F2 vert = 0.5625W =    28.97 kN.
$M_3$ vert = WL/16 =      32.188 kNm.      F3 vert = 0.5W =      25.75 kN.
Design moment, Mv =    **49.29 kNm.**      Design shear, Fv =    **28.97 kN.**

**Horz. moment and shear forces :**

W =   $w_{horz} \times L_{span}$ =    38.27 kN.

M1 horz = 0.0957WL =    37 kNm.      F1 horz = 0.4357W =    16.67 kN.
M2 horz = - WL/16 =      24 kNm.      F2 horz = 0.5625W =    21.53 kN.
M3 horz = WL/16 =       24 kNm.      F3 horz = 0.5W =       19.14 kN.
Design moment, Mh =    **37 kNm.**      Design shear, Fh =    **22 kN.**

**Longitudinal moment and shear forces :**

*a) Wind :*      proj    Shape    Force
          Nos    mm     factor    kN/m
col leg      2      140      2      0.84 (2x0.14x2x1.5)
lacing      2      100      2      0.60
*b) Pipe friction :*
     5% of vert load/col        2.58 (0.05 x 5.15 x 10)
                     $w_{horj\ L}$    4.00 kN/m run.

Moment at base =    4 x 4.5^2 / 2=      40.5 kNm      Shear =    18 kN

### 12.3.1.7 Design of member sections
### 12.3.1.7 (A) According to IS: 800 1984

**Member 1 :**   ISMC   100    Pipe support steel

**Member 2 :**   Twin beam
  Center distance =   d =   0.9 meter.
  Axial force per chord, P = Mh / d =   **41** kN.
  Bending Moment/Chord = Mv / 2 =   **25** kNm.

Try with   ISMB   300   Span =   10 m.   Laced @   0.9 m. c/c.

Sectional properties:
  wt/m =   44.2 kg.         B =   14 cm
  C/S area, A =   56.26 cm2.   T =   1.2 cm
  $r_x$ =   12.37 cm.        tw =   0.8 cm
  $r_y$ =   2.84 cm.         $I_y$ =   454 cm4.
  $Z_x$ =   574 cm3.         $I_x$ =   8604 cm4.
  $KL_x$ =   700 cm.   (0.7L)   $KL_y$ =   90 cm.

**Axial stress :**   (IS:800 1984 Table 5.1)
  $KL_x / r_x$ =   57
  $KL_y / r_y$ =   32    [KL/r] max =   57

  Permissible stress, $\sigma_{ac}$ =   169 kg/cm2.   (33 % increased for wind)
  Calculated stress,   $\sigma_c$ = P/A =   72 kg/cm2.   **Safe.**

**Bending stress :**                                            (IS:800 1962 10.2.2.1)
  $M_x / Z_x$ =   436 kg/cm2    D /T =   24.194    T =   12
  (250000 / 574)
  Permissible stress =   $P_{bc}$ =   880 kg/cm2.
  Eff. Lenth, l =   850 cm      $I_y.h / Z_x.l^2$ =   3E-05
            h =   27.52 cm     $K.l^2 / I_y.h^2$ =   45.535
            K =   21.7 cm4        $C_s$ =   880 kg/cm2
  Calculated stress =   $f_{bc}$ =   436 kg/cm2.   **Safe.**

**Stress ratio :**   $\sigma_c / \sigma_{ac}$ + $f_{bc} / P_{bc}$ =   **0.9**   **< 1.33**   **Safe.**

# Design of Industrial Structures

**_Deflection_**  [ Refer Steel designer's manual ELBS; 4th ed]
L = 10 m  N = 0.1465L = 1.465 m  n = N/L = 0.1465
$w_{vert}$ = 2.575 kN/m./per single beam
$\delta$ span = (w $L^4$ /384EI) x (5 - 24 $n^2$) = 17.48 mm
$\delta$ splice = (w $L^3$ N / 24EI) x (1 - 6$n^2$ - 3$n^3$) = 7.87 mm
$\delta$ allow = L / 325 = 10000 / 325 = 31 mm
$\delta$ max = 17.48 mm  **Safe.< L /325.**

*[Note: Allowable deflexion shall be limited according to piping designer's recommendation ]*

**Lacing :**  Transverse shear = 2.5 % of P (=Mh/d) = 1.02 kN.

Diagonal :  Force in lacing per face = 0.72 kN.  (1.02 x 1.414 / 2)

Try with  **ISA 50 50 6**  Sectional properties:
**2 faces**  wt/m = 4.5 kg.
C/S area = 5.68 cm2.
rx = 1.51 cm.  rvv = 0.96 cm.
ry = 1.51 cm.

Lx = 127 cm.  0.7* Lx/rx = 59  welded to member 2

Permissible stress, $\sigma_{ac}$ = 169 kg/cm2.  33.33 % stress increased.  (IS:800 1984 Table 5.1)
Calculated stress,  $\sigma_c$ = P/A = 13 kg/cm2.  **Safe.**

**Member 3 :**  *Twin columns laced together and designed as combined section*

Height = **4.5** m.
Center distance = d = 0.9 meter.
Axial Load = P = **54.72** kN.  (28.97 + 25.75)
Shear at base = F2 + F3 = 40.66 kN.  (21.526875 + 19.135)
Transverse Moment at base = Myy = (F2+F3)*H = **183** kNm.  (wind)
Longitudinal wind moment = Mxx = 40.5 kNm (less than 10% of Myy, hence ignored)

Try with  2 nos  ISMB  250  @  0.9 m  crs

**Sectional properties:** Twin legged column
A= 95 cm2    wt/m = 37.3 kg.    C/S area = 47.55 cm2.(single leg)
Ixx = 10263 cm3.    rxx = 10.39 cm    d/2 = 45 cm
                                      Iy = 335 cm4    (one leg)
                    Iyy = Iy +A(0.5d)2 = 193044 cm4
                                      ryy = 45 cm
                                      ry = 2.65 cm.    (one leg)
                    Zyy = Iyy /(d/2) = 4290 cm3

*Axial stress :*

Lx = 450 cm.    Ly = 90 cm.    Laced @    0.9 m. c/c.
Lx / rx = 43    Ly / ry = 34
Lyy (overall sec) = 1.5 L = 675 cm    Lyy /ryy = 15
[L/r ] max = 43

Permissible stress, $\sigma ac$ = 139 kg/cm2.    (IS:800 1984 10.2.3)
Calculated stress,    $\sigma c$ = P/A = 54.72 / 95 =    58 kg/cm2.    **Safe.**

*Bending stress :*

Permissible stress =    Pbc =    1650 kg/cm2.    (IS:800 1984 Table 10.2.3)

Calculated stress =    fbc =    P/A + My/Zyy
                            =    58    +    427
                            =    484 kg/cm2.    **Safe.**

*Stress Interaction :*
$\sigma c$ / $\sigma ac$ + fbc / Pbc = 58/139 + 484 / 1650 =    0.4173    +    0.2933    =    0.71
                                                                                                    **< 1; Safe.**

**Lacing :**

Axial load per chord = P / 2 + M / d =    231 kN
Transverse force = 2.5 % of axial =    6.00 kN.

Force / lacing per face =    4.24 kN.    (6 x 1.414 / 2)

Try with    **ISA 50 50 6**    Sectional properties:
            **2 faces**    wt/m =    4.5 kg.
                          C/S area =    5.68 cm2.
                          rx =    1.51 cm.    rvv =    0.96 cm.
                          ry =    1.51 cm.

Lx =    127 cm.    0.7* Lx/rvv =    93    welded to column flange

Permissible stress, $\sigma ac$ =    120 kg/cm2.    33.33 % stress increased.    (IS:800 1984 Table 5.1)
Calculated stress,    $\sigma c$ = P/A =    75 kg/cm2.    **Safe.**

## Deflexion :

$H = $ 4.5 m  $I_{yy} = $ 193044 cm4
Horz force at top, $F = F_2 + F_3 = $ 40.66 kN

$\delta$ top $= F H^3 / 3EI = $ 3.2 mm  **Safe.< H /150.**

$\delta$ allow $= $ H/150 $= $ 30 mm  (IS:800 1984 Table 10.2.3)

### 12.3.1.7 (B) According to IS 800 : 2007 [Working Stress Method]

**Member 2 :**   Twin beam
 Span  L= 10 meter.
 Center distance = d = 0.9 meter.
 Axial force per chord, **Ps** = Mh / d = **41** kN.
 Bending Moment/Chord = Mv / 2 = **25** kNm.

**Axial Comp:**
 Section provided :

| ISMB | 300 | (IS 800:2007 7.1.2.1) | |
|---|---|---|---|
| Ae = | 5626 | mm2 | E = | 200000 | Mpa |
| rx = | 123.7 | mm | fy = | 250 | Mpa |
| ry= | 28.4 | mm | $\gamma_{m0} = $ | 1.10 | |
| Lx (0.7L) | 7000 | mm | KL/r = | 57 | |
| Ly = | 900 | mm | fcc = | 616 | MPa |
| Zx = | 573600 | mm3 | $\lambda = $ | 0.6371 | |
| $\alpha = $ | 0.21 | | $\phi = $ | 0.75 | |
| $\chi = $ | 0.873 | | fcd = | 198 | Mpa |

 Allowable comp load  (IS 800:2007 -11.3.1)
 Pa = 0.6 fcd Ae =   0.6 x 198 x 5626 /1000  =  668 kN
  Ps =  41 kN  **> Ps ; Safe,**

**Flexural strength :** (IS 800:2007 -11.4.1)

Allowable stress, fabc or fabt = 0.66 fy = 0.66 x 250 = 165 Mpa
Ma = fbc . Zx = 165 x 573600 = 95 kNm
Mv = 25 kNm   > Mv ; Safe,

*Stress Interaction :*
Ps / Pa + Mv / Ma = 41 / 668 + 25 / 95 = 0.061 + 0.263 = 0.32
< 1; Safe.

**Lacings**

Force, **P** = 2.5% x 41 / 2          Section : **ISA 50 50 6**  on both faces
       = **0.51** kN                   b1 = b2 = 50 mm       t = 6 mm
L =0.7L= 0.89 m                        Ae = 568 mm2         $r_{vv}$ = 9.6 mm

Slenderness ratio, $\lambda = (k1 + k2\,\lambda_{vv}^{2} + k3\,\lambda_{\varphi}^{2})^{0.5}$    (IS 800:2007 -7.5.12)
    $\lambda_{vv}$ = 1.04        k1 = 0.75       k3 = 20
    $\lambda_{\varphi}$ = 0.09   k2 = 0.35
    $\lambda$ = 1.14
    fcd = $(fy/\gamma_{mo}) / [\phi + (\phi^{2} - \lambda^{2})]^{0.5}$     (IS 800:2007 -7.1.2.1)
    $\phi$ = 1.3849
    fcd = 161 Mpa

Allowable Comp strength, **Pa** = 0.6. fcd. Ae =   0.6 x 161 x 568/ 1000 =  **54.87** kN    > P; Safe.

**Member 3 :**  Two legged column
              Height              H= 4.5 meter.
              Center distance =   d= 0.9 meter.
              Axial Load at base = P = 54.72 kN.
              Moment at base , M = (F2+F3)*H = 183 kNm.

              Axial force per chord = P / 2 + M / d =  **231** kN.

**Axial Comp strength:** (IS 800:2007 7.1.2.1)

Section provided :     @  0.9 m    crs

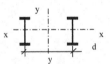

| | | | Single leg | |
|---|---|---|---|---|
| Ae = | 4755 | mm2 | D = | 250 mm |
| rxx = | 104 | mm | B = | 125 mm |
| ryy= | 27 | mm | T = | 12.5 mm |
| Lx = | 4500 | mm | tw = | 6.9 mm |
| Ly = | 900 | mm | | |
| Zxx = | 410500 | mm3 | | |
| $\alpha$ = | 0.21 | | | |
| E = | 200000 | Mpa | | |
| fy = | 250 | Mpa | | |
| $\gamma_{m0}$ = | 1.10 | | | |
| KL/r = | 43 | | | |
| fcc = | 1051 | MPa | | |
| $\lambda$ = | 0.4877 | | | |
| $\phi$ = | 0.65 | | | |
| $\chi$ = | 0.926 | | | |
| fcd = | 210 | Mpa | | |

# Design of Industrial Structures

<u>Allowable comp load</u>  (IS 800:2007 -11.3.1)
Pa = 0.6 fcd Ae =   0.6 x 210 x 4755 /1000   =   599 kN
                                         Ps =   231 kN   > Ps ; Safe,

## *Bending strength :*

Ms = 0.6 fy Zyy      Zyy =  ###### mm3        (IS:800 2007 - 11.4.1 )

Md =  0.6 x 250 x 4289867/1000000
    =  643 kNm
M =   183 kNm    Md > M; Safe.

*Stress Interaction :*
    Pa / Ps + M /Md  = 599/231 + 182.9784375 / 643 =   0.39   +   0.28   =   0.67
                                                                                    < 1; Safe.

## Lacings

Force, P =   4.24 kN                       Section : ISA 50 50 6   on both faces
                                           b1 = b2 =   50 mm       t =    6 mm
L = 0.7L =   0.89 m                        Ae =    568 mm2         $r_{vv}$ =  9.6 mm

Slenderness ratio, $\lambda = (k1 + k2\, \lambda_{vv}^2 + k3\, \lambda_{\varphi}^2)^{0.5}$      (IS 800:2007 -7.5.12)
        $\lambda_{vv}$ =   1.04       k1 =    0.75      k3 =     20
        $\lambda_{\varphi}$ =   0.09       k2 =    0.35
        $\lambda$ =    1.14
        fcd = $(fy/\gamma_{mo}) / [\phi + (\phi^2 - \lambda^2)]^{0.5}$                (IS 800:2007 -7.1.2.1)
        $\phi$ =   1.38

        fcd =   161 Mpa

Allowable Comp strength, **Pa** = 0.6. fcd. Ae =    0.6 x 161 x 568/1000 =       55 kN
                                                                                 > P; Safe.

End

## 12.3.2 Two tier Pipe rack for Ash Pipes and Cable racks

### 12.3.2.1 Description

The pipe rack is designed to carry ash pipes and cable trays from Power plants to Ash water pump house. There will be a maintenance walk way accessible by cage ladders from ground at suitable locations. The structural arrangement is made with long span lattice girder trusses spanning between portal frames. These portal frames will rest on trestles. The trestles are steel columns braced along transverse direction. All vertical loads from pipes, cables and walkways will be carried by cross beams at each tier. These cross beams will transfer the load to trusses along column lines. Plan bracings are provided at each tier for resisting wind/seismic forces, which will finaly transferred to foundation system. The pipe rack structure is a continuous structure according to piping layout drawing. There will be expansion joint or separtion gap with twin columns at intervals as recommended by codes of practice to avoid stresses from seasonal temperature variation. Longitudinal bracings are provided at intervals to resist longitudinal forces due to wind/seismic as well as to ensure frame stability. The design of longitudinal bracing has not been furnished in this workout example. Ash pipes are carrying cold fluid. Hence the friction load for temperature changes has not been considered in this workout example. However, designer shall consider piping friction and loads at corners and bends, wherever specified by piping designer.

### 12.3.2.2 Reference

1.0   IS 800 : 1962; 1984; 2007
2.0   IS 875 : 1987
3.0   Steel designer's manual - ELBS  4 Ed.

### 12.3.2.3 Material

| IS 2062 | | | | | | |
|---|---|---|---|---|---|---|
| Grade | | E 250 (Fe 410 W) A | | | | |
| Yield Stress | $f_y$ | 250 | 240 | 230 | MPa | (IS 800:2007 - Table 1) |
| d or t | | <20 | 20-40 | >40 | | |
| Tensile Stress | $f_u$ | 410 | 410 | 410 | MPa | |
| Elasticity | E | | | 200000 | MPa | (IS 800:2007 - 2.2.4.1) |
| Mat. safety fac | $\gamma_{m0}$ | 1.10 | $\gamma_{m1}$ | 1.25 | | (IS 800:2007 - Table 5) |

# Design of Industrial Structures

## 12.3.2.4 Sketch

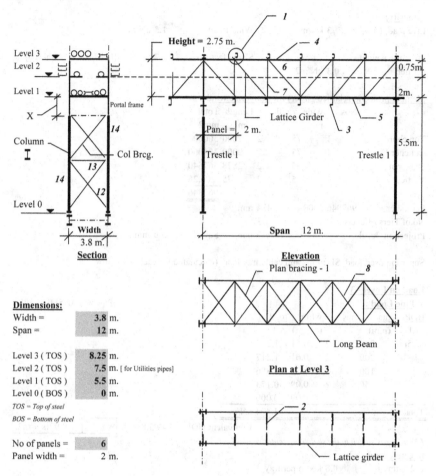

**Dimensions:**
Width = 3.8 m.
Span = 12 m.

Level 3 (TOS) = 8.25 m.
Level 2 (TOS) = 7.5 m. [for Utilities pipes]
Level 1 (TOS) = 5.5 m.
Level 0 (BOS) = 0 m.
*TOS = Top of steel*
*BOS = Bottom of steel*

No of panels = 6
Panel width = 2 m.

**Lattice:**
Member Projection : 200 mm/Level
*for transverse wind (see calc below)*
Depth of Lattice girder = 2.75 m.

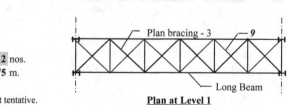

**Col Brcg:**
No of sets = 2 nos.
Height of panel = 2.75 m.
*Notes:*
1.0 No of pipes in the sketch is just tentative.
2.0 X = Minimum dimension required for connection.

## 12.3.2.5 Load

**Intensity:**
Live load, LL = **3** kN/m2          Wind Load   **1.5** kN/m2

Structure weight                     Shape factor   **2**   [IS 875 : 1987 Table 26]
per tier =  **0.3** kN/m2

Member projection (lattice girder) for transverse wind

|  | Nos | wide mm | sides | Length mm | Total mm2 |
|---|---|---|---|---|---|
| Top chord | 1 | 75 | 2 | 2000 | 300000 |
| Bot chord | 1 | 75 | 2 | 2000 | 300000 |
| Diagonal | 1 | 65 | 2 | 2828 | 367640 |
| Vertical | 0.5 | 75 | 2 | 2.75 | 206.25 |
|  |  |  |  |  | 967846 |

Per meter = 967846/2000 = 484 mm
No of Tiers / Levels = 3
Projection /level = 484/3 = 161.33 mm     Say,   **200** mm

Super imposed load, SL : Pipe and cable tray load (calculated for each tier below).

### Load on Level - 3
**a. Pipe Load :**                                                [Pipe wt : ANSI B 36.10]

| BOP (RL) | dia (mm) | nos | wt/m (kN/m) | total/m |
|---|---|---|---|---|
| 8.25m. | 200 | 2 | 0.61 | 1.217 |
|  | 200 | 2 | 0.61 | 1.217 |
|  | 100 | 4 | 0.22 | 0.886 |
|  | 50 | 2 | 0.09 | 0.173 |
|  |  |  | 0.00 | 0.000 |
| Total weight |  |  |  | 3.494 kN/m |

Equivalent UDL = 3.49/3.8 =   $W_{pipe}$  **0.92** kN/m2.

*[ No of pipes shown in the sketch are tentative only]*

**b. Cable load**
Cable tray =  **0.8** kN/m per tray

|  | No of trays | Wt/m (kN) |
|---|---|---|
| 1. Left Post | 0 | 0 |
| 2. Right Post | 3 | 2.4 |
| Total weight |  | 7.2 kN/m |

Equivalent UDL = 7.2/3.8 =   $W_{cable}$  **1.89** kN/m2.

# Design of Industrial Structures

**c. Live Load :**
1. Incidental Point Load at any location :
   Pinci    1.25 kN
   Equivalent UDL = 1.25/3.8 =    $P_{inci}$    0.33 kN/m2.
2. Live load on walkway :
   Width of walkway =    0.6 m.
   $W_{walkway}$    1.8 kN/m.
   Equivalent UDL = 1.8/3.8 =    $P_{wway}$    0.47 kN/m2.

Total UDL for Pipe + Cable + Walkway + Incidental =    **SL =**    **3.62** kN/m2.

### Design UDL on Level - 3 :

**Vert :** Dead load of Structure    =    0.3 kN/m2.
Super imposed load = Maximum of ( LL or SL)    =    3.62 kN/m2.
$W_{Lev\,3}$    **3.92** kN/m2.

**Horz : Wind load**
Projection - Pipe =    1300 mm.    (200x2+200x2+100x4+50x2)
Structure Projection =    400 mm.    ( mem proj 200 x shape factor 2)
sum =    1700 mm.
No of Cable tray =    3    (3x tray ht 100 mm )
Wind load per meter :
1.5 x [ (1700/1000) + (3 x 0.1) ] =    $W_{wind3}$    **3.00** kN/m.

### Load on Level 2

**a. Pipe Load :**

| BOP (RL) | dia (mm) | nos | wt/m (kN/m) | total/m |
|---|---|---|---|---|
| 7.5m. | 100 | 1 | 0.22 | 0.221 |
| | 80 | 1 | 0.15 | 0.150 |
| | | | 0.00 | 0.000 |
| Total weight | | | | 0.372 kN/m |

Equivalent UDL = 0.37/3.8 =    $W_{pipe}$    0.10 kN/m2.

**b. Cable load**
Cable tray = **0.4** kN/m per tray

|  | No of trays | Wt/m (kN) |
|---|---|---|
| 1. Left Post | 0 | 0 |
| 2. Right Post | 0 | 0 |
| Total weight |  | 0 kN/m |

Equivalent UDL = 0/3.8 =  $W_{cable}$  **0.00** kN/m2.

**c. Live Load :**
1. Incidental Point Load at any location :
$\qquad$ Pinci $\quad$ **0** kN

Equivalent UDL = 0/3.8 = $P_{inci}$ **0.00** kN/m2.

2. Live load on walkway :
Width of walkway = **0** m.

$\qquad w_{walkway}\qquad$ 0 kN/m.

Equivalent UDL = 0/3.8 = $P_{wway}$ **0.00** kN/m2.

Total UDL for Pipe + Cable + Walkway + Incidental = $\qquad$ **SL =** **0.10** kN/m2.

**Design UDL on Level - 2 :**

**Vert :** Dead load of Structure $\qquad$ = $\quad$ 0.3 kN/m2.
$\qquad$ Super imposed load = Maximum of ( LL, SL) $\quad$ = $\quad$ 3.00 kN/m2.
$\qquad\qquad\qquad\qquad\qquad\qquad\qquad\qquad$ $W_{Lev2}$ **3.30** kN/m2.

**Horz : Wind load**
$\qquad$ Projection - Pipe = $\quad$ 180 mm. $\qquad$ No of Cable tray = $\quad$ 0
$\qquad$ Structure Projection = $\quad$ 400 mm. $\qquad$ Tray height = $\quad$ 100 mm
$\qquad$ Wind load per meter :
$\qquad$ 1.5*[ (580/1000) + (0*0.1) ] = $\qquad$ $W_{wind2}$ **0.87** kN/m.

**Load on Level - 1**

**a. Pipe Load :**

| BOP (RL) | dia (mm) | nos | wt/m (kN/m) | total/m |
|---|---|---|---|---|
| 5.5m. | 200 | 2 | 0.61 | 1.217 |
|  | 100 | 4 | 0.22 | 0.886 |
|  | 150 | 2 | 0.43 | 0.855 |
|  | 50 | 4 | 0.09 | 0.347 |
|  |  |  | 0.00 | 0.000 |
| Total weight |  |  |  | 3.305 kN/m |

Equivalent UDL = 3.31/3.8 = $W_{pipe}$ **0.87** kN/m2.

# Design of Industrial Structures

**b. Cable load**
Cable tray = **0.8** kN/m per tray

| | No of trays | Wt/m (kN) |
|---|---|---|
| 1. Left Post | 0 | 0 |
| 2. Right Post | 0 | 0 |
| Total weight | | 0 kN/m |

Equivalent UDL = 0/3.8 =    $W_{cable}$    0.00 kN/m2.

**c. Live Load :**
1. Incidental Point Load at any location :
   Pinci    **1.25** kN
   
   Equivalent UDL = 1.25/3.8 =    $P_{inci}$    0.33 kN/m2.

2. Live load on walkway :
   Width of walkway =    **0.7** m.
   $W_{walkway}$    2.1 kN/m.  (0.7 x 3)
   
   Equivalent UDL = 2.1/3.8 =    $P_{wway}$    0.55 kN/m2.

Total UDL for Pipe + Cable + Walkway + Incidental =    SL =    1.75 kN/m2.

**Design UDL on Level - 1 :**

Vert : Dead load of Structure                              =    0.3 kN/m2.
       Super imposed load = Maximum of ( LL, SL )    =    3.00 kN/m2.
                                                    $W_{Lev\,1}$    3.30 kN/m2.

Horz : **Wind load**
       Projection - Pipe =       1300 mm.          No of Cable tray =    0
       Structure Projection =    400 mm.           Tray height =         100 mm
       Wind load per meter :
       1.5*[ (1700/1000) + (0*0.1) ] =             $W_{wind1}$    2.55 kN/m.

## 12.3.2.6 Member Forces in Cross Beams

| | member | | | | | | | spcg x $W_{Lev}$ |
|---|---|---|---|---|---|---|---|---|
| Lev 3 | 1 | Span = | 3.8 m. | Spcg = | 2 m. | UDL = | 7.8 kN/m | (2 x 3.92) |
| Lev 2 | 2 | Span = | 3.8 m. | Spcg = | 2 m. | UDL = | 6.6 kN/m | (2 x 3.3) |
| Lev 1 | 3 | Span = | 3.8 m. | Spcg = | 2 m. | UDL = | 6.6 kN/m | (2 x 3.3) |

| mem | Moment | kNm | End Shear | kN | |
|---|---|---|---|---|---|
| 1 | 7.8 x 3.8^2 / 8 = | 14.08 | 0.5 x 7.8 x 3.8 = | 14.82 | Simply supported |
| 2 | 6.6 x 3.8^2 / 8 = | 11.91 | 0.5 x 6.6 x 3.8 = | 12.54 | Simply supported |
| 3 | 6.6 x 3.8^2 / 8 = | 11.91 | 0.5 x 6.6 x 3.8 = | 12.54 | Simply supported |

## 12.3.2.7  Member Forces in Lattice Girder

Span = 12 m.  
No of Vertical panels = 6  
Depth of Girder = 2.75 m.  
Width of pipe rack = 3.8 m.

**Load:**

| | | | |
|---|---|---|---|
| Level 3 Vert : | W Lev 3 * 0.5* width. | 7.45 kN/m. | |
| Horz : | W wind3 | 3.00 kN/m. | H = **2.75** |
| Level 2 Vert : | W Lev 2 * 0.5* width. | 6.27 kN/m. | |
| Horz : | W wind2 | 0.87 kN/m. | Level 1 |
| Level 1 Vert : | W Lev 1 * 0.5* width. | 6.27 kN/m. | |
| Horz : | W wind1 | 2.55 kN/m. | |

Level 3  
Lv = 3.40 m.  
ELEVATION

**Force/ Top chord:**   *member 4*

$w_{vert}$ = 7.45 + 6.27 + 6.27 = 19.99 kN/m.  
$M_{vert}$ = 19.99 × 12^2 / 8 = 359.82 kNm.  
$M_{vert}$ / H = 359.82 / 2.75 = 131 kN (+) .

$w_{horz}$ = 3 + 0.5 × 0.87 = 3.44 kN/m.  
$M_{horz}$ = 3.44 × 12^2 / 8 = 61.92 kNm.  
$M_{horz}$ / W = 61.92 / 3.8 = 16.00 kN (+ -) .

Lh = 4.29 m.  
PLAN

Total Force/ Chord =   131 + 16 =   **147** kN (+) compression.

**Force/ bottom chord:**   *member 5*

$M_{vert}$ = 359.82 kNm.  
$M_{vert}$ / H = 359.82 / 2.75 = 131 kN ( - ).

$w_{horz}$ = 2.55 + 0.5 × 0.87 = 2.99 kN/m.  
$M_{horz}$ = 2.985 × 12^2 / 8 = 53.73 kNm.  
$M_{horz}$ / W = 53.73 / 3.8 = 14 kN ( - ).

Total Force/ Chord =   131 + 14 =   **145** kN ( - ). tension.

## Design of Industrial Structures

**Force in diagonal member:**      *member 6*
panel width =    12 / 6 =      2 m.
height =           2.75 m.
Length, Lv =      **3.40** m.
End Reaction =   19.99 x 12/2 =      120 kN.
Force in Diagonal =     120*3.4/2.75=    **148.38** kN (-)    tension.

**Force in vertical member :**      *member 7*
Axial =        120 - 19.99 x 2 =      80.02 kN.

**Cable Tray loading on vertical members :**
A. Gravity load has been included in level 3 floor load.

B. Moment due to eccentric joint with vertical , $M_c$ :
Weight of Cable tray =     7.2 kN/m    ( see loading at level 3)
P =     7.2 x 2(spcg) =     14.4 kN.
Eccentricity =    0.1 + 1/2 of tray width =     0.4 m
$M_C$ =     14.4 x 0.4 =      **5.76** kNm,

*[Pipe load has been considered in Level 2 floor load. Moment from pipe rack support will counteract the moment from cable rack, hence it is ignored to get the worse effect.]*

0.75 m.    P =    14.4
pipe    cable    kN.
     C      0.4
2 m.          m
     A
Cable tray load

### 12.3.2.8  Member Forces in Plan bracing

Length of diagonals, Lh =     **4.29** m     W =     3.8 m

Level 3     Rh =    3.44   x      6   =     20.64 kN.
mem 8     Force in diagonal =     20.64 x 4.29 / 3.8 =     **23.30** kN.     Tension / compresion
mem 1     Force in Vertical =                              **20.64** kN.     Compression

Level 1     Rh =    2.99   x      6   =     17.91 kN.
mem 9     Force in diagonal =                         **20.22** kN.     Tension / compresion
mem 3     Force in Vertical =                              **17.91** kN.     Compression

## 12.3.2.9 Member Forces in Trestle

**Portal frame at top:**     **Member 10 and 11**

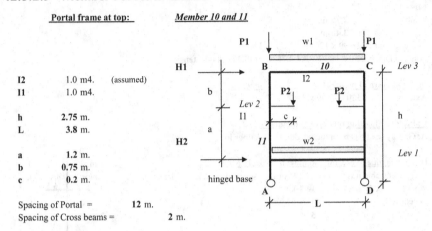

| | | |
|---|---|---|
| I2 | 1.0 m4. | (assumed) |
| I1 | 1.0 m4. | |
| h | 2.75 m. | |
| L | 3.8 m. | |
| a | 1.2 m. | |
| b | 0.75 m. | |
| c | 0.2 m. | |

Spacing of Portal = 12 m.
Spacing of Cross beams = 2 m.

**Loading and frame analysis :**
w1 =  7.8 kN/m   P1 =  (7.448+6.27+6.27) x (12 - 2) =   200 kN.   (load from lattice girder).
w2 =  6.6 kN/m   P2 =  6.6 x 3.8/2 =   13 kN ( from lev 2 pipe support ).
H1 =  (3 + 0.5x0.87) x 12 =  41.22 kN.   H2 =  (2.55 + 0.5x0.87) x 12=  35.82 kN.

*Using Formulae for rigid portal frames*           ( Steel designer's manual - ELBS - 4 ed )
k =   0.7237        N =   4.4474       a1 =   0.4364
     $(I_2/I_1).(h/L)$           (2k+3)              (a/h)

**Forces in Frame members (10 & 11)**

| | UDL | + | H1 | + | P2 | + | P1 | + | H2 | = | Sum |
|---|---|---|---|---|---|---|---|---|---|---|---|
| $M_B = M_C =$ | -6.33 | + | -56.68 | + | 0.18 | + | | + | | = | -63 |
| Mspan = | 7.75 | + | | + | | + | | + | | = | 8 |
| Va = | 14.82 | + | -29.83 | + | 13 | + | 200 | + | | = | 198 |
| Vd = | 14.82 | + | 29.83 | + | 13 | + | 200 | + | | = | 258 |
| Ha = Hd = | -2.30 | + | -20.61 | + | 1.01 | + | | + | -20.61 | = | -43 |

( Note: Load due to Cable trays on outside face of lattice girder of added separately in member design.)

## 12.3.2.10 Member Forces in Braced Columns—Bottom Part of Trestle

**a. DL + LL(Pipe full)+ WL**
Wind moment :  H1    41.22    x    8.25    =    340 kNm
               H2    35.82    x    5.5     =    197
                                                 537 kNm

| Axial load / Column leg : | | | | | | Leg 1 Comp kN | Leg 2 Tension kN | Any Shear kN |
|---|---|---|---|---|---|---|---|---|
| w1 | | 7.8 | x | 1.9 | = | 14.82 | 14.82 | |
| w2 | | 6.6 | x | 1.9 | = | 12.54 | 12.54 | |
| P1 | | | | | = | 200 | 200 | |
| P2 | | | | | = | 13 | 13 | |
| Wind | | 537 | / | 3.8 | = | 141 | -141 | 43 |
| | | | | | | 382 | 99 | 43 |

Design of Industrial Structures

**b. DL+ WL+ % LL (Pipe empty)**

| | | | | | | | Leg 1 | Leg 2 | Any |
|---|---|---|---|---|---|---|---|---|---|
| | | | | | | | kN | kN | kN |
| w1 | 0.7 | x | 7.8 | x | 1.9 | = | 10.37 | 10.37 | |
| w2 | 0.7 | x | 6.6 | x | 1.9 | = | 8.78 | 8.78 | |
| P1 | 0.7 | x | 200 | | | = | 140 | 140.00 | |
| P2 | 0.7 | x | 13 | | | = | 9 | 9.10 | |
| Wind | | | 537 | / | 3.8 | = | 141 | -141 | 43 |
| | | | | | | | 310 | 27 | 43 |

### 12.3.2.11 Member Forces in Column bracings

Diag: Base Shear = 43 kN
Height of panel = 2.75 m.
width = 3.8 m.
Diagonal length = 4.69 m.    Force/member =    53 kN.

Horz: Axial force = Base shear + 2.5 % of Col load = 53.00 kN

### 12.3.2.12 Summary of Member Forces and Designed Section

| Description | Mem mkd | Force and Moment | | | | | | Span / Length | | Member section |
|---|---|---|---|---|---|---|---|---|---|---|
| | | Comp | Ten | Shear | UDL | Mspan | Msup | Lx | Ly | |
| | | kN | kN | kN | kN/m | kNm | kNm | m | m | |
| | | | | | | | | (IS 800 2007 Table 11) | | |
| Cross beam: | | | | | | | | | | |
| Lev 3 | 1 | 20.64 | | 14.82 | 7.8 | 14.08 | | 3.8 | 3.8 | ISMC 200 |
| Lev 2 | 2 | | | 12.54 | 6.6 | 11.91 | | 3.8 | 3.8 | ISMC 200 |
| Lev 1 | 3 | 17.91 | | 12.54 | 6.6 | 11.91 | | 3.8 | 3.8 | ISMC 200 |
| Member forces in Lattice Girder | | | | | | | | | | |
| Top chord | 4 | 147 | | | | | | 2.0 | 2.0 | 2 ISA 75 75 8 |
| Bot chord | 5 | | 145 | | | | | 2.0 | 2.0 | 2 ISA 75 75 6 |
| Diagonal | 6 | | 148 | | | | | 3.40 | 3.40 | 2 ISA65 65 6 |
| Vertical | 7 | 80.02 | | | | 5.76 ** | | 2.75 | 2.75 | ISMC 200 |
| Plan Brcg. | | | | | | | | | | |
| Level 3 | | | | | | | | | | |
| Diagonal | 8 | 23.30 | 23.30 | | | | | 4.29 | 2.15 | ISA 75 75 8 |
| Vertical | 1 | 20.64 | | | | | | 3.80 | 3.80 | |
| Level 1 | | | | | | | | | | |
| Diagonal | 9 | 20.22 | 20.22 | | | | | 4.29 | 2.15 | ISA 75 75 8 |
| Vertical | 3 | 17.91 | | | | | | 3.80 | 3.80 | |
| Portal frame | | | | | | | | | | |
| Top Beam | 10 | | | | 7.8 | 8 | 63 | 3.8 | 3.8 | ISMB 250 |
| Column | 11 | 258 | 0 | 43 | | | 63 | 5.5 (2L) | 2.75 | Built up section |
| Col Brcg. | | | | | | | | | | |
| Diagonal | 12 | 53.00 | 53.00 | | | | | 4.69 | 4.69 | 2 ISA 75 75 8 |
| Horizontal | 13 | 53.00 | | | | | | 3.8 | 3.8 | 2 ISA 75 75 8 |
| Main Col | | | | | | | | | | |
| | 14 | 382 | 0 | 43 | | | | 2.75 | 5.5 | ISMB 250 |
| | | 310 | 0 | 43 | | | | | | |

(** from cable tray)    Mazor axix = x-x    Minor axis = y-y

## 12.3.2.13 Member Design According to IS 800 : 2007 [By Working Stress Method]

**Member 1**    **Cross beam**    Bending about mazor axis x-x

| | | | | | | |
|---|---|---|---|---|---|---|
| Span | 3.8 m | $L_x =$ | 3.8 m | UDL | 7.8 kN/m |
| | | $L_y =$ | 3.8 m | Mspan | 14.08 kNm |
| | | | | Shear | 14.82 kNm |

Try with    **ISMC 200**    wt/m    22.1 KG

**Flexural strength:**
*Lateraly unsupported beam ( susceptible to Lateral-torsional buckling )*

| Section | ISMC 200 | |
|---|---|---|
| Laterally unsupported beam | | |
| (IS 800:2007  8.2.2) | | |
| $L_{LT} =$ | 2660 | mm |
| B = | 75 | mm |
| T (tf) = | 11.4 | mm |
| D = | 200 | mm |
| tw = | 6.1 | mm |
| Zp = | 215497 | mm3 |
| Zx (Ze) = | 181900 | mm3 |
| $\gamma_{m0} =$ | 1.10 | |
| $\beta =$ | 1 | |
| E = | 200000 | Mpa |
| fy = | 250 | Mpa |
| hf = | 188.6 | mm |
| Iy = | 1404000 | mm4 |
| ry = | 22.3 | mm |
| Ix = | 18193000 | mm4 |
| $\alpha_{LT}$ | 0.21 | |
| $f_{bd} =$ | 170 | Mpa |
| Md = $\beta$ . Zp .$f_{bd}$ | | |
| **Md =** | 37 | Mpa |

$L_{LT} =$  0.7 Ly    (IS 800:2007 - Table 15)
$M_{cr} = ( \pi^2 EI_y hf / 2 L_{LT}2 ) . [ 1 + (1/20) . ((L_{LT} / r_y) / (hf/tf))^2 ]^{0.5}$
$\pi^2 E.I_y .hf / 2. L_{LT}2 =$  36898169
$((L_{LT} / r_y) / (hf/tf))^2 =$  51.99
$M_{cr} =$  70004491  Nmm

$\lambda_{LT} = (\beta Z_p fy / M_{cr})^{0.5}$  <= $(1.2 Z_e fy / M_{cr})^{0.5}$
$(\beta Z_p fy / M_{cr})^{0.5} =$  0.877
$(1.2 Z_e fy / M_{cr})^{0.5} =$  0.883
$\lambda_{LT} =$  0.877

$\phi_{LT} = 0.5 [ 1 + \alpha_{LT} (\lambda_{LT} - 0.2) + \lambda_{LT}2 ]$
$\phi_{LT} =$  0.96

$\chi_{LT} =$  $1 / [ \phi_{LT} + (\phi_{LT}2 - \lambda_{LT}2)^{0.5} ]$  <= 1
    =  0.749

$f_{bd} = \chi_{LT} fy / \gamma_{mo}$    =    170 Mpa

# Design of Industrial Structures

**Deflection:** (IS 800:2007 - 5.6.1)

$d = 5 w L^4 / 384 EI = $ 5 x 7.8 x 3800 ^4 / 384 x 200000 x 18193000 = 5.82 mm
Allowable deflection limit L /300 = 12.67 mm
OK.

**Flexural strength, Ma :** *(IS 800-11.4.1)*
For laterally Unsupported beam:

fabc = fabt = 0.6 Md/Zx          (Zx top = Zx bot)     Md =      37 knM

$\quad$ Ma = fabc. Zx = (0.6Md/Zx) Zx     (fabc = fabt)
$\qquad\quad$ = 0.6 Md
$\qquad\quad$ = 0.6 x 37 =          22 kNm

$\quad$ Allowable Flexural mom, **Ma** =   **22** kNm.
$\quad$ Actual Design Moment, Ms =   14 kNm.        **Ma > Ms; Safe.**

**Shear strength, Vs :** (IS 800-11.4.2)
Pure shear
$\quad$ Va = $\tau_{ab}$ . Av ,  where Av shear area
$\qquad$ = 0.4 fy x Av
$\qquad$ = 0.4 x 250 x 200 x 6.1/1000
$\qquad$ = 122 kN

$\quad$ Allowable Shear, **Va** =   **122** kN.
$\quad$ Actual design shear, Vs =   15 kN.          **Va > Vs; Safe.**

**Axial compression and bending:**

$\quad$ Length   3.8 m       Lx (Lz) =   3.8 m        Axial Comp =   21 kN
$\qquad\qquad\qquad\qquad\quad$ Ly =       3.8 m        Mspan =        14.08 kNm

| Section properties: | | | | Member dimensions | | | | | | |
|---|---|---|---|---|---|---|---|---|---|---|
| Section Provided: | s | | | wt/m (kg) | $A_g$ (mm$^2$) | $r_x (r_z)$ (mm) | $r_y$ (mm) | $Z_x(Z_e)$ (mm$^3$) | $I_x$ (mm$^4$) | $I_y$ (mm$^4$) | $C_{yy}$ (mm) |
| 1 | ISMC 200 | 0 | mm b/b | 22.1 | 2821 | 80.30 | 22.0 | 181900 | 18193000 | 1404000 | 21.7 |
| z----[-----z | | | | b mm | d mm | tf mm | tw mm | Connection bolt | | | |
| | | | | | | | | nos | $\phi$dia | $\phi$hole | rows |
| | | | | 75 | 200 | 11.4 | 6.1 | 4 | 16 | 18 | 2 |

**Buckling class**

$\qquad\qquad$ Buckling class =   c   for buckling about any axis   (IS 800:2007 - Table 10)
$\qquad\quad$ Imperfaction factor, α =   0.49                          (IS 800:2007 - Table 7)

**Proportioning limit**                                              (IS 800:2007 - Table 2)

$\quad$ ε = (250 / fy ) ^0.5 =   (250 / 250)^0.5 =    1.00

Outstanding leg
$\quad$ **b / t =**       75 / 11.4 =            7       Plastic.         (IS 800:2007 - 3.7.2)
$\quad$ *Limiting width to thickness ratio:*
$\quad$ Plastic, λr =           9.4ε = 9.4 x 1 =           9
$\quad$ Compact, λp =        10.5e = 10.5 x 1 =         11
$\quad$ Semi compact, λsp =  15.7e = 15.7 x 1 =         16

### Effective Lengths (L) and Slenderness ratio ( L/r ) :      (IS 800:2007 - Table 11)

| Laterally Unsupported length | Lz | 3.8 M | Kz | 1 | KLz | 3.80 M |
|---|---|---|---|---|---|---|
|  | Ly | 3.8 M | Ky | 0.7 | Kly | 2.66 M |

(IS 800:2007 - Table 15)

Maxm slenderness ratio (L/r) shall not exceed     **250**     (IS 800:2007 - Table 3)

Here, $KLz/r_z$ =    3800/80.3 =    47
       $KLy/r_y$ =    2660/22 =    121
             L/r =    **121**    <250, OK.

### Compression strength
*Compression due to buckling about minor axis, y-y (KLy / ry)*      (IS 800:2007 7..5.1.1)

Design compression strength, $P_{dy}$ = $A_e \cdot f_{cd}$     where, $A_e$ = effective sectional area
                                                               $f_{cd}$ = design compressive stress

$f_{cd}$ = $\chi f_y / \gamma_{mo}$    $\leq f_y / \gamma_{mo}$      (IS 800:2007 7.1.2.1)
$\chi$ = $1 / [\phi + (\phi^2 - \lambda^2)^{0.5}]$
$\phi$ = $0.5 [1 + \alpha(\lambda - 0.2) + \lambda^2]$
$\lambda$ = $(f_y / f_{cc})^{0.5}$

$f_{cc}$ = $\pi^2 E / (KLz/r_z)^2$      Euler buckling stress

Now, putting following values into above equation

| $\alpha$ = | 0.49 | E = | 200000 Mpa | $f_y$ = | 250 Mpa |
|---|---|---|---|---|---|
| $\gamma_{m0}$ = | 1.10 | $KLy/r_y$ = | 121 | $A_e$ = | 2821 mm2 |

$f_{cc}$ =   $3.14^2 \times 200000 / 121^2$ =        135 MPa
$\lambda$ =   $(250 / 135)^{0.5}$ =        1.36
$\phi$ =   $0.5[1 + 0.49 \times (1.36 - 0.2) + 1.36^2]$ =    1.71
$\chi$ =   $1 / [1.71 + (1.71^2 - 1.36^2)^{0.5}]$ =    0.364
$f_{cd}$ =   $0.364 \times 250 / 1.1$ =    83 Mpa    $\leq f_y / \gamma_{mo}$    $\leq$    250 / 1.1
                                                                                                   $\leq$    227 Mpa

$f_{cd}$ =    83 Mpa    (=$f_{cdy}$)

**$P_{dy}$** = $A_e \cdot f_{cd}$ =    $2821 \times 83 / 1000$ =    **234** kN

*Axial compression by working stress method:*      (IS 800:2007 11.3.1)

Permissible comp. stress, $f_{ac}$ = $0.6 f_{cd}$      $f_{cd}$ =    83 Mpa (see calc. above.)
                                                                  $A_g$ =    2821 mm2

Permissible Comp strength, $P_d$ = $0.6 f_{cd} \cdot A_g$    =    $0.6 \times 83 \times 2821$    =    140486 N
                                                                                             $P_d$ =    140 kN
                                                                            Actual load, P =    21 kN
                                                                                      P < Pd; Safe.

# Design of Industrial Structures

*Interaction factor :*

$P / Pd + Ms / Ma =$   $20.64 / 140 + 14.079 / 22.2 =$   $0.15 + 0.63 =$   0.78 < 1; Safe.

**Combined Axial bending and shear stress check:**   (IS 800:2007 11.5.4)
*axial stress*
fac = 0.6 x fcd =    0.6 x 83=    50 Mpa
*bending stress*
fabc = 0.6 x fbd =   0.6 x 170=   102 Mpa
*shear stress*
τb = Vs / Av =   14820 / (200x 6.1) =   12 Mpa

The equivalent stress, $fe = (fac^2 + fabc^2 + fac \cdot fabc + 3 \tau b^2)^{0.5}$  < 0.9 fy
$fe = (50^2 + 102^2 + 50 \times 102 + 3 \times 12^2)^{0.5} =$   136
$0.9 \times fy = 0.9 \times 250 =$   225

fe < 0.9fy; Safe.

## Member 2 & 3
Provide same section as in member 1

## Lattice Girder
**Member 4**   Lattice Girder top chord
Length   2.0 m   Lx (Lz) =   2.0 m   Axial Comp =   147.0 kN
Ly =   2.0 m

| Section properties: | | | Built-up member dimensions | | | | | | |
|---|---|---|---|---|---|---|---|---|---|
| Section Provided: | | s | wt/m (kg) | $A_g$ (mm$^2$) | $r_x$ (mm) | $r_y$ (mm) | $r_{vv}$ (mm) | $I_x$ (mm$^4$) | $I_y$ (mm$^4$) | $C_{yy}$ (mm) |
| 2 | ISA 75 75 8 | 12 mm b/b | 17.8 | 2276 | 22.80 | 36.0 | 14.5 | 1180000 | 2888730 | 21.4 |
| | | | Single Leg dimensions | | | | | | |
| | | | b | d | t | Connection bolt | | | An |
| | | | mm | mm | mm | nos | φdia | φhole | rows | mm2 |
| | | | 75 | 75 | 8 | 4 | 20 | 22 | 1 | 2100 |

where, An = net area after deduction of bolt holes.
An =   Ag - (no of rows x φ hole x t)
Distance between tacking plate (welded spacer plate) = **a =**   256 mm   (IS 800:2007 - 10.2.5.2)
(the value of ' a ' shall not exceed 32 t or 300 mm, whichever is less)

**Buckling class**
Buckling class =   c   for buckling about any axis   (IS 800:2007 - Table 10)
Imperfaction factor, α =   0.49   (IS 800:2007 - Table 7)

**Proportioning limit**   (IS 800:2007 - Table 2)
$\varepsilon = (250 / fy)^{0.5} =$   $(250 / 250)^{0.5} =$   1.00

## Design of Industrial Structures

Outstanding leg
b / t =        75 / 8 =          9        Plastic.        (IS 800:2007 - 3.7.2)
*Limiting width to thickness ratio:*
Plastic, $\lambda r$ =      $9.4\varepsilon = 9.4 \times 1$ =        9
Compact, $\lambda p$ =    $10.5e = 10.5 \times 1$ =       11

Semi compact, $\lambda sp$ =   $15.7e = 15.7 \times 1$ =    16
**Effective Lengths (L) and Slenderness ratio ( L/r ) :**     (IS 800:2007 - Table 11)

Laterally Unsupported length    Lz    2.0 M    Kz    0.85    KLz    1.70 M
                                Ly    2.0 M    Ky    1       Kly    2.00 M

Maxm slenderness ratio (L/r) shall not exceed        250        (IS 800:2007 - Table 3)
Here,   KLz/rz =         1700/22.8 =         75
        KLy/ry =         2000/36 =           56
                         L/r =               75        <250, OK.

**Sectional capacity / Design strength:**

**Axial compression**
*(i) Compression due to buckling about major axis, z-z [Lz / rz (=rx)]*     (IS 800:2007 7.5.2.1)

Design compression strength, Pdz = Ae. fcd            (IS 800:2007 7.1.2)
                                            where, Ae = effective sectional area
                                                   fcd = design compressive stress

fcd =    $\chi$ fy / $\gamma_{mo}$    <= fy / $\gamma_{mo}$
$\chi$ =    $1 / [\phi + (\phi^2 - \lambda^2)^{0.5}]$
$\phi$ =    $0.5 [ 1 + \alpha(\lambda - 0.2) + \lambda^2]$
$\lambda$ =    $(fy / fcc)^{0.5}$

fcc =    $\pi^2 E / (KLz/rz)^2$        Euler buckling stress

Now, putting following values into above equation

$\alpha$ =      0.49       E =       200000 Mpa       fy =      250 Mpa
$\gamma_{m0}$ = 1.10       KLz/rz =  75               Ae =      2276 mm2   =Ag

fcc =       $3.14^2 \times 200000 / 75^2$ =               351 MPa
$\lambda$ =  $(250 / 351)^{0.5}$=                          0.84
$\phi$ =     $0.5[1 + 0.49 \times (0.84 - 0.2) + 0.84^2$=  1.01
$\chi$ =     $1 /[1.01 + (1.01^2 - 0.84^2)^{0.5}]$=        0.637

fcd =    $0.637 \times 250 / 1.1$ =    145 Mpa    <= fy / $\gamma_{mo}$    <=  250 / 1.1
                                                                          <=  227 Mpa

fcd =       145 Mpa     (=fcdz)

**Pdz** = Ae . fcd =     $2276 \times 145 / 1000$=    **330 kN**

# Design of Industrial Structures

*(ii) Compression due to buckling about minor axis, y-y (KLy / ryy)*     (IS 800:2007 7.5.2.1)

Design compression strength, Pdy = Ae. fcd     where. Ae = effective sectional area
fcd = design compressive stress

$$fcd = \chi \, fy / \gamma_{mo} \quad <= fy / \gamma_{mo}$$     (IS 800:2007 7.1.2.1)
$$\chi = 1 / [\phi + (\phi^2 - \lambda^2)^{0.5}]$$
$$\phi = 0.5 [1 + \alpha(\lambda - 0.2) + \lambda^2]$$
$$\lambda = (fy / fcc)^{0.5}$$

$$fcc = \pi^2 E / (KLz/rz)^2 \quad \text{Euler buckling stress}$$

Now, putting following values into above equation

| | | | | | | |
|---|---|---|---|---|---|---|
| $\alpha =$ | 0.49 | E = | 200000 Mpa | fy = | 250 Mpa | |
| $\gamma_{m0} =$ | 1.10 | KLy/ryy= | 56 | Ae = | 2276 mm2 | |

fcc = 3.14^2 x 200000 / 56^2 =           629 MPa
$\lambda =$ (250 / 629)^0.5=              0.63
$\phi =$ 0.5[1 + 0.49 x (0.63 - 0.2) + 0.63^2 =   0.8
$\chi =$ 1 /[0.8 + (0.8^2 - 0.63^2)^0.5]=         0.773

fcd = 0.773 x 250 / 1.1 =     176 Mpa     <= fy / $\gamma_{mo}$     <=   250 / 1.1
                                                                    <=   227 Mpa

fcd =     176 Mpa     (=fcdy)

**Pdy** = Ae . fcd =     2276 x 176 /1000=     **401 kN**

*The design compressive strength,* **Pd** *will be lowest of above calculated strengths of the member.*

    **Pd =**     **330 kN**     and     fcd =     145 Mpa


**Axial compression :**     (IS 800:2007 11.3.1)

Permissible comp. stress, fac = 0.6 fcd                    fcd =     145 Mpa
                                                           Ag =      2276 mm2

Permissible Comp strength, Pd = 0.6 fcd .Ag
                              = 0.6 x 145 x 2276
                              = 198012 N
                              = 198 kN
                Actual load, P = 147 kN
                                 P < Pd; Safe.

**Member 5**  **Lattice Girder bottom chord**
　　Length　2.0 m　　Lx (Lz) = 　2.0 m　　Axial tension = 　145 kN
　　　　　　　　　　Ly = 　　　2.0 m

| Section properties: | | | | Built-up member dimensions | | | | | | |
|---|---|---|---|---|---|---|---|---|---|---|
| Section Provided: | s | | | wt/m (kg) | $A_g$ (mm$^2$) | $r_x$ (mm) | $r_y$ (mm) | $r_{vv}$ (mm) | $I_x$ (mm$^4$) | $I_y$ (mm$^4$) | $C_{yy}$ (mm) |
| 2　ISA 75 75 6 | 12 | mm b/b | | 13.6 | 1732 | 23.00 | 35.0 | 14.6 | 914000 | 2139494 | 20.6 |
| | | | | Single Leg dimensions | | | | | | |
| | | | | b | d | t | Connection bolt | | | An |
| | s | | | mm | mm | mm | nos | φdia | φhole | rows | mm2 |
| | | | | 75 | 75 | 6 | 4 | 20 | 22 | 1 | 1600 |

where,  An = net area after deduction of bolt holes.
　　　An =　Ag - (no of rows x φ hole x t)
Distance between tacking plate (welded spacer plate) = **a** = 　　192　mm　　(IS 800:2007 - 10.2.5.2)
(the value of ' a ' shall not exceed 32 t or 300 mm, whichever is less)

**Buckling class**
　　　　　　　　Buckling class = 　c　　for buckling about any axis　　(IS 800:2007 - Table 10)
　　　　　　Imperfaction factor, $\alpha$ = 　0.49　　　　　　　　　　　　(IS 800:2007 - Table 7)

**Proportioning limit**　　　　　　　　　　　　　　　　　　　　　　(IS 800:2007 - Table 2)

　　$\varepsilon = (250 / f_y)^{0.5}$ = 　$(250/250)^{0.5}$ = 　　1.00

Outstanding leg
　　**b / t =** 　　75 / 6 = 　　13　　Semi compact　　(IS 800:2007 - 3.7.2)
　　_Limiting width to thickness ratio:_
　　Plastic, $\lambda r$ = 　　　9.4$\varepsilon$ = 9.4 x 1 = 　　　9
　　Compact, $\lambda p$ = 　　10.5e = 10.5 x 1 = 　　11
　　Semi compact, $\lambda sp$ = 　15.7e = 15.7 x 1 = 　　16

**Effective Lengths (L) and Slenderness ratio ( L/r ) :**　　(IS 800:2007 - Table 11)

　Laterally Unsupported length　　Lz　2.0 M　　Kz　0.85　　KLz　1.70 M
　　　　　　　　　　　　　　　　 Ly　2.0 M　　Ky　1　　　Kly　2.00 M

　　　Maxm slenderness ratio (L/r) shall not exceed　　250　　(IS 800:2007 - Table 3)
　　　Here,　KLz/rz = 　1700/23 = 　　74
　　　　　　KLy/ry = 　2000/35 = 　　57
　　　　　　　　　　**L/r =**　　　74　　<250, OK.

# Design of Industrial Structures

**Sectional capacity / Design strength:**

**Axial Tension**
Design tensile strength of member = Td

*(i) Design tensile strength due to yielding of gross section, Tdg*  (IS 800:2007 - 6.2)

$$Tdg = Ag\ fy\ /\ \gamma_{mo} = \quad 1732 \times 250\ /\ 1.1\ /\ 1000 = \quad 394\ kN$$

*(ii) Design tensile strength due to rupture of critical section, Tdn*

$Tdn = 0.9\ Anc.\ fu\ /\ \gamma_{m1} + \beta.\ Ago.\ fy\ /\ \gamma_{mo}$   (IS 800:2007 - 6.3.3)
$\beta = 1.4 - 0.076\ (w/t)\ (fy/fu)\ (bs/Lc) \ <=\ 0.9\ fu\ \gamma_{mo}\ /\ fy\ \gamma_{m1} >= 0.7$

where,  Anc = net area of connected leg =  $2 \times [75 \times 6 - (1 \times 22 \times 6)] =$   636 mm2
         Ago = gross area of outstanding leg =  $2 \times (75 \times 6) =$   900 mm2

         w = width of outstanding leg = d =                                   75 mm
         bs = shear lag width = w + w1 (= Cyy) - t =   75 + 20.6 - 6 =        90 mm
         Lc = length of end connection = (nos of bolt -0.5) x 3 φ hole =     231 mm

         f u =   410 Mpa        $\gamma_{m0} =$   1.10      t =    6 mm
         fy =    250 Mpa        $\gamma_{m1} =$   1.25

$\beta = 1.4 - 0.076\ (w/t)\ (fy/fu)\ (bs/Lc)\ <=\ 0.9\ fu\ \gamma_{mo}\ /\ fy\ \gamma_{m1} >= 0.7$
  =  1.4 - 0.0076 x (75/6) x (250/410) x (90/231)  <=  0.9 x 410 x 1.1 / 250 x 1.25 > = 0.7
  =        1.17         <=      1.30       >=     0.7        Ok.

Tdn = 0.9 Anc. fu / $\gamma_{m1}$ + β. Ago. fy / $\gamma_{mo}$
   =  0.9 x 636 x 410 / 1.25 + 1.17 x 900 x 250 / 1.1
   =  427065 N
   =  427 kN

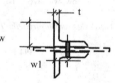

*(iii) Design tensile strength due to block shear, Tdb*   (IS 800:2007 - 6.4.1)

a)   Tdb = [ Avg. fy / ( 1.732 $\gamma_{mo}$) + 0.9 Atn. fu / $\gamma_{m1}$]      OR

b)   Tdb = (0.9 Avn. fu / (1.732 $\gamma_{m1}$ ) + Atg. fy / $\gamma_{mo}$ )     whichever is smaller

$A_{vg}$ = gross shear area along bolt line parallel to line (1-2) of external force
= 2 . Lc . t
= 2 x 231 x 6   =   2772 mm2
$A_{vn}$ = net shear area along bolt line parallel to the line (1-2) of external force
= 2 [Lc . t - ( no of bolts x $\phi$ hole x thickness)]
= 2 x [ 231 x 6 - (4x22x6) ] =    1716 mm2

$A_{tg}$ = gross area in tension from bolt hole to toe of the angle, perpendicular to line (2-3) of force
= 2 . (d - Cyy) . t
= 2 x (75 - 20.6) x 6 ) =    653 mm2
$A_{tn}$ = net area in tension from bolt hole to toe of the angle, perpendicular to line (2-3) of force
= 2 [ (d - Cyy) . t - ( no of rows x $\phi$ hole x thickness)]
= 2 [ (75 - 20.6) x 6 - (4 x 22 x 6) ] =    389 mm2

a)   $T_{db}$ = [ $A_{vg}$ . fy / ( 1.732 $\gamma_{mo}$) + 0.9 $A_{tn}$ . fu / $\gamma_{m1}$]
   = 2772 x 250 /(1.732 x 1.1) + 0.9 x 388.8 x 410 / 1.25
   = 478515 N
   = 479 kN
   OR
b)   $T_{db}$ = (0.9 $A_{vn}$. fu / (1.732 $\gamma_{m1}$ ) + $A_{tg}$. fy / $\gamma_{mo}$ )    whichever is smaller
   = 0.9 x 1716 x 410 / (1.732 x 1.25) + 652.8 x 250 /1.1
   = 440837 N
   = 441 kN

So, Tensile strength due to block shear, $T_{db}$ =    441 kN

now, we have   $T_{dg}$ =   394 kN,   $T_{dn}$ =   427 kN and   $T_{db}$ =   441 kN

The design tensile strength, **Td** will be lowest of above calculated strengths of the member.

**Td =   394 kN**


**Axial Tension:**                                           (IS 800:2007  11.2.1)

Tensile stresses for :
a) yielding    fat = 0.6 fy =    0.6 x 250    =    150 MPa

b) rupture    fat = 0.69 Tdn/Ag    =    0.69 x 427000 / 1732    =    170 MPa

c) block shear    fat = 0.69 Tdb/Ag    =    0.69 x 441000 / 1732    =    176 MPa
Permissible value of fat =    150 MPa

Permissible Tensile strength, Td = fat .Ag    =    150 x 1732
                                                =    259800 N
                                                =    260 kN
                         Actual load, T =    145 kN   T < Td; Safe.

# Design of Industrial Structures

**Member 6**     Lattice Girder diagonals

      Length     3.40 m          Lx (Lz) =  3.40 m          Axial tension =   148 kN
                                 Ly =       3.40 m

| Section properties: | | | | Built-up member dimensions | | | | | | |
|---|---|---|---|---|---|---|---|---|---|---|
| Section Provided: | | s | | wt/m (kg) | $A_g$ (mm$^2$) | $r_x$ (mm) | $r_y$ (mm) | $r_{vv}$ (mm) | $I_x$ (mm$^4$) | $I_y$ (mm$^4$) | $C_{yy}$ (mm) |
| 2 | ISA 65 65 6 | 12 | mm b/b | 11.6 | 1488 | 19.80 | 31.0 | 12.6 | 582000 | 1446245 | 18.1 |
| | | | | Single Leg dimensions | | | | | | |
| | | | | b | d | t | Connection bolt | | | An |
| | | | | mm | mm | mm | nos(n) | $\phi$dia | $\phi$hole | rows | mm2 |
| | | | | 65 | 65 | 6 | 3 | 16 | 18 | 1 | 1380 |

where, An = net area after deduction of bolt holes.
    An =   Ag - (no of rows x $\phi$ hole x  t)
Distance between tacking plate (welded spacer plate) =  **a** =     192 mm     (IS 800:2007 - 10.2.5.2)
(the value of ' a ' shall not exceed 32 t or 300 mm, whichever is less)

| Axial Tensile Strength, Td : | | |
|---|---|---|
| (IS 800:2007 - 6.2; 6.3 & 6.4) | | |
| fy | 250 | Mpa |
| fu | 410 | Mpa |
| E | 200000 | Mpa |
| $\gamma_{m0}$ | 1.1 | |
| $\gamma_{m1}$ | 1.25 | |
| Tdg = Ag fy / $\gamma_{mo}$ = | 338 | kN |
| Anc = net area of  b = | 564 | mm2 |
| Ago = gr. area of d = | 780 | mm2 |
| w = d = | 65 | mm |
| bs = shear lag   = w +  $C_{yy}$ - t | | |
| bs = | 77 | mm |
| Lc | 135 | mm |
| x     > β >      0.7 | | |
| β =  1.4 - 0.076 (w/t) (fy/fu) (bs/Lc) | | |
| x = 0.9 fu $\gamma_{mo}$ / fy  $\gamma_{m1}$ | 1.30 | |
| β = | 1.11 | Ok. |
| Tdn = 0.9 Anc. fu / $\gamma_{m1}$ + β. Ago. fy / $\gamma_{mo}$ | | |
| Tdn = | 363 | kN |
| Avg = 2 Lc t = | 1620 | mm2 |
| Avn = 2(Lc.t - n. $\phi$hol.t) = | 972 | mm2 |
| Atg = 2 (d - $C_{yy}$) . t = | 562.8 | mm2 |
| Atn = 2 [(d - $C_{yy}$)t  - (n x $\phi$ hol x t)] | | |
| Atn = | 346.8 | mm2 |
| Tdb1=[Avg. fy/( 1.732 $\gamma_{mo}$)+0.9Atn.fu/$\gamma_{m1}$] | | |
| Tdb1= | 315 | kN |
| Tdb2 =(0.9 Avn.fu/(1.732 $\gamma_{m1}$)+Atg.fy/$\gamma_{mo}$) | | |
| Tdb2 = | 294 | |
| Tdb =min(Tdb1, Tdb2) = | 294 | kN |
| Tdb = min(Tdg, Tdn,Tdb) | | |
| Td = | 294 | kN |

**Axial Tension:**          (IS 800:2007 11.2.1)

Tensile stresses for :
a) yielding       fat = 0.6 fy =     0.6 x 250
                                          =        150 MPa
b) rupture         fat = 0.69 Tdn/Ag
                              =       0.69 x 363000 / 1488
                              =        168 MPa
c) block shear    fat = 0.69 Tdb/Ag
                             =        0.69 x 294000 / 1488
                             =        136 MPa

Permissible value of fat =         136 MPa

Permissible Tensile strength, Td = fat .Ag
                              =        136 x 1488
                              =        202368 N
                              =        202 kN
             Actual load, T =       148 kN

                                    T < Td; Safe.

**Member 7**  **Lattice Girder verticals**

   Length   2.8 m       Lx (Lz) =   2.8 m       Axial Comp =   80 kN
                         Ly =        2.8 m       Mspan =        5.76 kNm

| Section properties: | | | | Member dimensions | | | | | | |
|---|---|---|---|---|---|---|---|---|---|---|
| Section Provided: | | s | | wt/m (kg) | $A_g$ (mm$^2$) | $r_x (r_z)$ (mm) | $r_y$ (mm) | $Z_x(Z_e)$ (mm) | $I_x$ (mm$^4$) | $I_y$ (mm$^4$) | $C_{yy}$ (mm) |
| 1 | ISMC 200 | 0 | mm b/b | 22.1 | 2821 | 80.30 | 22.0 | 181900 | 18193000 | 1404000 | 21.7 |
| | | | | b | d | tf | tw | Connection bolt | | | |
| | | | | mm | mm | mm | mm | nos | φdia | φhole | rows |
| | | | | 75 | 200 | 11.4 | 6.1 | 4 | 16 | 18 | 2 |

**Buckling class**

   Buckling class =   c    for buckling about any axis    (IS 800:2007 - Table 10)
   Imperfaction factor, α =   0.49                        (IS 800:2007 - Table 7)

**Proportioning limit**                                    (IS 800:2007 - Table 2)

   ε = (250 / fy )^0.5 =   (250 / 250)^0.5 =   1.00

# Design of Industrial Structures

Outstanding leg
b / t =     75 / 11.4 =     7     Plastic.     (IS 800:2007 - 3.7.2)

*Limiting width to thickness ratio:*
Plastic, $\lambda r$ =     9.4$\varepsilon$ = 9.4 x 1 =     9
Compact, $\lambda p$ =     10.5e = 10.5 x 1 =     11
Semi compact, $\lambda sp$ =     15.7e = 15.7 x 1 =     16

**Effective Lengths (L) and Slenderness ratio ( L/r) :**     (IS 800:2007 - Table 11)

Laterally Unsupported length    Lz   2.8 M    Kz   1    KLz   2.75 M
                                  Ly   2.8 M    Ky   0.85    Kly   2.34 M

Maxm slenderness ratio (L/r) shall not exceed     250     (IS 800:2007 - Table 3)
Here,    KLz/rz =    2750/80.3 =    34
         KLy/ry =    2337.5/22 =    106
                     L/r =    106     <250, OK.

**Compression strength**
*Compression due to buckling about minor axis, y-y (KLy / ry)*     (IS 800:2007 7..5.1.1)

Design compression strength, Pdy = Ae. fcd     where,    Ae =    effective sectional area
                                                   fcd =   design compressive stress

                 fcd =    $\chi$ fy / $\gamma_{mo}$    <= fy / $\gamma_{mo}$     (IS 800:2007 7.1.2.1)
                 $\chi$ =    1 / [ $\phi$ + ( $\phi^2 - \lambda^2$)^0.5]
                 $\phi$ =    0.5 [ 1 + $\alpha$($\lambda$ - 0.2) + $\lambda^2$]
                 $\lambda$ =    (fy / fcc)^0.5

                 fcc =    $\pi^2$ E / (KLz/rz)^2     Euler buckling stress

Now, putting following values into above equation

         $\alpha$ =    0.49     E =    200000 Mpa     fy =    250 Mpa
         $\gamma_{m0}$ =    1.10     KLy/ry =    106                Ae =    2821 mm2

fcc =    3.14^2 x 200000 / 106^2 =     176 MPa
$\lambda$ =    (250 / 176)^0.5=     1.19
$\phi$ =    0.5[1 + 0.49 x (1.19 - 0.2) + 1.19^2]=     1.45
$\chi$ =    1 /[1.45 + (1.45^2 - 1.19^2)^0.5]=     0.439
fcd =    0.439 x 250 / 1.1 =     100 Mpa    <= fy / $\gamma_{mo}$    <=    250 / 1.1
                                                                                     <=    227 Mpa
fcd =    100 Mpa    (=fcdy)

**Pdy** = Ae . fcd =     2821 x 100 /1000=     **282** kN

*Axial compression by working stress method:* (IS 800:2007 11.3.1)

Permissible comp. stress, fac = 0.6 fcd  fcd = 100 Mpa (see calc. above.)
  Ag = 2821 mm2

Permissible Comp strength, Pd = 0.6 fcd .Ag = 0.6 x 100 x 2821 = 169260 N
  Pd = 169 kN
  Actual load, P = 80 kN
  P < Pd; Safe.


**Flexural strength**
*Lateraly unsupported beam ( susceptible to Lateral-torsional buckling )*

Design flexural strength = Md = $\beta$ . Zp .$f_{bd}$  (IS 800:2007 - 8.2.2)

The expressions are
$f_{bd} = \chi_{LT}$ fy / $\gamma_{mo}$
$\chi_{LT}$ = 1 / [ $\phi_{LT}$ + ($\phi_{LT}^2 - \lambda_{LT}^2$)^0.5] <= 1
$\phi_{LT}$ = 0.5 [ 1 + $\alpha_{LT}$ ($\lambda_{LT}$ - 0.2) + $\lambda_{LT}^2$]
$\lambda_{LT}$ = ($\beta$ Zp fy / Mcr)^0.5  <= (1.2 Ze fy / Mcr)^0.5

Mcr = ( $\pi^2$ EIy hf / 2 $L_{LT}^2$) . [ 1 + (1/20) . (($L_{LT}$ / ry) / (hf/tf))^2 ]^0.5

Now, putting following values into above equation
hf = 188.6 mm  Iy = 1404000  mm4  $L_{LT}$ = 2750 mm
tf = 11.4 mm  E = 200000  Mpa
ry = 22 mm
Zp = plastic sectional modulus about the x axis = $\Sigma$ A y
  = 2 x 75 x 11.4 x 188.6/ 2 + 2 x (177.2/2) x 6.1 x (177.2/4)
  = 209138  mm4

$\beta$ = 1  $\alpha_{LT}$  0.49  (IS 800:2007 - 8.2.2)
fy = 250 Mpa  Zp = 209138  mm3
$\gamma_{m0}$ = 1.10  Ze = 181900  mm3

we get,

# Design of Industrial Structures

$Mcr = (\pi^2 EIy \, hf / 2 L_{LT}2) \cdot [1 + (1/20) \cdot ((L_{LT}/ry)/(hf/tf))^2]^{0.5}$

$\pi^2 EIy \, hf / 2 L_{LT}2 = $ 3.14^2 x 200000 x 1404000 x 188.6 / 2 x 2750^2 = 34522537

$((L_{LT}/ry)/(hf/tf))^2 = $ ((2750/22) / (188.6/11.4) )^2 = 57.09

Mcr = 34522537 x [1 + (1/20) x 57.09]^0.5 = 67777685 Nmm

$\lambda_{LT} = (\beta \, Zp \, fy / Mcr)^{0.5} \quad <= \quad (1.2 \, Ze \, fy / Mcr)^{0.5}$

$(\beta \, Zp \, fy / Mcr)^{0.5} = $ (1 x 209138 x 250 / 67777685)^0.5 = 0.8783

$(1.2 \, Ze \, fy / Mcr)^{0.5} = $ (1.2 x 181900 x 250 / 67777685)^0.5 = 0.8973

$\lambda_{LT} = $ 0.8783 = **0.88**

$\phi_{LT} = 0.5 [1 + \alpha_{LT}(\lambda_{LT} - 0.2) + \lambda_{LT}2] = 0.5[1+0.49(0.88-0.2)+0.88^2] = $ 1.0538

$\chi_{LT} = 1 / [\phi_{LT} + (\phi_{LT}2 - \lambda_{LT}2)^{0.5}] \quad <= 1$
$\quad = 1 / [1.0538 + (1.0538^2 - 0.88^2)^{0.5}]$
$\quad = 0.612$

$f_{bd} = \chi_{LT} \, fy / \gamma_{mo} \quad = 0.612 \times 250 / 1.1 = $ 139 Mpa

Design flexural strength = Md = $\beta \cdot Zp \cdot f_{bd}$ = 1 x 209138 x 139/10^6 = 29 kNm

*If the value of $\lambda_{LT}$ is less than 0.4, Lateral torsional buckling need not be checked and member shall be treated as Laterally supported beam (IS 800:2007 8.2.1.2)*

Hence, the Design flexural strength will be **Md =** **29** kNm

*Allowable flexural strength by working stress method, Ma :*

fabc = fabt = 0.6 Md/Zx    (Zx top = Zx bot)

Md = 29 knM

Ma = fabc. Zx = (0.6Md/Zx) Zx    (fabc = fabt)
$\quad = 0.6 \, Md \quad = 0.6 \times 29 = $ 17 kNm

Allowable Flexural mom, Ma = 17 kNm.

Actual Design Moment, Mspan = Ms = 6 kNm.

**Ma > Ms; Safe.**

*Interaction factor :*

$P / Pd + Ms / Ma =$   $80.02 / 169 + 5.76 / 17.4 =$   $0.47 + 0.33 =$   $0.805 < 1$; Safe.

## Member 8    Plan bracings at Level 3

This member is designed as compression member carrying wind load and tied with the cross member at mid point. The cross member will be effective when subjected to reversal of stress into compression resulting wind from opposite direction.

| Length | 4.29 m | Lx (Lz) = | 4.29 m | Axial Comp = | 23 kN |
|---|---|---|---|---|---|
| | | Ly = | 2.15 m | Axial tension = | 23 kN |

**Section properties:**   all dimensions are in mm

| Section Provided: | wt/m (kg) | $A_g$ (mm$^2$) | $r_x$ (mm) | $r_y$ (mm) | $r_{vv}$ (mm) | $I_x$ (mm$^4$) | $I_y$ (mm$^4$) | $C_{yy}$ (mm) |
|---|---|---|---|---|---|---|---|---|
| ISA 75 75 8 | 8.9 | 1138 | 22.80 | 22.8 | 14.5 | 590000 | 590000 | 21.4 |
| | Leg sizes (mm) | | | Connection bolt | | | | An |
| | b | d | t | nos | φdia | φhole | rows | mm2 |
| | 75 | 75 | 8 | 3 | 16 | 18 | 1 | 994 |

where, An = net area after deduction of bolt holes.    An =    Ag - (no of rows x φ hole x t)

**Materials**

| Yield Stress | $f_y$ | 250 | MPa |
|---|---|---|---|
| Tensile Stress | $f_u$ | 410 | MPa |
| Elasticity | E | 200000 | MPa |
| Mat. safety fac | $\gamma_{m0}$ | 1.10 | |
| | $\gamma_{m1}$ | 1.25 | |

**Buckling class**

| | | | |
|---|---|---|---|
| Buckling class = | c | for buckling about any axis | (IS 800:2007 - Table 10) |
| Imperfection factor, α = | 0.49 | | (IS 800:2007 - Table 7) |

**Proportioning limit**    (IS 800:2007 - Table 2)

ε = (250 / fy )^0.5 =    (250 / 250)^0.5 =    1.00

| For single angle (axial comp.) | Ratio | | limit | |
|---|---|---|---|---|
| b / t = 75 / 8 = | 9 | 15.7 ε = | 15.7 | Semi compact |
| d / t = 75 / 8 = | 9 | 15.7 e = | 15.7 | Semi compact |
| (b+d) / t = (75+75) / 8 = | 19 | 25 ε = | 25 | Semi compact |

# Design of Industrial Structures

**Effective Lengths (L) and Slenderness ratio ( L/r ) :** (IS 800:2007 - Table 11)

Laterally Unsupported length   Lx   4.3 M     Kx   0.85   KLx   3.65 M
                                Ly   2.1 M     Ky    1     Kly   2.15 M

Maxm slenderness ratio (L/r) shall not exceed   250   (IS 800:2007 - Table 3)
Here,  $KLx/rx$ =   3646.5/22.8 =   160
       $KLy/rvv$ =  2145/14.5 =     148    **L/r =**   **160**   < 250; OK.

## Sectional capacity / Design strength:
### Axial compression
*(i) Compression due to flexural-torsional buckling (Ly/rvv)*  (IS 800:2007 - 7.5.1.2)
*when loaded through one leg*

Design compression strength, $Pd = Ae \cdot fcd$     where, $Ae$ = effective sectional area
                                                            $fcd$ = design compressive stress

$$fcd = \chi\, fy / \gamma_{mo} \quad <= fy / \gamma_{mo}$$
$$\chi = 1 / [\, \phi + (\phi^2 - \lambda^2)^{\wedge}0.5\,]$$
$$\phi = 0.5\,[\,1 + \alpha(\lambda - 0.2) + \lambda^2\,]$$

$\lambda = \lambda e = (k1 + k2.\, \lambda_{vv}^2 + k3.\, \lambda\varphi^2)^{\wedge}0.5$   where,  $k1$ =   0.7   (IS 800:2007 Table 12)
                                                                 $k2$ =   0.6   partially fixed at end
                                                                 $k3$ =    5

$\lambda_{vv}$ =   $( L/r_{vv} ) / [\varepsilon\, (\pi^2\, E\, /250)]$         L = Ly =   215 mm
        =   (215/14.5) / [1x (3.14^2 x 200000 / 250)]
        =   0.002

$\lambda_{\varphi}$ =   $[(b + d)/2.t\,] / [\varepsilon\, (\pi^2\, E\, /250)]$
        =   ((75+75)/ (2x8)) / [1(3.14^2x200000 / 250)]
        =   0.001

$\lambda = \lambda e = (k1 + k2.\, \lambda_{vv}^2 + k3.\, \lambda\varphi^2)^{\wedge}0.5$
        =   (0.7+0.6x0.002^2+5x0.001^2)^0.5
        =   **0.837**

Now, putting following values into above equations

$\alpha =$ 0.49   $E =$ 200000 Mpa   $fy =$ 250 Mpa
$\gamma_{m0} =$ 1.10   $KLy/rvv =$ 148   $Ae =$ 1138 mm2

$\lambda =$   $\lambda e =$ 0.837
$\phi =$   $0.5[1 + 0.49 \times (0.837 - 0.2) + 0.837^2] =$ 1.01
$\chi =$   $1 /[1.01 + (1.01^2 - 0.837^2)^{0.5}] =$ 0.635
fcd = $0.635 \times 250 / 1.1 =$   144 Mpa   $<= fy/\gamma_{mo}$   $<=$ 250 / 1.1
     $<=$ 227 Mpa

fcd = **144** Mpa

**Pd** = Ae . fcd =   $1138 \times 144 /1000 =$   **164** kN

<u>(ii) Compression due to buckling about major axis, z-z [Lz / rz (=rx)]</u>   (IS 800:2007 7.5.1.1)

Design compression strength, Pdz = Ae. fcd   (IS 800:2007 7.1.2)

where, Ae = effective sectional area
fcd = design compressive stress

fcd = $\chi \, fy / \gamma_{mo}$   $<= fy / \gamma_{mo}$
$\chi =$ $1 / [\phi + (\phi^2 - \lambda^2)^{0.5}]$
$\phi =$ $0.5 [1 + \alpha(\lambda - 0.2) + \lambda^2]$
$\lambda =$ $(fy / fcc)^{0.5}$

fcc = $\pi^2 E / (KLz/rz)^2$   Euler buckling stress

Now, putting following values into above equation

$\alpha =$ 0.49   $E =$ 200000 Mpa   $fy =$ 250 Mpa
$\gamma_{m0} =$ 1.10   $KLz/rz =$ 160   $Ae =$ 1138 mm2

fcc = $3.14^2 \times 200000 / 160^2 =$   77 MPa
$\lambda =$ $(250 / 77)^{0.5} =$   1.8
$\phi =$ $0.5[1 + 0.49 \times (1.8 - 0.2) + 1.8^2] =$   2.51
$\chi =$ $1 /[2.51 + (2.51^2 - 1.8^2)^{0.5}] =$   0.235

fcd = $0.235 \times 250 / 1.1 =$   53 Mpa   $<= fy / \gamma_{mo}$   $<=$ 250 / 1.1
     $<=$ 227 Mpa

fcd = **53** Mpa   (=fcdz)

**Pdz** = Ae . fcd =   $1138 \times 53 /1000 =$   **60** kN

The design compressive strength, **Pd** will be lowest of above calculated strengths of the member.

**Pd** =   **60 kN**

# Design of Industrial Structures 427

**Axial compression :** (IS 800:2007 11.3.1)

Permissible comp. stress, fac = 0.6 fcd     fcd =    53 Mpa (see calc. above.)
                                        Ag =    1138 mm2
Permissible Comp strength, Pd = 0.6 fcd .Ag    =    0.6 x 53 x 1138    =    36188 N
                                                                                    Pd =    **36** kN
                                                          Actual load, P =    23 kN
                                                                                P < Pd; Safe.

**Axial Tension:** (IS 800:2007 11.2.1)
Tensile stresses for yielding :    fat = 0.6 fy =    0.6 x 250    =    150 MPa
                                   Td = fat. Ag =    150 x 1138/1000 =    171 kN
                                                           Actual load, T =    23 kN
                                                                      T < Td; Safe.

**Member 9**
                Provide same as member 8

**Member 10**     **Portal frame beam**

Span, L=    3.8 m           UDL     7.8 kN/m    Msup =    62.83 kNm    (DL + WL)
Lx =    3.8 m           (Ends restrained)         Shear =    14.82 kNm
Ly =    3.2 m           (unsupported length clearing w/way etc.)

| Try with | | ISMB 250 | | wt/m | | 37.3 KG | | | | |
|---|---|---|---|---|---|---|---|---|---|---|
| **Section properties:** | | | | Member dimensions | | | | | | |
| Section Provided: | s | | | wt/m (kg) | $A_g$ (mm$^2$) | $r_x(r_z)$ (mm) | $r_y$ (mm) | $Z_x(Z_e)$ (mm) | $I_x$ (mm$^4$) | $I_y$ (mm$^4$) | $C_{yy}$ (mm) |
| 1 | **ISMB 250** | 0 | mm b/b | 37.3 | 4755 | 104 | 27.0 | 410500 | 51316000 | 3345000 | 0 |
| | | | | b | d | tf | tw | Connection bolt | | | |
| | | | | mm | mm | mm | mm | nos | φdia | φhole | rows |
| | | | | 125 | 250 | 12.5 | 6.9 | | | | |

## Flexural strength:
*Lateraly unsupported beam ( susceptible to Lateral-torsional buckling )*

| Section | ISMB 250 | | |
|---|---|---|---|
| Laterally unsupported beam | | | |
| (IS 800:2007 8.2.2) | | | |
| $L_{LT} =$ | | 2240 | mm |
| B = | | 125 | mm |
| T (tf) = | | 12.5 | mm |
| D = | | 250 | mm |
| tw = | | 6.9 | mm |
| Zp = | 468395 | | mm3 |
| Zx (Ze) = | 410500 | | mm3 |
| $\gamma_{m0} =$ | | 1.10 | |
| $\beta =$ | | 1 | |
| E = | | 200000 | Mpa |
| fy = | | 250 | Mpa |
| hf = | | 237.5 | mm |
| Iy = | 3345000 | | mm4 |
| ry = | | 26.5 | mm |
| Ix = | 51316000 | | mm4 |
| $\alpha_{LT}$ | 0.21 | | |
| $f_{bd} =$ | | 190 | Mpa |
| $Md = \beta . Zp . f_{bd}$ | | | |
| **Md =** | | **89** | **Mpa** |

$L_{LT} =$  0.7 Ly     (IS 800:2007 - Table 15)

$Mcr = ( \pi^2 EIy \, hf / 2 \, L_{LT}2 ) . [ 1 + (1/20) . ((L_{LT} / ry) / (hf/tf))^2 ]^{0.5}$

$\pi^2 E.Iy .hf / 2 . L_{LT}2 =$ 1.56E+08

$((L_{LT} / ry) / (hf/tf))^2 =$ 19.79

Mcr = 220188669 Nmm

$\lambda_{LT} = (\beta \, Zp \, fy / Mcr)^{0.5}$  <=  $(1.2 \, Ze \, fy / Mcr)^{0.5}$

$(\beta \, Zp \, fy / Mcr)^{0.5} =$  0.729

$(1.2 \, Ze \, fy / Mcr)^{0.5} =$  0.748

$\lambda_{LT} =$  0.729

$\phi_{LT} = 0.5 [ 1 + \alpha_{LT} (\lambda_{LT} - 0.2) + \lambda_{LT}2 ]$

$\phi_{LT} =$  0.82

$\chi_{LT} =$  $1 / [ \phi_{LT} + (\phi_{LT}2 - \lambda_{LT}2)^{0.5} ]$  <= 1
     =  0.834

$f_{bd} = \chi_{LT} \, fy / \gamma_{mo}$     =     190 Mpa

### Flexural strength, Ma :     *(IS 800-11.4.1)*
For laterally Unsupported beam:

fabc = fabt = 0.6 Md/Zx     (Zx top = Zx bot)     Md =     89 knM

Ma = fabc. Zx = (0.6Md/Zx) Zx     (fabc = fabt)
     =  0.6 Md
     =  0.6 x 89 =     53 kNm
Increased 33% for wind combination, **Ma** =     71
Allowable Flexural mom, **Ma** =     **71** kNm.
Actual Design Moment, Ms =     63 kNm.     **Ma > Ms; Safe.**

### Shear strength, Vs :     *(IS 800-11.4.2)*
Pure shear

$Va = \tau_{ab} . Av$ , where Av shear area
     =  0.4 fy x Av
     =  0.4 x 250 x 250 x 6.9/1000
     =  172.5 kN

Allowable Shear, **Va** =     **173** kN.
Actual design shear, Vs =     15 kN.     **Va > Vs; Safe.**

# Design of Industrial Structures

## Deflection: (IS 800:2007 - 5.6.1)

$\delta = wL^4 / 384 EI =$  1 x 7.8 x 3800 ^4 / 384 x 200000 x 51316000 =   0.41 mm
Allowable deflection limit L /300 =  12.67 mm
**OK.**

## Member 11   Portal Column    Bending about mazor axis x-x

| | | | | | |
|---|---|---|---|---|---|
| L (h)= | 2.75 m | Axial | 258 kN | (from analysis Portal frame) |
| Lx = | 2.75 m | | 86 kN | (12 x 7.2) - Cable tray |
| Ly = | 2.75 m | | 344 kN | |

Msup    63 kNm          Shear    43 kN    (= H1)

Try with following section

## Axial compression:

| Section properties: | | | Member dimensions | | | | | |
|---|---|---|---|---|---|---|---|---|
| Built up section | | | wt/m (kg) | $A_g$ (mm$^2$) | $r_x (r_z)$ (mm) | $r_y$ (mm) | $I_x$ (mm$^4$) | $I_y$ (mm$^4$) |
| Flange plates | 125 | 16 | 51.94 | 6616 | 99 | 28 | 65201565 | 5239725 |
| Web Plate | 218 | 12 | | | | | | |
| z ---- I ---- z | | | b mm | d mm | tf mm | tw mm | Zx mm3 | Av mm2 |
| | | | 125 | 218 | 16 | 12 | 521613 | 2616 |

## Buckling class

Buckling class = c for buckling about any axis     (IS 800:2007 - Table 10)
Imperfaction factor, $\alpha$ = 0.49                (IS 800:2007 - Table 7)

## Proportioning limit                                (IS 800:2007 - Table 2)
$\varepsilon = (250 / f_y)^{0.5} =$   (250 / 250)^0.5 =     1.00

## Outstanding leg

b / t =    125 / 16 =    **8**    Plastic.    (IS 800:2007 - 3.7.2)

*Limiting width to thickness ratio:*

| | | |
|---|---|---|
| Plastic, $\lambda r =$ | $9.4\varepsilon = 9.4 \times 1 =$ | 9 |
| Compact, $\lambda p =$ | $10.5e = 10.5 \times 1 =$ | 11 |
| Semi compact, $\lambda sp =$ | $15.7e = 15.7 \times 1 =$ | 16 |

## Effective Lengths (L) and Slenderness ratio ( L/r ) :    (IS 800:2007 - Table 11)

| | | | | | | |
|---|---|---|---|---|---|---|
| Laterally Unsupported length | Lz | 2.8 M | Kz | 1.2 | KLz | 3.30 M |
| | Ly | 2.8 M | Ky | 0.85 | Kly | 2.34 M |

( Lz = 1.2 x h ; top free for translation)

Maxm slenderness ratio (L/r) shall not exceed    **250**    (IS 800:2007 - Table 3)

Here,    KLz/rz =    3300/99 =    33

         KLy/ry =    2340/28 =    84

                   L/r =    84    <250, OK.

## Compression strength

*Compression due to buckling about minor axis, y-y (KLy / ry)*    (IS 800:2007 7..5.1.1)

Design compression strength, Pdy = Ae. fcd    where.    Ae = effective sectional area

                                                                     fcd = design compressive stress

$fcd = \chi\, fy / \gamma_{mo}$    $<= fy / \gamma_{mo}$                                  (IS 800:2007 7.1.2.1)

$\chi = 1 / [\,\phi + (\phi^2 - \lambda^2)^{0.5}\,]$

$\phi = 0.5\,[\,1 + \alpha(\lambda - 0.2) + \lambda^2\,]$

$\lambda = (fy / fcc)^{0.5}$

$fcc = \pi^2 E / (KLz/rz)^2$    Euler buckling stress

Now, putting following values into above equation

| | | | | | | |
|---|---|---|---|---|---|---|
| $\alpha =$ | 0.49 | E = | 200000 Mpa | fy = | 250 Mpa | |
| $\gamma_{m0} =$ | 1.10 | KLy/ry = | 84 | Ae = | 6616 mm2 | |

| | | | | |
|---|---|---|---|---|
| fcc = | 3.14^2 x 200000 / 83.5714285714286^2 = | 282 MPa | | |
| $\lambda =$ | (250 / 282)^0.5= | 0.94 | | |
| $\phi =$ | 0.5[1 + 0.49 x (0.94 - 0.2) + 0.94^2= | 1.12 | | |
| $\chi =$ | 1 /[1.12 + (1.12^2 - 0.94^2)^0.5]= | 0.578 | | |
| fcd = | 0.578 x 250 / 1.1 =    131 Mpa | <= fy / $\gamma_{mo}$ | <= | 250 / 1.1 |
| | | | <= | 227 Mpa |

fcd =    131 Mpa    (=fcdy)

**Pdy** = Ae . fcd =    6616 x 131 /1000=    **867 kN**

# Design of Industrial Structures

*Axial compression by working stress method:* (IS 800:2007 11.3.1)

Permissible comp. stress, fac = 0.6 fcd    fcd = 131 Mpa (see calc. above.)
                                           Ag = 6616 mm2

Permissible Comp strength, Pd = 0.6 fcd .Ag  = 0.6 x 131 x 6616 = 520018 N
                                                                = 520 kN
                                    Increasing 33% stress for wind. Pd = 692 kN
                                                  Actual load, P = 344 kN
                                                  P < Pd; Safe.

**Lateral deflection:** (IS 800:2007 - 5.6.1)
P = H1 /2 =    21.5 kN per post.
d = P L3 / 3 EI =   21500 x 2750 ^3 / 3 x 200000 x 65201565 =
                                                          11.43 mm
                          Allowable deflection limit L /150 = 18.33 mm
                                                          OK.

## Flextural Strength
*Lateraly unsupported beam ( susceptible to Lateral-torsional buckling )*

| Section | Built up section | |
|---|---|---|
| Laterally unsupported beam | | |
| (IS 800:2007 8.2.2) | | |
| $L_{LT}$ = 0.7Ly | 1925 | mm |
| B = | 125 | mm |
| T (tf) = | 16 | mm |
| D = | 250 | mm |
| tw = | 12 | mm |
| Zp = | 632268 | mm3 |
| Zx (Ze) = | 521613 | mm3 |
| $\gamma_{m0}$ = | 1.10 | |
| β = | 1 | |
| E = | 200000 | Mpa |
| fy = | 250 | Mpa |
| hf = | 234 | mm |
| Iy = | 5239725 | mm4 |
| ry = | 28 | mm |
| Ix = | 65201565 | mm4 |
| $\alpha_{LT}$ | 0.49 | |
| $f_{bd}$ = | 182 | Mpa |
| Md = β . Zp .$f_{bd}$ | | |
| **Md =** | **115** | **Mpa** |

$L_{LT}$ = 0.7 Ly    (IS 800:2007 - Table 15)
$Mcr = ( \pi^2 EIy\, hf / 2\, L_{LT}2 ) . [ 1 + (1/20) . ((L_{LT} / ry) / (hf/tf))^2 ]^{0.5}$
$\pi^2 E.Iy\, .hf / 2.\, L_{LT}2$ = 3.26E+08
$((L_{LT} / ry) / (hf/tf))^2$ = 22.1
Mcr = 473312779  Nmm

$\lambda_{LT}$ = $(\beta\, Zp\, fy / Mcr)^{0.5}$ <= $(1.2\, Ze\, fy / Mcr)^{0.5}$
$(\beta\, Zp\, fy / Mcr)^{0.5}$ = 0.578
$(1.2\, Ze\, fy / Mcr)^{0.5}$ = 0.575
$\lambda_{LT}$ = 0.575

$\phi_{LT}$ = 0.5 [ 1 + $\alpha_{LT}$ ($\lambda_{LT}$ - 0.2) + $\lambda_{LT}$2 ]
$\phi_{LT}$ = 0.76

$\chi_{LT}$ = $1 / [ \phi_{LT} + (\phi_{LT}2 - \lambda_{LT}2)^{0.5} ]$ <= 1
    = 0.5

$f_{bd}$ = $\chi_{LT}$ fy / $\gamma_{mo}$    =  182 Mpa

## Flexural strength, Ma : *(IS 800-11.4.1)*
### For laterally Unsupported beam:

fabc = fabt = 0.6 Md/Zx        (Zx top = Zx bot)        Md =        115 knM

$$Ma = fabc \cdot Zx = (0.6Md/Zx) Zx \quad (fabc = fabt)$$
$$= 0.6 \, Md$$
$$= 0.6 \times 115 = \quad 69 \text{ kNm}$$

Increased 33% for wind combination, Ma =        92 kNm
Allowable Flexural strength, **Ma** =        **92** kNm.
Actual Design Moment, Ms =        63 kNm.        **Ma > Ms; Safe.**

## Shear strength, Vs : *(IS 800-11.4.2)*

$Va = \tau_{ab} \cdot Av$ , where Av shear area
= 0.4 fy × Av
= 0.4 × 250 × 2616/1000
= 261.6 kN

Allowable Shear, **Va** =        **262** kN.
Actual design shear, Vs =        43 kN.        **Va > Vs; Safe.**

## Interaction factor :
### at top
Axial load, $P_{top}$ = load from level 3 only =        P /(7.448/19.99) =        0.37   x P
$P_{top}$ / Pd  + Ms / Ma =    0.37 × 344.4 / 692 + 63 / 92 =        0.18 + 0.68 =        0.87   **< 1; Safe.**

### at base (hinged)
Moment =        0        Axial = P
P / Pd  + Ms / Ma =        344.4 / 692 + 0 =        0.5 + 0 =        0.50   **< 1; Safe.**

## Combined Axial bending and shear stress check:        *(IS 800:2007 11.5.4)*

**axial stress**
fac = 0.6 × fcd =        0.6 × 131=        79 Mpa
**bending stress**
fabc = 0.6 × fbd =        0.6 × =        0 Mpa
**shear stress**
$\tau b$ = Vs / Av =    43000 / (218× 12) =        16 Mpa

The equivalent stress, $fe = (fac^2 + fabc^2 + fac \cdot fabc + 3 \tau b^2)^{0.5} < 0.9 \, fy$
fe =        $(79^2 + 0^2 + 79 \times 0 + 3 \times 16^2)^{0.5}$ =        84
0.9 fy =        0.9 × 250 =        225

**fe < 0.9fy; Safe.**

# Design of Industrial Structures

**Member 12**  **Column bracings for Main column**

| Length | 4.69 m | Lx (Lz) = | 4.69 m |  | Axial Comp = | 53 kN |
|---|---|---|---|---|---|---|
|  |  | Ly = | 4.69 m |  | Axial tension = | 53 kN |

Try with     **2 ISA 75 75 8**

| Section properties: | | | | Built-up member dimensions | | | | | | | |
|---|---|---|---|---|---|---|---|---|---|---|---|
| Section Provided: | | s | | wt/m (kg) | $A_g$ (mm²) | $r_x$ (mm) | $r_y$ (mm) | $r_{vv}$ (mm) | $I_x$ (mm⁴) | $I_y$ (mm⁴) | $C_{yy}$ (mm) |
| 2 | ISA 75 75 8 | 12 | mm b/b | 17.8 | 2276 | 22.80 | 36.0 | 14.5 | 1180000 | 2888730 | 21.4 |
| | | | | Single Leg dimensions | | | | | | | |
| | | | | b | d | t | Connection bolt | | | | An |
| | | | | mm | mm | mm | nos | φdia | φhole | rows | mm2 |
| | | | | 75 | 75 | 8 | 4 | 16 | 18 | 1 | 2132 |

where, An = net area after deduction of bolt holes.
   An =    Ag - (no of rows x φ hole x t)
Distance between tacking plate (welded spacer plate) = **a** =    256 mm    (IS 800:2007 - 10.2.5.2)
(the value of ' a ' shall not exceed 32 t or 300 mm, whichever is less)

**Buckling class**

  Buckling class =    c    for buckling about any axis    (IS 800:2007 - Table 10)
  Imperfection factor, α =   0.49          (IS 800:2007 - Table 7)

**Proportioning limit**                                (IS 800:2007 - Table 2)

  ε = (250 / fy )^0.5 =   (250 / 250)^0.5 =    1.00

Outstanding leg
  **b / t =**     75 / 8 =        9       Plastic.         (IS 800:2007 - 3.7.2)
  *Limiting width to thickness ratio:*
  Plastic, λr =       9.4ε = 9.4 x 1 =           9
  Compact, λp =      10.5ε = 10.5 x 1 =         11
  Semi compact, λsp =  15.7ε = 15.7 x 1 =       16

### Effective Lengths (L) and Slenderness ratio ( L/r ) : (IS 800:2007 - Table 11)

| Laterally Unsupported length | Lz | 4.7 M | Kz | 0.85 | KLz | 3.99 M |
|---|---|---|---|---|---|---|
| | Ly | 4.7 M | Ky | 1 | Kly | 4.69 M |

Maxm slenderness ratio (L/r) shall not exceed **250** (IS 800:2007 - Table 3)

Here, $KL_z/r_z$ = 3990/22.8 = 175
$KL_y/r_y$ = 4690/36 = 130
L/r = 175 <250, OK.

### Sectional capacity / Design strength:

### Axial compression

*(i) Compression due to buckling about major axis, z-z [Lz / rz (=rx)]* (IS 800:2007 7.5.2.1)

Design compression strength, $P_{dz} = A_e \cdot f_{cd}$ (IS 800:2007 7.1.2)

where, $A_e$ = effective sectional area
$f_{cd}$ = design compressive stress

$f_{cd}$ = $\chi\, f_y / \gamma_{mo}$ <= $f_y / \gamma_{mo}$
$\chi$ = $1 / [\, \phi + (\, \phi^2 - \lambda^2\,)^{\wedge}0.5\,]$
$\phi$ = $0.5\,[\, 1 + \alpha(\lambda - 0.2) + \lambda^2\,]$
$\lambda$ = $(f_y / f_{cc})^{\wedge}0.5$

$f_{cc}$ = $\pi^2\, E / (KL_z/r_z)^{\wedge}2$    Euler buckling stress

Now, putting following values into above equation

$\alpha$ = 0.49       E = 200000 Mpa       $f_y$ = 250 Mpa
$\gamma_{m0}$ = 1.10       $KL_z/r_z$ = 175       $A_e$ = 2276 mm2       =$A_g$

$f_{cc}$ = 3.14^2 x 200000 / 175^2 =         64 MPa
$\lambda$ = (250 / 64)^0.5=         1.98
$\phi$ = 0.5[1 + 0.49 x (1.98 - 0.2) + 1.98^2=         2.9
$\chi$ = 1 /[2.9 + (2.9^2 - 1.98^2)^0.5]=         0.199

$f_{cd}$ = 0.199 x 250 / 1.1 =       45 Mpa       <= $f_y / \gamma_{mo}$       <= 250 / 1.1
                                                                          <= 227 Mpa

$f_{cd}$ =    45 Mpa    (=$f_{cdz}$)

**$P_{dz}$** = $A_e \cdot f_{cd}$ =    2276 x 45 /1000=       **102** kN

# Design of Industrial Structures

*(ii) Compression due to buckling about minor axis, y-y (KLy / ryy)* (IS 800:2007 7.5.2.1)

Design compression strength, Pdy = Ae. fcd 

where. Ae = effective sectional area
fcd = design compressive stress

$$fcd = \chi\, fy / \gamma_{mo} \quad \leq fy / \gamma_{mo}$$ (IS 800:2007 7.1.2.1)
$$\chi = 1 / [\phi + (\phi^2 - \lambda^2)^{0.5}]$$
$$\phi = 0.5\,[1 + \alpha(\lambda - 0.2) + \lambda^2]$$
$$\lambda = (fy / fcc)^{0.5}$$

$$fcc = \pi^2 E / (KLz/rz)^2 \quad \text{Euler buckling stress}$$

Now, putting following values into above equation

| | | | | | |
|---|---|---|---|---|---|
| $\alpha =$ | 0.49 | E = | 200000 Mpa | fy = | 250 Mpa |
| $\gamma_{m0} =$ | 1.10 | KLy/ryy= | 130 | Ae = | 2276 mm2 |

fcc = 3.14^2 x 200000 / 130^2 =     117 MPa
$\lambda$ = (250 / 117)^0.5=     1.46
$\phi$ = 0.5[1 + 0.49 x (1.46 - 0.2) + 1.46^2=     1.87
$\chi$ = 1 /[1.87 + (1.87^2 - 1.46^2)^0.5]=     0.329

fcd = 0.329 x 250 / 1.1 =     75 Mpa    $\leq$ fy / $\gamma_{mo}$    $\leq$    250 / 1.1
                                                                                                                                                                        $\leq$    227 Mpa

fcd =     75 Mpa (=fcdy)

**Pdy** = Ae . fcd =     2276 x 75 /1000=     **171 kN**

The design compressive strength, **Pd** will be lowest of above calculated strengths of the member.

        **Pd =**     **102 kN**    and    fcd =     45 Mpa

**Axial compression :**     (IS 800:2007 11.3.1)

Permissible comp. stress, fac = 0.6 fcd                          fcd =     45 Mpa
                                                                                                                                                       Ag =     2276 mm2

        Permissible Comp strength, Pd = 0.6 fcd .Ag
                                                     =     0.6 x 45 x 2276
                                                     =     61452 N
                                                     =     61 kN
                          Actual load, P =     53 kN
                                           P < Pd; Safe.

## Axial Tension

Design tensile strength of member = Td

**(i) Design tensile strength due to yielding of gross section, Tdg**  (IS 800:2007 - 6.2)

$Tdg = Ag\, fy / \gamma_{mo} =$    2276 x 250 / 1.1 / 1000 =    517 kN

**(ii) Design tensile strength due to rupture of critical section, Tdn**

$Tdn = 0.9\, Anc.\, fu / \gamma_{m1} + \beta.\, Ago.\, fy / \gamma_{m0}$   (IS 800:2007 - 6.3.3)
$\beta = 1.4 - 0.076\,(w/t)(fy/fu)(bs/Lc) <= 0.9\, fu\, \gamma_{mo} / fy\, \gamma_{m1} >= 0.7$

where,  Anc = net area of connected leg =   2 x [75x8-(1x18x8)] =   912 mm2
        Ago = gross area of outstanding leg =   2 x (75 x 8) =   1200 mm2

        w = width of outstanding leg = d =                              75 mm
        bs = shear lag width = w + w1 (= Cyy) - t =   75 + 21.4 - 8 =   88 mm
        Lc = length of end connection = (nos of bolt -0.5) x 3 φ hole =   189 mm

        f u =   410 Mpa        $\gamma_{m0}$ =   1.10        t =    8 mm
        fy =    250 Mpa        $\gamma_{m1}$ =   1.25

$\beta = 1.4 - 0.076\,(w/t)(fy/fu)(bs/Lc) <= 0.9\, fu\, \gamma_{mo} / fy\, \gamma_{m1} >= 0.7$
  =   1.4-0.0076x(75/8)x(250/410)x(88/189)   <=   0.9x410x1.1 / 250x1.25> =0.7
  =   1.2          <=    1.30         >=     0.7              Ok.

Tdn = 0.9 Anc. fu / $\gamma_{m1}$ + β. Ago. fy / $\gamma_{m0}$
   =   0.9 x 912x410 / 1.25 + 1.2 x 1200 x 250 / 1.1
   =   596495 N
   =   596 kN

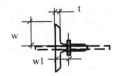

**(iii) Design tensile strength due to block shear, Tdb**  (IS 800:2007 - 6.4.1)

a)   Tdb = [ Avg. fy / ( 1.732 $\gamma_{mo}$) + 0.9 Atn. fu / $\gamma_{m1}$]     OR

b)   Tdb = (0.9 Avn. fu / (1.732 $\gamma_{m1}$ ) + Atg. fy / $\gamma_{mo}$ )    whichever is smaller

    Avg =  gross shear area along bolt line parallel to line (1-2) of external force
        =   2. Lc . t
        =   2 x 189 x 8      =    3024 mm2

Design of Industrial Structures

$A_{vn}$ = net shear area along bolt line parallel to the line (1-2) of external force
   = 2 [Lc . t - ( no of bolts x $\phi$ hole x thickness)]
   = 2 x [ 189 x 8 - (4x18x8) ] =   2000 mm2


$A_{tg}$ = gross area in tension from bolt hole to toe of the angle, perpendicular to line (2-3) of force
   = 2 . (d - Cyy) . t
   = 2 x (75 - 21.4) x 8 ) =   858 mm2
$A_{tn}$ = net area in tension from bolt hole to toe of the angle, perpendicular to line (2-3) of force
   = 2 [ (d - Cyy) . t - ( no of rows x $\phi$ hole x thickness)]
   = 2 [ (75 - 21.4) x 8 - (1 x 18 x 8) ] =   570 mm2

a) $T_{db}$ = [ $A_{vg}$ . fy / ( 1.732 $\gamma_{mo}$) + 0.9 $A_{tn}$ . fu / $\gamma_{m1}$]
   = 3024 x 250 /(1.732 x 1.1) + 0.9 x 569.6 x 410 / 1.25
   = 564955 N
   = 565 kN
   OR
b) $T_{db}$ = (0.9 $A_{vn}$. fu / (1.732 $\gamma_{m1}$ ) + $A_{tg}$. fy / $\gamma_{mo}$ )   whichever is smaller
   = 0.9 x 2000 x 410 / (1.732 x 1.25) + 857.6 x 250 /1.1
   = 535787 N
   = 536 kN

So, Tensile strength due to block shear, $T_{db}$ =   536 kN

now, we have   $T_{dg}$ =   517 kN,   $T_{dn}$ =   596 kN and   $T_{db}$ =   536 kN

*The design tensile strength, **Td** will be lowest of above calculated strengths of the member.*

Td =   517 kN


**Axial Tension:**                                                                                       (IS 800:2007 11.2.1)

Tensile stresses for :
a) yielding      fat = 0.6 fy =   0.6 x 250        =   150 MPa

b) rupture       fat = 0.69 Tdn/Ag       =   0.69 x 596000 / 2276       =   181 MPa

c) block shear   fat = 0.69 Tdb/Ag       =   0.69 x 536000 / 2276       =   162 MPa


Permissible value of fat =        150 MPa

| | | | |
|---|---|---|---|
| Permissible Tensile strength, Td = fat .Ag | = | 150 x 2276 | |
| | = | 341400 N | |
| | = | 341 kN | |
| Actual load, T = | | 53 kN | T < Td; Safe. |

**Member 13**   **Horz tie bracing between main column legs**

Provide same section as member 12

**Member 14**   **Main Column leg**

The total height of the column is divided in two segments along minor axis. Each segments are crossed braced and tied with horizontal ties. The column base plate will have gusset plates to distribute load on base plate. There will be anchor bolts to transfer shear and resist uplift. Column base plate will rest on Reinforced concrete pedestal.

Height  L =   5.5 m      Lx =   5.5 m     Ly =   2.75 m.

Axial comp, P =   382 kN      Tension, T =   0 kN     Shear =   43 kN

Try with     ISMB 250       wt/m     37.3 KG

**Axial compression:**

| Section properties: | | | | Member dimensions | | | | | | |
|---|---|---|---|---|---|---|---|---|---|---|
| Section Provided: | | s | | wt/m (kg) | $A_g$ (mm$^2$) | $r_x$ ($r_z$) (mm) | $r_y$ (mm) | $Z_x(Z_e)$ (mm) | $I_x$ (mm$^4$) | $I_y$ (mm$^4$) | $C_{yy}$ (mm) |
| 1 | ISMB 250 | 0 | mm b/b | 37.3 | 4755 | 104 | 27.0 | 410500 | 51316000 | 3345000 | 0 |
| y ----⊢⊣---- y  z | | | | b mm | d mm | tf mm | tw mm | Connection bolt | | | |
| | | | | | | | | nos | φdia | φhole | rows |
| | | | | 125 | 250 | 12.5 | 6.9 | | | | |

**Buckling class**

Buckling class =   c     for buckling about any axis       (IS 800:2007 - Table 10)
Imperfaction factor, α =   0.49                             (IS 800:2007 - Table 7)

**Proportioning limit**                                      (IS 800:2007 - Table 2)

ε = (250 / fy ) ^0.5 =    (250 / 250)^0.5 =    1.00

# Design of Industrial Structures

Outstanding leg
  0.5b / t =   62.5 / 12.5 =   **5**   Compact   (IS 800:2007 - 3.7.2)
  *Limiting width to thickness ratio:*
  Plastic, $\lambda r$ =   $9.4\varepsilon$ = 9.4 x 1 =   9
  Compact, $\lambda p$ =   10.5e = 10.5 x 1 =   11
  Semi compact, $\lambda sp$ =   15.7e = 15.7 x 1 =   16

### Effective Lengths (L) and Slenderness ratio ( L/r ) :   (IS 800:2007 - Table 11)

  Laterally Unsupported length   Lz   5.5 M   Kz   1   KLz   5.50 M
                                 Ly   2.8 M   Ky   0.85   Kly   2.34 M

  Maxm slenderness ratio (L/r) shall not exceed   **180**   (IS 800:2007 - Table 3)
  Here,   KLz/rz =   5500/104 =   53
          KLy/ry =   2337.5/27 =   87
                          L/r =   87   <180, OK.

### Compression strength
*Compression due to buckling about minor axis, y-y (KLy / ry)*   (IS 800:2007 7..5.1.1)
Design compression strength, Pdy = Ae. fcd   where,  Ae =   effective sectional area
                                                     fcd =   design compressive stress
  fcd =   $\chi$ fy / $\gamma_{mo}$   <= fy / $\gamma_{mo}$   (IS 800:2007 7.1.2.1)
  $\chi$ =   $1 / [\phi + (\phi^2 - \lambda^2)^{0.5}]$
  $\phi$ =   $0.5 [ 1 + \alpha(\lambda - 0.2) + \lambda^2]$
  $\lambda$ =   (fy / fcc)^0.5
  fcc =   $\pi^2$ E / (KLz/rz)^2   Euler buckling stress

Now, putting following values into above equation
  $\alpha$ =   0.49   E =   200000 Mpa   fy =   250 Mpa
  $\gamma_{m0}$ =   1.10   KLy/ry =   87   Ae =   4755 mm2

  fcc =   3.14^2 x 200000 / 86.5740740740741^2 =   263 MPa
  $\lambda$ =   (250 / 263)^0.5 =   0.97
  $\phi$ =   0.5[1 + 0.49 x (0.97 - 0.2) + 0.97^2 =   1.16
  $\chi$ =   1 /[1.16 + (1.16^2 - 0.97^2)^0.5]=   0.557
  fcd =   0.557 x 250 / 1.1 =   127 Mpa   <= fy / $\gamma_{mo}$   <=   250 / 1.1
                                                                  <=   227 Mpa

  fcd =   127 Mpa   (=fcdy)

  **Pdy** = Ae . fcd =   4755 x 127 /1000=   **604** kN

*Axial compression by working stress method:*   (IS 800:2007 11.3.1)
Permissible comp. stress, fac = 0.6 fcd   fcd =   127 Mpa  (see calc. above.)
                                          Ag =   4755 mm2

Permissible Comp strength, Pd = 0.6 fcd .Ag   =   0.6 x 127 x 4755   =   362331 N
                                                                      =   362 kN
                                 Increasing 33% stress for wind. Pd =   482 kN
                                                   Actual load, P =   382 kN
                                                                     P < Pd; Safe.

### Shear strength, Vs :

$V_a = \tau_{ab} \cdot A_v$ , where Av shear area
   = 0.4 fy x Av    = 0.4 x 250 x 1725/1000    = 173 kN
                                Actual shear =   43 kN    Safe.

### Column base design.

#### Base Plate

|  | mm | mm |
|---|---|---|
| Base Plate | 450 | 350 |
| Col size | 250 | 125 |

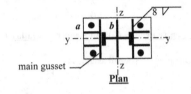

Plan

Comp =              382 kN
Tension =           0 kN
Base pressure =     382000/(450x350)=    2.43 Mpa         24.3 kg/cm2

*Panel - a :*                               *Panel - b :*
Maxm. projection = 100 mm.         span =    125 mm
Moment = 2.43x100^2/2               Moment = 2.43x125^2/8
       = 12150 Nmm                         = 4746 Nmm

M max =         12150 Nmm
fbs - bending stres = 0.75 fy =  0.75 x 250 =   188 Mpa        (IS 800:2007 11.3.3)
Thickness required, t = (12150 x 6 / 187.5)^0.5  =  20 mm
                                            provide   25 m thick

#### Provide
   Base Plate :   450 x 350 x 25 th
   Main gusset :  300 (Ht) x 12 th    and    Aux gusset   200 (Ht) x 12 th
   All welds are 6 mm fillet weld unless shown otherwise.

# Design of Industrial Structures

**Anchor bolt**

Tension = 141 kN  ( From  Wind load only)

Col base shear = 43 kN
$f_y$ = 240 MPA  $f_u$ = 410 MPA  $\gamma_{m0}$  1.10
$\gamma_{m1}$  1.25  $\gamma_{mb}$  1.25

Provide **4** nos.  **24** mm dia.
Anb, Net area of bolts = $4 \times 3.14 \times (0.87 \times 24)^2/4 =$  1369 mm2
Asb, Gross area of bolts = $4 \times 3.14 \times 24^2/4 =$  1809 mm2

*Nominal tension capacity, Tnb*
Tdn = $0.9 \times 1369 \times 410 / 1.25 / 1000 =$  404 kN  (IS 800:2007 6.3.2)
or,  Tdb = $f_{yb}$. Asb/$\gamma_{mo}$ = $240 \times 1809/1.1/1000 =$ 395 kN  (IS 800:2007 10.3.5)
Tdn = **395** kN

*Nominal shear capacity, Vnsb*
Vnb = [ ($f_u$/1.732) . (ns Asb)] / $\gamma_{mb}$  (IS 800:2007 10.3.3)
= **343** kN  ns = no of shear plane = 1

*Working stress method*  (IS 800:2007 11.6)
Permissble tension capacity, Tab = 0.6 Tdn = 0.6 x 395 = 237 kN
Permissble shear capacity, Vsb = 0.6 Vnb = 0.6 x 343 = 206 kN
Interaction factor:
(Tab / Tdn )^2  +  (Vsb / Vnb )^2  <= 1
(237/395) ^ 2 +  (205.8 / 343) ^2=  0.36  +  0.36  =  0.72
**< 1 ; Safe..**

End

## 12.3.3  Single Tier Pipe and Cable Racks Crossing Road

### 12.3.3.1  Description

This structure is used for pipes and cables crossing over the road. There are cable pull pits or manholes on either sides of the road. Pipe and cable racks rise up climbing on supports clamped on the column structures. The clear height above road is kept 5 meter considering movement of plant vehicles and construction equipment. The structure is designed to bear its own weight, cables and pipes in combination with wind or seismic load, whichever is worst. A walkway has been provided for maintenance work on the top of the bridge structure. The field connections will be bolted structure using 8.8 grade high tensile bolt. Painting work shall be done with synthetic enamel or epoxy based paint as per environment condition.

### 12.3.3.2  Reference

1.0  IS 800 : 1962; 1984; 2007
2.0  IS 875 : 1987
3.0  Steel designer's manual - ELBS  4 Ed.

### 12.3.3.3 Material

| IS 2062 | | | | | | |
|---|---|---|---|---|---|---|
| Grade | E 250 (Fe 410 W) A | | | | | |
| Yield Stress | $f_y$ | 250 | 240 | 230 | MPa | (IS 800:2007 - Table 1) |
| d or t | | <20 | 20-40 | >40 | | |
| Tensile Stress | $f_u$ | 410 | 410 | 410 | MPa | |
| Elasticity | E | | | 200000 | MPa | (IS 800:2007 - 2.2.4.1) |
| Mat. safety fac | $\gamma_{m0}$ | 1.10 | $\gamma_{m1}$ | 1.25 | | (IS 800:2007 - Table 5) |

### 12.3.3.4 Sketch

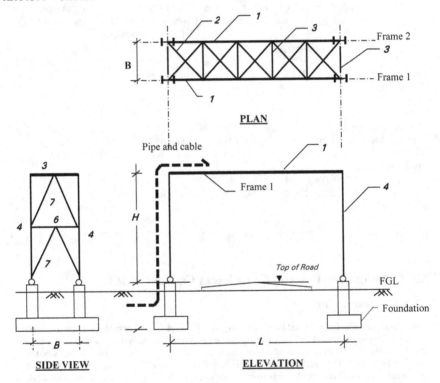

# Design of Industrial Structures

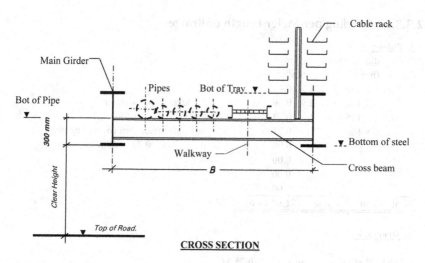

**CROSS SECTION**

## 12.3.3.5 Design Parameters

**General.**
| | | | |
|---|---|---|---|
| Finished grade level (RL) | FGL = | 26.5 | M |
| Natural grade level (RL) | NGL = | 26.5 | M |
| Ground water level (RL) | GWL = | 26.5 | M |
| Top Elevation of Road (RL) | Road = | 26.75 | M |

**Load:**
| | | | |
|---|---|---|---|
| Design Wind pressure | pwind = | 2 | kN/m2 |
| Ground Surcharge | gsurch = | 10 | kN/m2 |

**Bridge Structure.**
| | | | |
|---|---|---|---|
| Span | L = | 12 | M |
| Width | B = | 3.5 | M |
| Clear height | CLH = | 5 | M |
| Bot. of Base Plate (RL) | BOS = | 26.75 | M |
| Maxm. Depth of Girder | Md = | 0.6 | M |
| Bot. of Pipe (RL) | BOP = | 34.20 | M |
| Heigth of Column | H = | 5 | M |
| Bot. Of Bridge (RL) : | BOB = | 31.75 | M |

Bridge structure :
No of panels and width        nP = 5 nos        wP = 2.4 M

Column structure :
No of vertical panels and width

nP1 = 2        wP1 = 2.5 M

**Load Intensity:**
| | | | | |
|---|---|---|---|---|
| Live Load on Walk-way | | LLI = | 3 | kN/m2 |
| Wind Load | | WLI = | 2 | kN/m2 |
| Wind force coeffecient | | cf = | 2 | |
| Solidity Ratio : | Bridge | SR = | 1 | |
| | Column | SRC = | 1 | |

Horz. Seismic coefficient        $\alpha h$ =        0.04

## 12.3.3.6 Loading per Meter Length of Bridge

**A. Pipes :**

| BOP (RL) | dia (mm) | nos | wt/m (kn/m) |
|---|---|---|---|
| 34.2m | 200 | 1 | 0.40 |
| | 150 | 2 | 0.39 |
| | 125 | 1 | 0.29 |
| | 100 | 1 | 0.21 |
| | | | 0.00 |
| | | | 0.00 |
| | | | 0.00 |
| | | | 0.00 |
| Total weight | **wpipe** | | **1.69** kn/m |

Maxm. Pipe diameter :
Pdia = 0.2 m
Total surface area covered by pipes
= 0.725 m2/m

**B. Structure :**

Width of Walk-way    way    0.75 M

Unit weight of structure including columns, walk way etc =    0.5 kN/m2 (assumed).

Self Weight of Structure :    **wstruc** = 0.5 x 3.5 =    **1.75** kN/m

**C. Cable racks :**

| Type : | No of trays | Wt/m (kn) |
|---|---|---|
| 1.0 Power cables | 3 | 1.2 |
| 2.0 Control cables | 2 | 1 |
| 3.0 Instrument cables | 1 | 0.8 |
| Total weight | wcable | **6.40** kn/m |

Bot.(RL) of 1st tray            BOTra =   34.5 M
Vertical Dist. Betn. Trays :    Htra =    0.3 M

Maxm. No. of rows in a single Col :    nTra =    5 nos

**D. Incidental Point Load :**

Vertical Point Load at any point    Pinc =    2.5 kN

# Design of Industrial Structures

## 12.3.3.7 Analysis

**To determine equivalent UDL on each frame ( frame 1 or 2)**

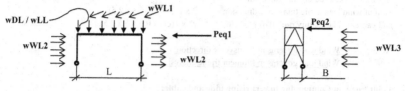

**Loading diagram for Wind / Seismic**

| | | | | |
|---|---|---|---|---|
| Dead Load : | wDL = | 17.22 kN/m | $(1.69 + 6.4 + 1.75) \times B/2$ | |
| Live Load on walk way ; | wLL = | 1.125 kN/m | $(0.75 \times 3) / 2$ | |
| | | | | |
| Transverse wind : | wWL1 = | 7.6 kN/m | $(2 \times 2 \times 1 \times 1.9)$ | per girder |
| | wWL3 = | 3 kN/m | $(2 \times 2 \times 1 \times 0.75)$ | per column frame |
| Along wind : | wWL2 = | 7.72 kN/m | $(2 \times 2 \times 1 \times 1.93)$ | per column/side |

*( Wind load = intensity x force coefficient x solidity ratio x projected area.)*

**Projected area against wind:**

(i) Transverse wind - wWL1

| mem | nos | L (mm) | W (mm) | Proj (m2) |
|---|---|---|---|---|
| Girder | 2 | 12 | 0.6 | 14.4 |
| Cable | 5 | 12 | 0.1 | 6 |
| pipes | 1 | 12.00 | 0.2 | 2.4 |
| | | sum total of bridge = | | 22.8 m |

hp = 22.8/12/2 = **1.9** m/m
( per girder )

(ii) Long wind - wWL2

| mem | nos | L (mm) | W (mm) | Proj (m2) |
|---|---|---|---|---|
| Col | 2 | 5 | 0.25 | 2.5 |
| pipes | 1 | 5 | 0.725 | 3.63 |
| cable | 1 | 5 | 0.7 | 3.50 |
| | | Total gable wind = | | 9.63 sqm |

hp = 9.63/5 = **1.93** m/m ht
( per column )

(iii) Transv. Wind on col - wWL3

| mem | nos | L (mm) | W (mm) | Proj (m2) |
|---|---|---|---|---|
| Col | 1 | 5 | 0.45 | 2.25 |
| pipe | 1 | 5 | 0.2 | 1.00 |
| cable | 1 | 5.0 | 0.1 | 0.50 |
| | | sum per col = | | 3.75 sqm |

hp = 3.75/5 = **0.75** m/m ht
( per column )

Seismic Load :
Along Bridge -          $P_{eq1}$ =    167 kN       (1.69 + 6.4 + 1.75) kN/m x (12 + 2 x (5/2))
Transverse dircn. -     $P_{eq2}$ =     84 kN       (167 / 2)
(for DL - pipe, structure and cable only)
Sum of wind load along transverse direction =    2 x (7.6 x 12 + 3 x 2.5) =           197 kN
and gable wind ( solidity ratio 0.6) =           [2 x 1(cf) x 0.6 x 5 x 3.5] =        221 kN

Note:   1.0  Wind load governs in Transverse direction.
        2.0  Wind load governs in Moment frame analysis.

**Point Load on Column due to vert rising pipe and cable:**

$P_{DL}$ =    43 kN      (17.22 x 5/2)
$P_{LL}$ =   2.5 kN      (incidental load)

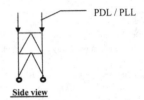

Side view

# Design of Industrial Structures

## Loading Diagram on Frame - 1 :

**Load case - 1**   Dead Load         **Load Case - 2**   Live Load

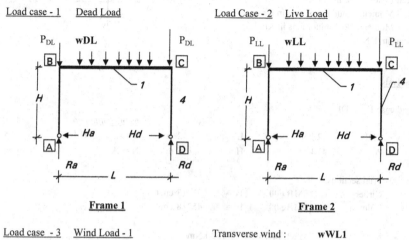

Frame 1                                Frame 2

**Load case - 3**   Wind Load - 1     Transverse wind :   wWL1

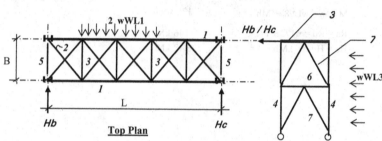

Top Plan

**LOAD CASE -4**   Wind Load - 2     Along wind :        wWL2

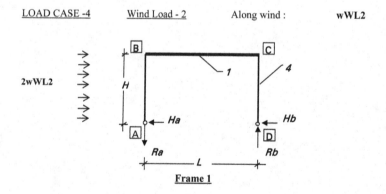

Frame 1

**Member forces: Unit - KN ,Kn-m**
Axial Load , P in KN
Mx = Moment about mazor axis in Kn-m
My = Moment about minor axis in Kn-m
Shear in KN.

## Analysis

<u>Load case 1</u>    DL

|  |  |  |  |  | *frame cofficients* |  |
|---|---|---|---|---|---|---|
| wDL = | 17.22 kN/m (w) | L = | 12 m | k = $I_{beam}/I_{col}$. H/L= | 0.85 |
| $P_{DL}$ = | 43 kN | H = | 5 m | N = 2k + 3 = | 4.69 |

Trial sections:
Girder    **ISMB 600**    Ix =    91813 cm4
Column    **ISMB 500**    Ix =    45218 cm4

Mb = Mc =    ( -) $wL^2/4N$ =    -132 kNm
Mspan = $wL^2/8$ + Mb =    178 kNm
Ra = Rd = wL/2 + $P_{DL}$    =    146 kN
Ha = Hd = (-) Mb / H =    26.42 kN

# Design of Industrial Structures

Load case 2    LL

    $wLL =$  1.13 kN/m  (w)  $L =$  12 m  
    $P_{LL} =$  2.5 kN  $H =$  5 m  

*frame cofficients*  
$k = Ix/Iy . H/L =$  1  
$N = 2k + 3 =$  5

Trial sections:  
Girder    ISMB 600    $Ix =$    91813 cm4  
Column    ISMB 500    $Ix =$    45218 cm4

$Mb = Mc =$    $(-) wL^2/4N =$    -8.63 kNm  
$Mspan = wL^2/8 + Mb =$    12 kNm

$Ra = Rd = wL/2 + P_{DL}$  =  9 kN  
$Ha = Hd = (-) Mb / H =$  1.73 kN

Load case 3    wind load  
*Top plan*  
$w = 2wWL1 =$  7.6 kN/m  Transverse wind acting on bridge deck  
$Mspan = w L^2 /8 =$  137 kNm  End shear, $Hb = Hc =$  45.6 kN ( for 2 wWL1))  
Force/ chord = Axial load (P / T) =  $Mspan / B =$  39.1 kN

*member 1:*  Axial Comp / Tension =  39.1 kN  
*member 5 :*  Axial Comp = 45.6 / 2 =  22.8 kN  
*member 2 :*  Length =  4.24 m  
    Axial Comp / Tension =  27.6 kN  
*member 3 :*  Axial Comp =  4.56 kN

*Side view*

    wWL3 =  3 kN/m height  $Hb =$  45.6 kN  
    Moment at base =  3 x 5^2/2 + 45.6 x 5 =  266 kNm.

*member 4 :*  Axial Comp / Tension =  75.86 kN  ( = 265.5 / 3.5)  
*member 6 :*  Axial Comp= 0.5 x 45.6+3 x 2.5 =  30.3 kN  
*member 7 :*  Length =  3.05 m  
    Base shear/leg =  3 x 5 / 2 + 45.6 / 2 =  30.3 kN  
    Axial Comp / Tension =  52.81 kN  ( = 30.3 x 3.05 /1.75)

Load case 4    LL

w/unit height = 15.44 kN/m    L = 12 m    *frame cofficients*
                              H = 5 m     $k = I_x/I_y$. H/L= 1
                                          $N = 2k + 3 =$  5

Trial sections:
Girder    ISMB 600    $I_x =$  91813 cm4
Column    ISMB 500    $I_x =$  45218 cm4

$M_b = wH^2/4 [-k/2N + 1] =$    88 kNm
$M_c = wH^2/4 [-k/2N - 1] =$    -105 kNm

$R_a = (-) R_d = (-) w H^2/2L =$    -16 kN
$H_d = (-) M_c / H =$               21 kN
$H_a = -1 (w.H - H_d) =$            -56.16 kN

## Member forces: Unit - KN, Kn-m
Axial Load, P in KN
Mx = Moment about mazor axis in Kn-m
My = Moment about minor axis in Kn-m
Shear in KN.

| Item | Mem. mkd | P Mx My | Load cases: 1 DL | 2 LL | 3 WL1 | 4 WL2 | load Combinations Comb1 1+2 | comb2 1+3 | Comb3 1+4 |
|---|---|---|---|---|---|---|---|---|---|
| Girder | mem1 | P | | | 39.1 | | 0 | 39 | 0 |
| | | Mxsup | -132 | -8.63 | | -105 | -141 | -132 | -237 |
| | | Mxsp | 178 | 11.62 | | | 189 | 178 | 178 |
| | | Shear | 103 | 7 | | -1.4 | 110 | 103 | 102 |
| | | My | | | | | 0 | 0 | 0 |
| Plan | mem2 | P | | | 28 | | 0 | 28 | 0 |
| Brcg. | mem3 | P | | | 5 | | 0 | 5 | 0 |
| Column | mem4 | P | 146 | 9 | 75.86 | 16 | 156 | 222 | 162 |
| | | T | | | -75.86 | | 0 | -76 | 0 |
| | | MxTop | -132 | -8.63 | | -105 | -141 | -132 | -237 |
| | | Mxsp | | | | | 0 | 0 | 0 |
| | | MxBot | 0 | 0 | | | 0 | 0 | 0 |
| | | Shear | 26.42 | 1.73 | | 56.2 | 28 | 26 | 83 |
| Col.Tie | mem5 | P | | | 22.80 | | 0 | 23 | 0 |
| | | | | | | | 0 | 0 | 0 |
| Vert. col. | mem6 | P | | | 30.30 | | 0 | 30 | 0 |
| | | | | | | | 0 | 0 | 0 |
| Brcg. | mem7 | P | | | 52.81 | | 0 | 53 | 0 |
| | | T | | | -52.81 | | 0 | -53 | 0 |

# Design of Industrial Structures

## 12.3.3.8  Member Design

*Member 1*          Girder          bending about major axis

Span L =   12 m          Mx (Mz) =   189 kNm          UDL =   18.0 kN/m   DL+LL
                         V =         110 kN
                         P =         39 kN

**Section properties:**

| Section Provided: | s | Member dimensions | | | | | |
|---|---|---|---|---|---|---|---|
| | | wt/m (kg) | $A_g$ (mm²) | $r_x (r_z)$ (mm) | $r_y$ (mm) | $Z_x (Z_e)$ (mm) | |
| 1  ISMB 600 | | 122.6 | 15621 | 242 | 41 | 3060400 | |
| | | b | d | tf | tw | Ix | Iy |
| | | mm | mm | mm | mm | mm4 | mm4 |
| | | 210 | 600 | 20.8 | 12 | 918130000 | 26510000 |

**Buckling class**

        Buckling class =   b    for buckling about minor axis   (IS 800:2007 - Table 10)
        Imperfection factor, α =   0.34                                     (IS 800:2007 - Table 7)

**Proportioning limit**                                                                 (IS 800:2007 - Table 2)

    ε = $(250 / f_y)^{0.5}$ =   $(250 / 250)^{0.5}$ =     1.00

Outstanding leg
    0.5b / t =     105 / 20.8 =     **5**       Compact       (IS 800:2007 - 3.7.2)
    *Limiting width to thickness ratio:*
    Plastic, $\lambda r$ =         9.4ε = 9.4 x 1 =          9
    Compact, $\lambda p$ =      10.5e = 10.5 x 1 =        11
    Semi compact, $\lambda sp$ = 15.7e = 15.7 x 1 =    16

**Effective Lengths (L) and Slenderness ratio ( L/r ) :**                    (IS 800:2007 - Table 11)

Laterally Unsupported length    Lz =   12.0 M    Kz =   1    KLz =   12.00 M
                                Ly =   2.4 M     Ky =   1    Kly =   2.40 M

    Maxm slenderness ratio (L/r) shall not exceed        180       (IS 800:2007 - Table 3)
    Here,   KLz/rz =    12000/242.4 =     50
            KLy/ry =    2400/41 =          59
                   L/r =           59         < 180, OK.

**Deflection:**                                                                         (IS 800:2007 - 5.6.1)
d = $5 w L^4 / 384 EI$ =   5 x 18 x 12000 ^4 / 384 x 200000 x 918130000 =     26.47 mm
                                     Allowable deflection limit L /300 =     40.00 mm
                                                                             OK.

**Laterally supported beam** (*web not susceptible for shear buckling; design shear ≤ 0.6Vd*)

Nominal flexural strength, Mn = Mp = fy Zp = Plastic moment

$f_y$ = Yield stress = 250 Mpa
$Z_p$ = plastic sectional modulus about the x axis = Σ A y
= 2 x 210 x 20.8 x 579.2/ 2 + 2 x (558.4/2) x 12 x (558.4/4)
= 3465377    mm4

Mn =   250 x 3465377/1000000 =      866 kNm.    (= Mp)

Design flexural strength = Md = β . Mn / $\gamma_{mo}$ < 1.2 Ze fy /$\gamma_{mo}$      (IS 800:2007 - 8.2.1.2)

Md =   1 x 866 / 1.1    =    787 kNm.   where, β = 1 for plastic and compact section

1.2 Ze fy /$\gamma_{mo}$   =   1.2 x 3060400 x 250 / 1.1/1000000 =      835 kNm    > Md; Okay.

Design flextural strength, **Md** =    **787** kNm

### Nominal Shear strength of web , Vn                   (IS 800:2007 - 8.4)

The nominal shear strength of web, Vn is goverered by plastic shear resistance.

Vn =    Vp = Av fyw / (3)^0.5                       (IS 800:2007 - 8.4.1)
Shear Area, Av = d x tw =    600 x 12 =    7200 mm2
Yield strength of web, $f_{yw}$ = fy =              250 Mpa

Vn = Vp =    7200 x 250 /3^0.5/1000 =    1039 kN.

### Design Shear strength of web , Vd
Vd = Vn / $\gamma_{m0}$                                      (IS 800:2007 - 8.4)
**Vd** =    1039 / 1.1=    **945** kN

# Design of Industrial Structures

## Strength design

| Limit state design<br>(IS 800:2007 - Section 8) | Working stress design<br>(IS 800:2007 - Section 11) |
|---|---|
| Factored design moment = M<br>Design flexural strength of the section = Md<br>Factored design shear = V<br>Design shear strength of the section = Vd<br><br>$M \leq Md$<br>$V \leq Vd$<br>Load factor : 1.5<br><br>M = 1.5 x 189 = 284 kNm<br>Md = 787 kNm<br>**Md > M; Safe.** | Actual design moment = Ms<br>Permissible flexural strength = Ma<br>Actual design shear = Vs<br>Permissible shear strength = Va<br><br>$Ms \leq Ma$<br>$Vs \leq Va$<br><br>_Allowable flexural strength, Ma :_ (IS 800-11.4.1)<br>_For laterally supported beam:_<br>$fabc = fabt = Ma/Zx$ (Zx top = Zx bot)<br>$Ma = fabc \cdot Zx = 0.66 f_y Zx$ (fabc = fabt)<br>= 0.66x250x3060400/10^6<br>= 504.97 kNm<br>Allowable Moment, **Ma** = **505** kNm.<br>Actual Design Moment, Ms = 189 kNm.<br>**Ma > Ms; Safe.** |
| V = 1.5 x 110 = 165 kN<br>Vd = 945 kN<br>**Vd > V; Safe.** | V = 110 kN<br><br>_Shear strength, Vs :_ (IS 800-11.4.2)<br>Pure shear<br>$Va = \tau_{ab} \cdot Av$, where Av shear area<br>= $0.4 f_y \times Av$<br>= 0.4 x 250 x 600 x 12/1000<br>= 720 kN<br>Allowable Shear, **Va** = **720** kNm.<br>Actual design shear, Vs = 110 kNm.<br>**Va > Vs; Safe.** |

**Axial stress**
The magnitude of axial force P is less than 10% of section capacity, hence ignored.

## Member 2    Plan Bracing    Axial Compresion / Tension.

L =    4.24 m         P =    28 kN
                      T =    28 kN

**Section properties:**

| Section Provided: | s | | wt/m (kg) | $A_g$ (mm$^2$) | $r_x$ (mm) | $r_y$ (mm) | $r_{vv}$ (mm) | $I_x$ (mm$^4$) | $I_y$ (mm$^4$) | $C_{yy}$ (mm) |
|---|---|---|---|---|---|---|---|---|---|---|
| 2  ISA 75 75 8 | 8 | mm b/b | 17.8 | 2276 | 22.80 | 34.0 | 14.5 | 1180000 | 2648384 | 21.4 |

| | | | Single Leg dimensions | | | | | | | |
|---|---|---|---|---|---|---|---|---|---|---|
| | | | b | d | t | Connection bolt | | | | An |
| | | | mm | mm | mm | nos | φdia | φhole | rows | mm2 |
| | | | 75 | 75 | 8 | 3 | 16 | 18 | 1 | 2132 |

where, An = net area after deduction of bolt holes.
    An =    Ag - (no of rows x φ hole x t)
Distance between tacking plate (welded spacer plate) = **a** =    256 mm    (IS 800:2007 - 10.2.5.2)
(the value of ' a ' shall not exceed 32 t or 300 mm, whichever is less)

### Buckling class

            Buckling class =   c    for buckling about any axis    (IS 800:2007 - Table 10)
            Imperfection factor, α =    0.49                       (IS 800:2007 - Table 7)

### Proportioning limit                                             (IS 800:2007 - Table 2)

ε = (250 / fy ) ^0.5  =    (250 / 250)^0.5=    1.00

Outstanding leg
    b / t =    75 / 8 =        9        Plastic.                   (IS 800:2007 - 3.7.2)
    *Limiting width to thickness ratio:*
    Plastic, λr =        9.4ε = 9.4 x 1 =     9
    Compact, λp =        10.5e = 10.5 x 1 =   11
    Semi compact, λsp =  15.7e = 15.7 x 1 =   16

### Effective Lengths (L) and Slenderness ratio ( L/r ) :           (IS 800:2007 - Table 11)

Laterally Unsupported length    Lz =    4.24 M    Kz =    1    KLz =    4.24 M
                                Ly =    2.12 M    Ky =    1    Kly =    2.12 M

    Maxm slenderness ratio (L/r) shall not exceed       250      (IS 800:2007 - Table 3)
        Here,   KLz/rz =    4240/22.8 =    186
                KLy/ry =    2120/34 =       62
                            L/r =           186      <250, OK.

# Design of Industrial Structures

**Sectional capacity / Design strength:**

**Axial compression**
*(i) Compression due to buckling about major axis, z-z [Lz / rz (=rx)]*  (IS 800:2007 7.5.2.1)

Design compression strength, $P_{dz} = A_e \cdot f_{cd}$  (IS 800:2007 7.1.2)

where.  $A_e$ = effective sectional area
$f_{cd}$ = design compressive stress

$f_{cd} = \chi\, f_y / \gamma_{mo}  \quad <= f_y / \gamma_{mo}$
$\chi = 1 / [\phi + (\phi^2 - \lambda^2)^{0.5}]$
$\phi = 0.5\,[1 + \alpha(\lambda - 0.2) + \lambda^2]$
$\lambda = (f_y / f_{cc})^{0.5}$

$f_{cc} = \pi^2 E / (KL_z/r_z)^2$  Euler buckling stress

Now, putting following values into above equation

$\alpha$ = 0.49  $E$ = 200000 Mpa  $f_y$ = 250 Mpa
$\gamma_{m0}$ = 1.10  $KL_z/r_z$ = 186  $A_e$ = 2276 mm2  $=A_g$

$f_{cc}$ = 3.14^2 x 200000 / 186^2 =  57 MPa
$\lambda$ = (250 / 57)^0.5 = 2.09
$\phi$ = 0.5[1 + 0.49 x (2.09 - 0.2) + 2.09^2]= 3.15
$\chi$ = 1 /[3.15 + (3.15^2 - 2.09^2)^0.5]= 0.182

$f_{cd}$ = 0.182 x 250 / 1.1 = 41 Mpa  $<= f_y / \gamma_{mo}$  $<=$  250 / 1.1
 $<=$  227 Mpa

$f_{cd}$ = 41 Mpa  (=$f_{cdz}$)

**$P_{dz}$** = $A_e \cdot f_{cd}$ = 2276 x 41 /1000= 93 kN

*(ii) Compression due to buckling about minor axis, y-y (KLy / ryy)*  (IS 800:2007 7.5.2.1)

Design compression strength, $P_{dy} = A_e \cdot f_{cd}$  where.  $A_e$ = effective sectional area
$f_{cd}$ = design compressive stress

$f_{cd} = \chi\, f_y / \gamma_{mo} \quad <= f_y / \gamma_{mo}$  (IS 800:2007 7.1.2.1)
$\chi = 1 / [\phi + (\phi^2 - \lambda^2)^{0.5}]$
$\phi = 0.5\,[1 + \alpha(\lambda - 0.2) + \lambda^2]$
$\lambda = (f_y / f_{cc})^{0.5}$

$f_{cc} = \pi^2 E / (KL_z/r_z)^2$  Euler buckling stress

Now, putting following values into above equation

| | | | | | |
|---|---|---|---|---|---|
| $\alpha =$ | 0.49 | $E =$ | 200000 Mpa | $fy =$ | 250 Mpa |
| $\gamma_{m0} =$ | 1.10 | KLy/ryy= | 62 | $Ae =$ | 2276 mm2 |

$fcc =$  3.14^2 x 200000 / 62^2 =     513 MPa
$\lambda =$   (250 / 513)^0.5=    0.7
$\phi =$   0.5[1 + 0.49 x (0.7 - 0.2) + 0.7^2]=   0.87
$\chi =$   1 /[0.87 + (0.87^2 - 0.7^2)^0.5]=    0.721

$fcd =$   0.721 x 250 / 1.1 =   164 Mpa   <= $fy / \gamma_{mo}$   <=   250 / 1.1
                                                                <=   227 Mpa

$fcd =$   164 Mpa   (=fcdy)

**Pdy** = Ae . fcd =   2276 x 164 /1000=   **373 kN**

The design compressive strength, **Pd** will be lowest of above calculated strengths of the member.

**Pd =**   **93 kN**

### Axial Tension

The factored design tension, T  <  Design tensile strength of member, Td

*(i) Design tensile strength due to yielding of gross section, Tdg*    (IS 800:2007 - 6.2)

Tdg = Ag fy / $\gamma_{mo}$ =    2276 x 250 / 1.1/ 1000  =    517 kN

*(ii) Design tensile strength due to rupture of critical section, Tdn*

Tdn = 0.9 Anc. fu / $\gamma_{m1}$ + β. Ago. fy / $\gamma_{m0}$    (IS 800:2007 - 6.3.3)

β = 1.4 - 0.076 (w/t) (fy/fu) (bs/Lc) <= 0.9 fu $\gamma_{mo}$ / fy $\gamma_{m1}$ >= 0.7

where,  Anc = net area of connected leg =    2 x [75x8-(1x18x8) ] =    912 mm2
        Ago = gross area of outstanding leg =   2 x (75 x 8 ) =    1200 mm2

w = width of outstanding leg = d =                                   75 mm
bs = shear lag width = w + w1 (= Cyy) - t =    75 + 21.4 - 8 =       88 mm
Lc = length of end connection = (nos of bolt -0.5) x 3 $\phi$ hole =    135 mm

# Design of Industrial Structures

$fu =$ 410 Mpa  $\gamma_{m0} =$ 1.10  $t =$ 8 mm
$fy =$ 250 Mpa  $\gamma_{m1} =$ 1.25

$\beta = 1.4 - 0.076 \,(w/t)\,(fy/fu)\,(bs/Lc) <= 0.9\, fu\, \gamma_{mo} / fy\, \gamma_{m1} >= 0.7$
$= 1.4 - 0.076 \times (75/8) \times (250/410) \times (88/135) <= 0.9 \times 410 \times 1.1 / (250 \times 1.25) >= 0.7$
$= 1.12 \quad <= \quad 1.30 \quad >= \quad 0.7 \quad\quad$ Ok.

$Tdn = 0.9\, Anc.\, fu / \gamma_{m1} + \beta.\, Ago.\, fy / \gamma_{mo}$
$\quad = 0.9 \times 912 \times 410 / 1.25 + 1.12 \times 1200 \times 250 / 1.1$
$\quad = 574677\ N$
$\quad = 575\ kN$

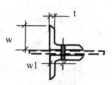

### *(iii) Design tensile strength due to block shear, Tdb*    (IS 800:2007 - 6.4.1)

a) $\quad Tdb = [\ Avg.\, fy / (1.732\, \gamma_{mo}) + 0.9\, Atn.\, fu / \gamma_{m1}]$    OR

b) $\quad Tdb = (0.9\, Avn.\, fu / (1.732\, \gamma_{m1}) + Atg.\, fy / \gamma_{mo})$    whichever is smaller

$Avg =$ gross shear area along bolt line parallel to line (1-2) of external force
$\quad\ \ = 2.\, Lc.\, t$
$\quad\ \ = 2 \times 135 \times 8 \quad = \quad 2160\ mm2$
$Avn =$ net shear area along bolt line parallel to the line (1-2) of external force
$\quad\ \ = 2\,[Lc.\,t - (\text{no of bolts} \times \phi\, \text{hole} \times \text{thickness})]$
$\quad\ \ = 2 \times [\,135 \times 8 - (3 \times 18 \times 8)\,] = \quad 1296\ mm2$

$Atg =$ gross area in tension from bolt hole to toe of the angle, perpendicular to line (2-3) of force
$\quad\ \ = 2.\,(d - Cyy).\,t$
$\quad\ \ = 2 \times (75 - 21.4) \times 8\,) = \quad 857.6\ mm2$
$Atn =$ net area in tension from bolt hole to toe of the angle, perpendicular to line (2-3) of force
$\quad\ \ = 2\,[\,(d - Cyy).\,t - (\text{no of rows} \times \phi\, \text{hole} \times \text{thickness})]$
$\quad\ \ = 2\,[\,(75 - 21.4) \times 8 - (1 \times 18 \times 8)\,] = \quad 569.6\ mm2$

a) $\quad Tdb = [\ Avg.\, fy / (1.732\, \gamma_{mo}) + 0.9\, Atn.\, fu / \gamma_{m1}]$
$\quad\quad = 2160 \times 250 / (1.732 \times 1.1) + 0.9 \times 569.6 \times 410 / 1.25$
$\quad\quad = 451581\ N$
$\quad\quad = 452\ kN$
OR

b) $\quad Tdb = (0.9\, Avn.\, fu / (1.732\, \gamma_{m1}) + Atg.\, fy / \gamma_{mo})$    whichever is smaller
$\quad\quad = 0.9 \times 1296 \times 410 / (1.732 \times 1.25) + 857.6 \times 250 / 1.1$
$\quad\quad = 415798\ N$
$\quad\quad = 416\ kN$

So, Tensile strength due to block shear, $Tdb = $ 416 kN
now, we have $Tdg =$ 517 kN,  $Tdn =$ 575 kN and $Tdb =$ 416 kN

*The design tensile strength, **Td** will be lowest of above calculated strengths of the member.*
$\quad Td =\quad 416\ kN$

| Strength design | Compression /Tension | |
|---|---|---|
| | Limit state method | Working stress method |
| property | Plastic. | Plastic. |
| Load | Axial Comp, P = 28 kN<br>Axial Tension, T = 28 kN | Axial Comp, P = 28 kN<br>Axial Tension, T = 28 kN |
| | Load factor = 1.5<br>Factored load :<br>P = 1.5 x 27.645449  41.468 kN<br>T = 1.5 x 27.645449  41.468 kN | **Axial compression :**  (IS 800:2007 11.3.1)<br>Permissible comp. stress, fac = 0.6 fcd<br>fcd = 41 Mpa (see calc. above.)<br>Ag = 2276 mm2 |
| | Design Comp strength, Pd = 93 kN<br>Factor load , P = 41 kN<br>P < Pd; Safe. | Permissible Comp strength, Pd = 0.6 fcd .Ag<br>= 0.6 x 41 x 2276<br>= 55990 N<br>= 56 kN<br>Actual load, P = 28 kN<br>P < Pd; Safe. |
| | Design Tensile strength, Td = 416 kN<br>Factor load , T = 41 kN<br>T < Td; Safe. | **Axial Tension:**  (IS 800:2007 11.2.1)<br><br>Tensile stresses for :<br>a) yielding   fat = 0.6 fy = 0.6 x 250<br>= 150 MPa<br>b) rupture   fat = 0.69 Tdn/Ag<br>= 0.69 x 575000 / 2276<br>= 174 MPa<br>c) block shear   fat = 0.69 Tdb/Ag<br>= 0.69 x 416000 / 2276<br>= 126 MPa<br><br>Permissible value of fat = 126 MPa<br><br>Permissible Tensile strength, Td = fat .Ag<br>= 126 x 2276<br>= 286776 N<br>= 287 kN<br>Actual load, T = 28 kN<br>T < Td; Safe. |

# Design of Industrial Structures

### Member 3    Cross beam    bending about major axis

Span L = 3.5 m    UDL = 18.0 kN/m   (wDL17.22 + wLL1.125)    DL+LL
           $M_x$ ($M_z$) = 18 x 3.5^2 / 8 = 28 kNm
           V = 18 x 3.5 x 0.5 = 32 kN

| Section properties: | | | Member dimensions | | | | | | |
|---|---|---|---|---|---|---|---|---|---|
| Section Provided: | | s | wt/m (kg) | $A_g$ (mm$^2$) | $r_x$ ($r_z$) (mm) | $r_y$ (mm) | $Z_x$($Z_e$) (mm) | | |
| 1 | ISMC 250 | | 30.4 | 3867 | 99 | 24 | 305300 | | |
| | | | b | d | tf | tw | Ix | | Iy |
| | | | mm | mm | mm | mm | mm4 | | mm4 |
| | | | 80 | 250 | 14.1 | 7.1 | 38168000 | | 2191000 |

### Buckling class
                    Buckling class =   c    for buckling about any axis    (IS 800:2007 - Table 10)
                Imperfaction factor, $\alpha$ =   0.49                                        (IS 800:2007 - Table 7)

### Proportioning limit                                                                                (IS 800:2007 - Table 2)

    $\varepsilon$ = (250 / fy )^0.5 =    (250 / 250)^0.5 =    1.00

### Outstanding leg
    b / t =    80 / 14.1 =    6    Plastic.    (IS 800:2007 - 3.7.2)
    *Limiting width to thickness ratio:*
    Plastic, $\lambda r$ =    9.4$\varepsilon$ = 9.4 x 1 =    9
    Compact, $\lambda p$ =    10.5e = 10.5 x 1 =    11
    Semi compact, $\lambda sp$ =    15.7e = 15.7 x 1 =    16

### Effective Lengths (L) and Slenderness ratio ( L/r ) :                    (IS 800:2007 - Table 11)

Laterally Unsupported length    Lz =    3.5 M    Kz =    0.85    KLz =    2.98 M
                                                  Ly =    3.5 M    Ky =    1           Kly =    3.50 M

    Maxm slenderness ratio (L/r) shall not exceed           180       (IS 800:2007 - Table 3)
    Here,    KLz/rz =    2975/99.4 =    30
            KLy/ry =    3500/24 =    146
                    L/r =    146      < 180, OK.

**Deflection:** (IS 800:2007 - 5.6.1)
$d = 5wL^4 / 384 EI$ = 5 x 18 x 3500 ^4 / 384 x 200000 x 38168000 = 4.61 mm
Allowable deflection limit L /300 = 11.67 mm   OK.

**Laterally supported beam** ( *web not susceptible for shear buckling; design shear $\leq 0.6Vd$* )

Nominal flexural strength, $M_n = M_p = f_y Z_p$ = Plastic moment

$f_y$ = Yield stress =      250 Mpa
$Z_p$ = plastic sectional modulus about the x axis = $\Sigma A y$
   =    2 x 80 x 14.1 x 235.9/ 2 + 2 x (221.8/2) x 7.1 x (221.8/4)
   =       353417    mm4

$M_n$ =   250 x 353417/1000000 =         88 kNm.    (= $M_p$)

Design flexural strength = $M_d = \beta \cdot M_n / \gamma_{mo}$  < 1.2 $Z_e f_y /\gamma_{mo}$     (IS 800:2007 - 8.2.1.2)

$M_d$ =   1 x 88 / 1.1     =     80 kNm.   where, $\beta$ = 1 for plastic and compact section

1.2 $Z_e f_y /\gamma_{mo}$   =    1.2 x 305300 x 250 / 1.1/1000000 =       83 kNm    > $M_d$; Okay.

Design flextural strength, **Md** =     **80** kNm

### Nominal Shear strength of web , Vn                              (IS 800:2007 - 8.4)

The nominal shear strength of web, Vn is goverered by plastic shear resistance.

$V_n$ =   $V_p = A_v f_{yw} / (3)^{0.5}$                                        (IS 800:2007 - 8.4.1)
Shear Area, $A_v = d \times t_w$ =      250 x 7.1 =        1775 mm2
Yield strength of web, $f_{yw} = f_y$ =                    250 Mpa

$V_n = V_p$ =     1775 x 250 /3^0.5/1000 =       256 kN.

### Design Shear strength of web , Vd
$V_d = V_n / \gamma_{m0}$                                                 (IS 800:2007 - 8.4)
**Vd** =   256 / 1.1=      **233** kN

# Design of Industrial Structures

**Strength design**

| Limit state design<br>(IS 800:2007 - Section 8) | Working stress design<br>(IS 800:2007 - Section 11) |
|---|---|
| Factored design moment = M<br>Design flexural strength of the section = Md<br>Factored design shear = V<br>Design shear strength of the section = Vd<br><br>$\quad$ M <= Md<br>$\quad$ V <= Vd<br>Load factor :  $\quad$ 1.5<br><br>M = 1.5 x 28 = $\quad$ 42 $\quad$ kNm<br>$\quad\quad\quad$ Md = $\quad$ 80 $\quad$ kNm<br>$\quad\quad\quad$ **Md > M; Safe.** | Actual design moment = Ms<br>Permissible flexural strength = Ma<br>Actual design shear = Vs<br>Permissible shear strength = Va<br><br>$\quad$ Ms <= Ma<br>$\quad$ Vs <= Va<br><br><u>Allowable flexural strength, Ma :</u> $\quad$ (IS 800-11.4.1)<br>For laterally supported beam:<br>$fabc = fabt = Ma/Zx$ $\quad$ (Zx top = Zx bot)<br>$Ma = fabc . Zx = 0.66\, fy\, Zx$ $\quad$ (fabc = fabt)<br>$\quad$ = $\quad$ 0.66x250x305300/10^6<br>$\quad$ = $\quad$ 50.375 kNm<br>Allowable Moment, **Ma** = $\quad\quad$ **50** kNm.<br>Actual Design Moment, Ms = $\quad\quad$ 28 kNm.<br>$\quad\quad\quad\quad\quad\quad\quad\quad$ **Ma > Ms; Safe.** |
| V = $\quad$ 1.5 x 32 = $\quad$ 48 kN<br>Vd = $\quad\quad\quad\quad\quad\quad$ 233 kN<br>$\quad\quad\quad$ **Vd > V; Safe.** | V = $\quad\quad$ 32 kN<br><br><u>Shear strength, Vs :</u> $\quad$ (IS 800-11.4.2)<br>Pure shear<br>$Va = \tau_{ab} . Av$ , $\quad$ where Av shear area<br>$\quad$ = $\quad$ 0.4 fy x Av<br>$\quad$ = $\quad$ 0.4 x 250 x 250 x 7.1/1000<br>$\quad$ = $\quad$ 178 kN<br>Allowable Shear, **Va** = $\quad\quad$ **178** kNm.<br>Actual design shear, Vs = $\quad\quad$ 32 kNm.<br>$\quad\quad\quad\quad\quad\quad\quad\quad$ **Va > Vs; Safe.** |

**Axial stress**
   The magnitude of axial force is negligible, hence ignored.

**Member 4    Column leg**

The total height of the column is divided in two segments along minor axis. Each segments are crossed braced and tied with horizontal tie. The column base plate will have gusset plates to distribute load on base plate. There will be anchor bolts to transfer shear and resist uplift. Column base plate will rest on Reinforced concrete pedestal.

Height  L = 5 m      Lx = 5 m      Ly = 2.5 m.

| DL + WL | Axial | | Tension | | Moment | | Shear | |
|---|---|---|---|---|---|---|---|---|
| Base | P = | 222 kN | T = | -76 kN | M = | 0 kNm | V = | 26 kN |
| Top | | 110 kN* | | 0 kN | | -237 kNm | | 23 kN** |

[ (*) Reaction from top girder      **(Hb/2) ]

Try with    **ISMB 500**    wt/m    86.9 kg

**Axial compression:**

| Section properties: | | | | Member dimensions | | | | | | | |
|---|---|---|---|---|---|---|---|---|---|---|---|
| Section Provided: | | s | | wt/m (kg) | $A_g$ (mm$^2$) | $r_x(r_z)$ (mm) | $r_y$ (mm) | $Z_x(Z_e)$ (mm) | $I_x$ (mm$^4$) | $I_y$ (mm$^4$) | $C_{yy}$ (mm) |
| 1 | ISMB 500 | 0 | mm b/b | 86.9 | 11074 | 202 | 35.0 | 1808700 | 4.52E+08 | 13698000 | 0 |
| | | | | b | d | tf | tw | Connection bolt | | | |
| | | | | mm | mm | mm | mm | nos | φdia | φhole | rows |
| | | | | 180 | 500 | 17.2 | 10.2 | | | | |

**Buckling class**

Buckling class =   b    for buckling about minor axis    (IS 800:2007 - Table 10)
Imperfaction factor, α =   0.34                          (IS 800:2007 - Table 7)

# Design of Industrial Structures

**Proportioning limit** (IS 800:2007 - Table 2)

$\varepsilon = (250 / f_y)^{0.5} =$  $(250/250)^{0.5} =$   1.00

**Outstanding leg**
  $b/t =$   180 / 17.2 =   **10**   Compact   (IS 800:2007 - 3.7.2)
  *Limiting width to thickness ratio:*
  Plastic, $\lambda r =$   $9.4\varepsilon = 9.4 \times 1 =$   9
  Compact, $\lambda p =$   $10.5e = 10.5 \times 1 =$   11
  Semi compact, $\lambda sp =$   $15.7e = 15.7 \times 1 =$   16


**Effective Lengths (L) and Slenderness ratio ( L/r ) :**   (IS 800:2007 - Table 11)

Laterally Unsupported length   $L_z =$  5.0 M   $K_z =$  0.85   $KL_z =$  4.25 M
                               $L_y =$  2.5 M   $K_y =$  1      $KL_y =$  2.50 M

  Maxm slenderness ratio (L/r) shall not exceed   250   (IS 800:2007 - Table 3)
  Here,  $KL_z/r_z =$   4250/202 =   21
         $KL_y/r_y =$   2500/35 =    71
                        **L/r =**    71   <250, OK.

**Compression strength**
*(i) Compression due to buckling about minor axis, y-y (KLy / ry)*   (IS 800:2007 7..5.1.1)

Design compression strength, $P_{dy} = A_e \cdot f_{cd}$   where,  $A_e =$  effective sectional area
                                                                   $f_{cd} =$  design compressive stress
              $f_{cd} =$   $\chi f_y / \gamma_{mo}$  $\leq f_y / \gamma_{mo}$   (IS 800:2007 7.1.2.1)
              $\chi =$     $1 / [\phi + (\phi^2 - \lambda^2)^{0.5}]$
              $\phi =$     $0.5 [1 + \alpha(\lambda - 0.2) + \lambda^2]$
              $\lambda =$  $(f_y / f_{cc})^{0.5}$
              $f_{cc} =$   $\pi^2 E / (KL_z/r_z)^2$   Euler buckling stress

Now, putting following values into above equation
    $\alpha =$      0.34      $E =$      200000 Mpa      $f_y =$    250 Mpa
    $\gamma_{m0} =$ 1.10      $KL_y/r_y =$   71           $A_e =$    11074 mm2

    $f_{cc} =$   $3.14^2 \times 200000 / 71.4285714285714^2 =$   386 MPa
    $\lambda =$  $(250 / 386)^{0.5} =$                           0.8
    $\phi =$     $0.5[1 + 0.34 \times (0.8 - 0.2) + 0.8^2] =$    0.92
    $\chi =$     $1 / [0.92 + (0.92^2 - 0.8^2)^{0.5}] =$         0.728
    $f_{cd} =$   $0.728 \times 250 / 1.1 =$   165 Mpa   $\leq f_y/\gamma_{mo}$   $\leq$   250 / 1.1
                                                                              $\leq$   227 Mpa

              $f_{cd} =$   165 Mpa   (=fcdy)

    **$P_{dy} = A_e \cdot f_{cd} =$**   $11074 \times 165 / 1000 =$   **1827** kN

## (ii) Compression due to buckling about major axis, z-z ($KL_z/r_z$)

Design compression strength, $P_{dz} = A_e \cdot f_{cd}$      where, $A_e$ = effective sectional area
                                                         $f_{cd}$ = design compressive stress

$f_{cd} = \chi f_y / \gamma_{mo} \quad <= f_y / \gamma_{mo}$          (IS 800:2007 7.1.2.1)

$\chi = 1 / [\phi + (\phi^2 - \lambda^2)^{0.5}]$

$\phi = 0.5 [1 + \alpha(\lambda - 0.2) + \lambda^2]$

$\lambda = (f_y / f_{cc})^{0.5}$

$f_{cc} = \pi^2 E / (KL_z/r_z)^2$      Euler buckling stress

Now, putting following values into above equation

| | | | | | | |
|---|---|---|---|---|---|---|
| $\alpha =$ | 0.34 | E = | 200000 Mpa | $f_y =$ | 250 Mpa | |
| $\gamma_{m0} =$ | 1.10 | $KL_z/r_z$ : | 21 | $A_e =$ | 11074 mm2 | |

$f_{cc} = 3.14^2 \times 200000 / 21.039603960396^2 =$    4455 MPa

$\lambda = (250 / 4455)^{0.5} =$    0.24

$\phi = 0.5[1 + 0.34 \times (0.24 - 0.2) + 0.24^2] =$    0.54

$\chi = 1 /[0.54 + (0.54^2 - 0.24^2)^{0.5}] =$    0.977

$f_{cd} = 0.977 \times 250 / 1.1 =$      222 Mpa    $<= f_y / \gamma_{mo}$    $<=$    250 / 1.1
                                                                          $<=$    227 Mpa

$f_{cd} =$    222 Mpa    $(= f_{cdz})$

$P_{dz} = A_e \cdot f_{cd} =$    11074 × 222 /1000 =    **2458** kN

       **Pd =**    **1827 kN**      **fcd =**    **165 Mpa**

## Moment Capacity:

### (iii) Flexural strength due to buckling about major axis, z-z

Design flexural strength about major axis = $M_{dz}$                  (IS 800:2007 - 8.2)

### a) Laterally supported beam ( web not susceptible for shear buckling; design shear $\leq 0.6V_d$ )

Nominal flexural strength, $M_{nz} = M_{pz} = f_y Z_{pz}$ = Plastic moment

# Design of Industrial Structures

$f_y$ = Yield stress = 250 Mpa

$Z_p$ = plastic sectional modulus about the x axis = $\Sigma A \bar{y}$
   = 2 x 180 x 17.2 x 482.8/ 2 + 2 x (465.6/2) x 10.2 x (465.6/4)
   = 2047546   mm4

$M_{nz}$ = 250 x 2047546/1000000 = 512 kNm.   (= $M_{pz}$)

Design flexural strength = $M_{dz}$ = β . $M_{nz}$ / $\gamma_{mo}$ < 1.2 $Z_z$ $f_y$ /$\gamma_{mo}$   (IS 800:2007 - 8.2.1.2)

$M_{dz}$ = 1 x 512 / 1.1   = 465 kNm.   where, β = 1 for plastic and compact section

1.2 $Z_z$ $f_y$ /$\gamma_{mo}$   =   1.2 x 1808700 x 250 / 1.1/1000000 =   493 kNm   > Mdz; Okay.

Design flextural strength, **Md** =   **465** kNm

*b} Lateral torsional buckling need NOT be considered, when bending is about minor axis of the section.*
*(IS 800:2007 - 8.2.2)*

## Nominal Shear strength of web, Vn   (IS 800:2007 - 8.4)

The nominal shear strength of web, Vn is goverered by plastic shear resistance.

$V_n$ =   $V_p$ = $A_v$ $f_{yw}$ / $(3)^{0.5}$   (IS 800:2007 - 8.4.1)

Shear Area, $A_v$ = d x $t_w$ =   500 x 10.2 =   5100 mm2
Yield strength of web, $f_{yw}$ = $f_y$ =   250 Mpa

$V_n$ = $V_p$ =   5100 x 250 /$3^{0.5}$/1000 =   736 kN.

## Design Shear strength of web, Vd

$V_d$ = $V_n$ / $\gamma_{m0}$   (IS 800:2007 - 8.4)

**Vd** =   736 / 1.1=   **669** kN

## Strength design for combined axial force and flexure

| property | Limit state method | | | | Working stress method | | | | |
|---|---|---|---|---|---|---|---|---|---|
| | Flange: | Compact | | | Flange: | Compact | | | |
| | Web : | a) yielding | | | Web : | a) yielding | | | |
| | DL + WL | | | | DL + WL | | | | |
| Load | Base | Factored load | | | Base | Unfactored load | | | |
| | Axial | $M_{vertical}$ | $M_{horz}$ | $V_{vertical}$ | $V_{horz}$ | Axial | $M_{vertical}$ | $M_{horz}$ | $V_{vertical}$ | $V_{horz}$ |
| | P | $M_z$ | $M_y$ | $V_z$ | $V_y$ | P | $M_z$ | $M_y$ | $V_z$ | $V_y$ |

*[contd.]*

| Load factor | 222 | 0 | 0 | 26 | 0 | 222 | 0 | 0 | 26 | 0 |
|---|---|---|---|---|---|---|---|---|---|---|
| | 1.5 | 1.5 | 1.5 | 1.5 | 1.5 | | | | | |
| Factored load | 333 | 0 | 0 | 40 | 0 | 222 | 0 | 0 | 26 | 0 |

**Combined Axial compression and bending**
*Section Strength*  (IS 800:2007 - 9.3.1)

P = 333 kN   Pd = 1827 kN

Mz = 0 kNm   Mdz = 465 kNm

$P/P_d + M_y/M_{dy} + M_z/M_{dz} \leq 1$

= 333/1827 + 0/465
= 0.18 + 0 = 0.18
         < 1 ; Safe.

**Shear check**  (IS 800:2007 - 9.2.1)

Factored shear, V = 40 kN
Design shear strength, Vd = 669 kN
         Safe.

60% of Vd = 401 kN
*V < 60% of Vd; No reduction in Moment capacity, Md.*

**Axial Comp.**  (IS 800:2007 - 11.3.1)

Ps = 222 kN   Ae = 11074 mm2

Actual comp stress = fc = Ps/Ae
 = 222.23 x 1000 / 11074
 = 20 Mpa
Permissible comp. stress, fac = 0.60 fcd
 = 0.6 x 165
 = 99 Mpa
         fc<fac; Safe.
where, fcd = 165 Mpa ( see above)

**Bending.**  (IS 800:2007 - 11.4.1)
Mz = 0 kNm
My = 0 kNm
Ms = (0^2 + 0^2)^0.5 = 0.00 kNm
Ze = 1808700 mm3

Actual bending stress, fbc or fbt = Ms / Ze
 = 0 / 1808700
 = 0 Mpa

Permissible bending stress, fabc or fabt =
 = 0.60 Md / Ze
 = 0.6x465x10^6 / 1808700
 = 154 Mpa   fbc<fabc; Safe.

where, Md = 465 kNm  (see calc.above)

**Combined Axial compression and bending**

*Member strength requirement:*

$f_c/0.6f_y + f_{bcz}/f_{abcz}$    <= 1.0
= 20/ (0.6x250) + 0/154
 = 0.1333  +  0
 = 0.1333    < 1 ; Safe.

*[contd.]*

**Combined Axial compression, bending and shear**
(IS 800:2007 - 11.5.4) & ( IS 800 -1962 10.5.4)

Shear stress, $\tau_b$ = Vs/Av
= 26424/(500x10.2)
= 5.18 Mpa

fe =$(fc^2 + fbc^2 + fc.fbc + 3\tau b^2)$^0.5 < 0.9fy
= $(20^2 + 0^2 + 20 \times 0 + 3 \times 5.18^2)$^0.5
< 0.9 x 250
= 22 Mpa < 225 Mpa
                                   Safe.

| Load | DL + WL Top | | Factored load | | |
|---|---|---|---|---|---|
| | Axial | $M_{vertical}$ | $M_{horz}$ | $V_{vertical}$ | $V_{horz}$ |
| | P | $M_z$ | $M_y$ | $V_z$ | $V_y$ |
| | 110 | 237 | 0 | 23 | 0 |
| Load factor | 1.5 | 1.5 | 1.5 | 1.5 | 1.5 |
| Factored load | 165 | 356 | 0 | 34 | 0 |

**Combined Axial compression and bending**
*Section Strength*   (IS 800:2007 - 9.3.1)

P= 165 kN    Pd = 1827 kN

Mz = 356 kNm    Mdz = 465 kNm

P/Pd + Mz/Mdz <=1

=165/1827 + 356/465
=0.09 + 0.77 =    0.86
                  < 1 ; Safe.

**Shear check**     (IS 800:2007 - 9.2.1)
Factored shear , V =    34 kN

| DL + WL Top | | Unfactored load | | |
|---|---|---|---|---|
| Axial | $M_{vertical}$ | $M_{horz}$ | $V_{vertical}$ | $V_{horz}$ |
| P | $M_z$ | $M_y$ | $V_z$ | $V_y$ |
| 110 | 237 | 0 | 23 | 0 |
| | | | | |
| 110 | 237 | 0 | 23 | 0 |

**Axial Comp.**    (IS 800:2007 - 11.3.1)

Ps = 110 kN    Ae = 11074 mm2

Actual comp stress = fc = Ps/Ae
= 110 x 1000 / 11074
= 10 Mpa
Permissible comp. stress, fac = 0.60 fcd
= 0.6 x 165
= 99 Mpa
                          fc<fac; Safe.
where, fcd = 165 Mpa ( see above)

**Bending.**     (IS 800:2007 - 11.4.1)

Mz = 237 kNm

*[contd.]*

| | |
|---|---|
| Design shear strength, $V_d$ =    669 kN   Safe. <br><br> 60% of $V_d$ =   401 kN <br> *V < 60% of Vd; No reduction in Moment capacity, Md.* | $M_s$ =   237.00 kNm <br> $Z_e$ =   1808700   mm3 <br><br> Actual bending stress, $f_{bc}$ or $f_{bt}$ = $M_s / Z_e$ <br>        = 237000000 / 1808700 <br>        = 131 Mpa <br><br> Permissible bending stress, $f_{abc}$ or $f_{abt}$ = <br>      = 0.60 $M_d / Z_e$ <br>      = 0.6x465x10^6 / 1808700 <br>      = 154 Mpa    $f_{bc} < f_{abc}$; Safe. <br><br> where, $M_d$ =    465 kNm   (see calc.above) <br><br> **Combined Axial compression and bending** <br> *Member strength requirement:* <br> $f_c/0.6f_y + f_{bcz}/f_{abcz}$      <= 1.0 <br> =10/ (0.6x250) + 131/154 <br>      = 0.0667 + 0.851 <br>      = 0.92    < 1 ; Safe. <br> **Combined Axial compression, bending and shear** <br> (IS 800:2007 - 11.5.4) & ( IS 800 -1962 10.5.4) <br><br> Shear stress, $\tau_b$ = $V_s/A_v$ <br>      = 22800/(500x10.2) <br>      = 4.47 Mpa <br><br> $f_e = (f_c^2 + f_{bc}^2 + f_c.f_{bc} + 3\tau b^2)^{0.5}$ < 0.9$f_y$ <br> = (10^2 + 131^2 + 10 x 131 + 3 x 4.47^2)^0.5 <br>      < 0.9 x 250 <br>      = 136 Mpa   <    225 Mpa <br>                                 Safe. |

### Column base design.

**Base Plate**

|  | mm |  | mm |  | mm |
|---|---|---|---|---|---|
| Base Plate | 750 | x | 400 | x | 20 |
| Col size | 500 |  | 180 |  |  |
| Anchor bolts | 4 | nos | 24 | dia |  |

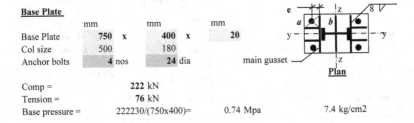

main gusset

Plan

| | | |
|---|---|---|
| Comp = | 222 kN | |
| Tension = | 76 kN | |
| Base pressure = | 222230/(750x400)= | 0.74 Mpa      7.4 kg/cm2 |

## Design of Industrial Structures

*Panel - a :*
Maxm. projection = 125 mm.
Moment = 0.74x125^2/2
       = 5781.3 Nmm

*Panel - b :*
span = 250 mm
Moment = 0.74x250^2/8
       = 5781 Nmm

M max = 5781 Nmm
fbs - bending stres = 0.75 fy = 0.75 x 250 = 188 Mpa          (IS 800:2007 11.3.3)
Thickness required, t = (5781.25 x 6 / 187.5)^0.5 = 14 mm

Bolt tension = 76 / 4 = 19 kN
Distance from col face, e = (750 - 500) / 4 = 62.5 mm
Moment = 19 x 62.5 = 1188 kN-mm                fy = 185 Mpa
Total width of support, b = (400 / 3) + (750 - 500) / 2 = 258 mm
Thickness required, t = ( 6 M / b . Fy)^0.5 = 12 mm
                                        Required thickness = 14 mm
                                        Thickness provided = 20 mm    **Safe.**

### Provide
Base Plate : **750 x 400 x 20 th**
Main gusset : **300 (Ht) x 12 th**   and   Aux gusset   **200 (Ht) x 12 th**
All welds are 6 mm fillet weld unless shown otherwise.

### Anchor bolt
Tension =        19 kN    ( From  Wind load only)
Col base shear = 83 kN
fy =  250 MPA      fu =  410 MPA          $\gamma_{m0}$   1.10
$\gamma_{m1}$  1.25      $\gamma_{mb}$  1.25

Provided        **4 nos.**         **24 mm dia.**
Anb, Net area of bolts =   4 x 3.14x(0.87x24)^2/4=     1369 mm2
Asb, Gross area of bolts = 4 x 3.14 x 24^2/4=          1809 mm2

*Nominal tension capacity, Tnb*
Tdn = 0.9 x 1369 x 410 / 1.25/ 1000 =   404 kN       (IS 800:2007 6.3.2)
or, Tdb = fyb. Asb/$\gamma$mo = 250 x 1809/1.1/1000 =  411 kN   (IS 800:2007 10.3.5)
                                    Tdn =  **404** kN

*Nominal shear capacity, Vnsb*
Vnb = [ (fu/1.732) . (ns Asb)] / $\gamma$mb              (IS 800:2007 10.3.3)
    = **343** kN       ns = no of shear plane =  1

*Working stress method*                                   (IS 800:2007 11.6)
Permissble tension capacity, Tab =  0.6 Tdn =  0.6 x 404  =  242.4 kN
Permissble shear capacity, Vsb =    0.6 Vnb =  0.6 x 343  =  206 kN
Interaction factor:

(Tab / Tdn )^2  +  (Vsb / Vnb )^2   <= 1
(242.4/404) ^ 2 + (205.8 / 343) ^2=    0.36   +   0.36  =   0.72
                                           **< 1 ; Safe..**

## Member 5, 6 and 7    Column Bracing    Axial Compresion / Tension.

L =   3.05 m        P =   53 kN
                    T =   53 kN

**Section properties:**

| Section Provided: | | s | | wt/m (kg) | $A_g$ (mm$^2$) | $r_x$ (mm) | $r_y$ (mm) | $r_{vv}$ (mm) | $I_x$ (mm$^4$) | $I_y$ (mm$^4$) | $C_{yy}$ (mm) |
|---|---|---|---|---|---|---|---|---|---|---|---|
| 2 | ISA 100 100 10 | 8 | mm b/b | 29.8 | 3806 | 30.50 | 44.0 | 19.4 | 3540000 | 7535387 | 28.4 |
| | | | | Single Leg dimensions | | | | | | | |
| | | | | b | d | t | Connection bolt | | | | An |
| | | s | | mm | mm | mm | nos | φdia | φhole | rows | mm2 |
| | | | | 100 | 100 | 10 | 3 | 16 | 18 | 1 | 3626 |

where, An = net area after deduction of bolt holes.
  An =   Ag - (no of rows x φ hole x  t)
Distance between tacking plate (welded spacer plate) = **a =**   300 mm   (IS 800:2007 - 10.2.5.2)
(the value of ' a ' shall not exceed 32 t or 300 mm, whichever is less)

**Buckling class**

  Buckling class =   c   for buckling about any axis      (IS 800:2007 - Table 10)
  Imperfection factor, α =   0.49                          (IS 800:2007 - Table 7)

**Proportioning limit**                                    (IS 800:2007 - Table 2)

  ε = (250 / fy ) ^0.5 =    (250 / 250)^0.5=   1.00

Outstanding leg
  **b / t =**    100 / 10 =    10        Compact          (IS 800:2007 - 3.7.2)
  *Limiting width to thickness ratio:*
  Plastic, λr =         9.4ε = 9.4 x 1 =         9
  Compact, λp =         10.5e = 10.5 x 1 =      11
  Semi compact, λsp =   15.7e = 15.7 x 1 =      16

**Effective Lengths (L) and Slenderness ratio ( L/r) :**           (IS 800:2007 - Table 11)

Laterally Unsupported length    Lz =  3.05 M    Kz =  0.85    KLz =  2.59 M
                                Ly =  3.05 M    Ky =  1       Kly =  3.05 M

  Maxm slenderness ratio (L/r) shall not exceed      250       (IS 800:2007 - Table 3)

        Here,   KLz/rz =   2592.5/30.5 =     85
                KLy/ry =   3050/44 =         69
                            **L/r =**         85      <250, OK.

# Design of Industrial Structures

**Sectional capacity / Design strength:**

**Axial compression**

*(i) Compression due to buckling about major axis, z-z [Lz / rz (=rx)]*  (IS 800:2007 7.5.2.1)

Design compression strength, $P_{dz}$ = Ae . fcd   (IS 800:2007 7.1.2)

where.  Ae = effective sectional area
fcd = design compressive stress

$$fcd = \chi \, fy / \gamma_{mo} \quad <= fy / \gamma_{mo}$$
$$\chi = 1 / [\, \phi + (\phi^2 - \lambda^2)^{0.5}\,]$$
$$\phi = 0.5\,[\,1 + \alpha(\lambda - 0.2) + \lambda^2\,]$$
$$\lambda = (fy / fcc)^{0.5}$$

$$fcc = \pi^2 E / (KLz/rz)^2 \quad \text{Euler buckling stress}$$

Now, putting following values into above equation

| | | | | | | |
|---|---|---|---|---|---|---|
| $\alpha$ = | 0.49 | E = | 200000 Mpa | fy = | 250 Mpa | |
| $\gamma_{m0}$ = | 1.10 | KLz/rz = | 85 | Ae = | 3806 mm2 | =Ag |

fcc = 3.14^2 x 200000 / 85^2 =     273 MPa
$\lambda$ = (250 / 273)^0.5=       0.96
$\phi$ = 0.5[1 + 0.49 x (0.96 - 0.2) + 0.96^2]=   1.15
$\chi$ = 1 /[1.15 + (1.15^2 - 0.96^2)^0.5]=   0.561

fcd = 0.561 x 250 / 1.1 =     128 Mpa   <= fy / $\gamma_{mo}$   <=   250 / 1.1
                                                        <=   227 Mpa

fcd =  128 Mpa   (=fcdz)

**Pdz** = Ae . fcd =   3806 x 128 /1000=   **487** kN

*(ii) Compression due to buckling about minor axis, y-y (KLy / ryy)*  (IS 800:2007 7.5.2.1)

Design compression strength, $P_{dy}$ = Ae. fcd     where.  Ae = effective sectional area
fcd = design compressive stress

$$fcd = \chi \, fy / \gamma_{mo} \quad <= fy / \gamma_{mo} \qquad \text{(IS 800:2007 7.1.2.1)}$$
$$\chi = 1 / [\, \phi + (\phi^2 - \lambda^2)^{0.5}\,]$$
$$\phi = 0.5\,[\,1 + \alpha(\lambda - 0.2) + \lambda^2\,]$$
$$\lambda = (fy / fcc)^{0.5}$$

$$fcc = \pi^2 E / (KLz/rz)^2 \quad \text{Euler buckling stress}$$

Now, putting following values into above equation

$\alpha$ = 0.49    E = 200000 Mpa    fy = 250 Mpa
$\gamma_{m0}$ = 1.10    KLy/ryy = 69    Ae = 3806 mm2

fcc = $3.14^2 \times 200000 / 69^2$ =    414 MPa
$\lambda$ = $(250/414)^{0.5}$ =    0.78
$\phi$ = $0.5[1 + 0.49 \times (0.78 - 0.2) + 0.78^2]$ =    0.95
$\chi$ = $1/[0.95 + (0.95^2 - 0.78^2)^{0.5}]$ =    0.67

fcd = $0.67 \times 250 / 1.1$ =    152 Mpa    <= fy / $\gamma_{mo}$    <=    250 / 1.1
                                                                                                        <=    227 Mpa

fcd = 152 Mpa    (=fcdy)

**Pdy** = Ae . fcd =    $3806 \times 152 / 1000$ =    **579 kN**

The design compressive strength, **Pd** will be lowest of above calculated strengths of the member.

**Pd =**    **487 kN**

### Axial Tension

The factored design tension, T < Design tensile strength of member, Td

*(i) Design tensile strength due to yielding of gross section, Tdg*      (IS 800:2007 - 6.2)

Tdg = Ag fy / $\gamma_{mo}$ =    $3806 \times 250 / 1.1 / 1000$ =    865 kN

*(ii) Design tensile strength due to rupture of critical section, Tdn*

Tdn = 0.9 Anc. fu / $\gamma_{m1}$ + $\beta$. Ago. fy / $\gamma_{m0}$      (IS 800:2007 - 6.3.3)

$\beta$ = 1.4 - 0.076 (w/t) (fy/fu) (bs/Lc) <= 0.9 fu $\gamma_{mo}$ / fy $\gamma_{m1}$ >= 0.7

where, Anc = net area of connected leg =    $2 \times [100 \times 10 - (1 \times 18 \times 10)]$ =    1640 mm2
         Ago = gross area of outstanding leg =    $2 \times (100 \times 10)$ =    2000 mm2

# Design of Industrial Structures

      w = width of outstanding leg = d =                                         100 mm
      bs = shear lag width = w + w1 (= Cyy) - t =      100 + 28.4 - 10 =          118 mm
      Lc = length of end connection = (nos of bolt -0.5) x 3 φ hole =         135 mm

      f u =        410 Mpa          $\gamma_{m0}$ =      1.10          t =          10 mm
      fy =          250 Mpa          $\gamma_{m1}$ =      1.25

β = 1.4 - 0.076 (w/t) (fy/fu) (bs/Lc) <= 0.9 fu $\gamma_{mo}$ / fy $\gamma_{m1}$ >= 0.7
  =    1.4-0.076x(100/10)x(250/410)x(118/135) <= 0.9x410x1.1 / (250x1.25) >= 0.7
  =      0.99         <=        1.30         >=        0.7           Ok.

Tdn = 0.9 Anc. fu / $\gamma_{m1}$ + β. Ago. fy / $\gamma_{m0}$
    =   0.9 x 1640x410 / 1.25 + 0.99 x 2000 x 250 / 1.1
    =   934128 N
    =      934 kN

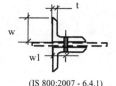

<u>(iii) Design tensile strength due to block shear, Tdb</u>              (IS 800:2007 - 6.4.1)

a)      Tdb = [ Avg. fy / ( 1.732 $\gamma_{mo}$) + 0.9 Atn. fu / $\gamma_{m1}$]         OR

b)      Tdb = (0.9 Avn. fu / (1.732 $\gamma_{m1}$ ) + Atg. fy / $\gamma_{mo}$ )         whichever is smaller

       Avg =    gross shear area along bolt line parallel to line (1-2) of external force
             =    2. Lc . t
             =    2 x 135 x 10    =      2700 mm2
       Avn =    net shear area along bolt line parallel to the line (1-2) of external force
             =    2 [Lc . t - ( no of bolts x φ hole x thickness)]
             =    2 x [ 135 x 10 - (3x18x10) ] =      1620 mm2

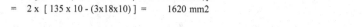

       Atg =    gross area in tension from bolt hole to toe of the angle, perpendicular to line (2-3) of force
             =    2 . (d - Cyy) . t
             =    2 x (100 - 28.4) x 10 ) =        1432 mm2
       Atn =    net area in tension from bolt hole to toe of the angle, perpendicular to line (2-3) of force
             =    2 [ (d - Cyy) . t - ( no of rows x φ hole x thickness)]
             =    2 [ (100 - 28.4) x 10 - (1 x 18 x 10) ] =      1072 mm2

a)      Tdb = [ Avg . fy / ( 1.732 $\gamma_{mo}$) + 0.9 Atn . fu / $\gamma_{m1}$]
          =    2700 x 250 /(1.732 x 1.1) + 0.9 x 1072 x 410 / 1.25
          =    670748 N
          =       671 kN

b) $T_{db} = (0.9 \, A_{vn} \, f_u / (1.732 \, \gamma_{m1}) + A_{tg} \, f_y / \gamma_{mo})$ whichever is smaller
   $= 0.9 \times 1620 \times 410 / (1.732 \times 1.25) + 1432 \times 250 / 1.1$
   $= 601565 \text{ N}$
   $= 602 \text{ kN}$

So, Tensile strength due to block shear, $T_{db}$ = 602 kN

now, we have $T_{dg}$ = 865 kN, $T_{dn}$ = 934 kN and $T_{db}$ = 602 kN

*The design tensile strength, **$T_d$** will be lowest of above calculated strengths of the member.*
   $T_d$ = 602 kN

# Design of Industrial Structures

**Strength design**     **Compression /Tension**

| property | Limit state method | Working stress method |
|---|---|---|
| | Compact | Compact |
| Load | Axial Comp, P = 53 kN<br>Axial Tension, T = 53 kN<br><br>Load factor = 1.5<br>Factored load :<br>P = 1.5 x 52.81 = 79.215 kN<br>T = 1.5 x 52.81 = 79.215 kN<br><br>Design Comp strength, Pd = 487 kN<br>Factor load , P = 79.215 kN<br>**P < Pd; Safe.**<br><br>Design Tensile strength, Td = 602 kN<br>Factor load , T = 79 kN<br>**T < Td; Safe.** | Axial Comp, P = 53 kN<br>Axial Tension, T = 53 kN<br><br>**Axial compression :** (IS 800:2007 11.3.1)<br><br>Permissible comp. stress, fac = 0.6 fcd<br><br>fcd = 128 Mpa (see calc. above.)<br>Ag = 3806 mm2<br><br>Permissible Comp strength, Pd = 0.6 fcd .Ag<br>= 0.6 x 128 x 3806<br>= 292301 N<br>= 292 kN<br>Actual load, P = 52.81 kN<br>**P < Pd; Safe.**<br><br>**Axial Tension:** (IS 800:2007 11.2.1)<br><br>Tensile stresses for :<br>a) yielding   fat = 0.6 fy = 0.6 x 250<br>= 150 MPa<br>b) rupture   fat = 0.69 Tdn/Ag<br>= 0.69 x 934000 / 3806<br>= 169 MPa<br>c) block shear   fat = 0.69 Tdb/Ag<br>= 0.69 x 602000 / 3806<br>= 109 MPa<br><br>Permissible value of fat = 109 MPa<br><br>Permissible Tensile strength, Td = fat .Ag<br>= 109 x 3806<br>= 414854 N<br>= 415 kN<br>Actual load, T = 53 kN<br>**T < Td; Safe.** |

End

## 12.4 DESIGN OF RCC PIPE SLEEPERS WITH CABLE RACKS FOR AN ASH HANDLING SYSTEM

### 12.4.1 Description

The Pipe sleeper is used for carrying pipe lines and cable racks over ground but at low height. It is an cost economic solution compared to overhead pipe and cable racks The sleepers in this example are designed for dead load and wind/seismic combination. The pipes are not carrying hot water, hence friction load for temperature is ignored . Strength design is done by working stress method in accordance with IS 456 - 2000. The pipes are clamped on support steel, which is fixed by welding to embedded plate on top of pedestal. Cable tray steel posts are also welded to embedded plate at sides The input data and results of analysis & design are furnished in tabular format. The break ups are furnished separately for understanding the embedded formula and values in cells of excel sheet.

### 12.4.2 Reference

1 IS 456 - 2000 ; 1978.
2 IS 875 (Part 3) - 1987

### 12.4.3 Design Parameters

| | | | | | |
|---|---|---|---|---|---|
| Finished grade level (RL) | FGL = | 3.8 M | | | |
| Ground water level (RL) | GWL = | 2.8 M | | | |
| Top of Pedestal (RL) | TOP = | 4.3 M | | | |
| Bottom of foundation (RL) | BOF = | 3.1 M | | | |

Material:

| | | | | | |
|---|---|---|---|---|---|
| Concrete Grade | Conc = | M20 | fck = | 20 | Mpa |
| Unit weight of concrete | $\gamma_{conc}$ = | 25 kN/cu.m | $\sigma cb$ = | 7 | Mpa |
| Cover to Reinforcement | cover = | 50 mm | | | |
| Max bar dia | $\phi dia$ = | 16 mm | | | |

Foundation Soil:

| | | | |
|---|---|---|---|
| Gross Bearing pressure | pgross = | 150 | kN/m2 |
| Shearing resistance | $\phi$ = | 34 | degree |
| Cohesion | coh = | 0 | kg /cm2 |
| Unit weight of Soil backfill | $\gamma soil$ = | 18 | kN/cu.m |
| Coeff. of Friction | $\mu$ = | 0.3 | granular |

Load:

| | | | |
|---|---|---|---|
| Design Wind pressure | $p_{wind}$ = | 1.28 | kN/m2 |
| Ground Surcharge | $\gamma surch$ = | 10 | kN/m2 |
| Horz. Seismic coefficient | $\alpha h$ = | 0.06 | |
| Wind force coeffecient | cf = | 2.05 | |
| Weight of Cable tray | Tray = | 1.2 | kN/m |

# Design of Industrial Structures

Vertical distance betn. trays    Htray = 0.3 m
Solidity Ratio :                SR = 0.5

## 12.4.4 SKETCH

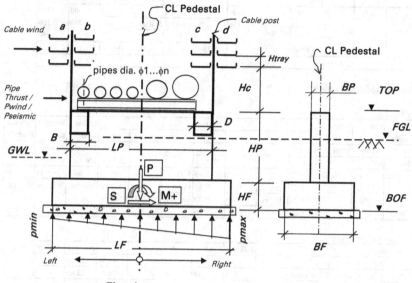

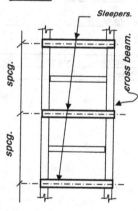

LF, BF, HF = Length, width and thickness of sleeper base.
LP, BP, HP = Length, width and thickness of sleeper pedestal.

**Top plan**

Wind force = Projection x Spcg. x intensity x co-eff. in kN
Seismic force, F = Pipe wt/m x Spcg. x Seismic coefficients in kN.

Wpl = Total wt of pipes on left in kN/m     Epl =   Ecc. of pipe load from CL at left (-) in m.
Wpr = Total wt of pipe on right in kN/m.     Epr =   Ecc. of pipe load from CL at right (+) in m
Wpipe = (Wpl + Wpr) x spcg of sleeper in kN.    $e_{pipe}$ = ( -Epl + Epr) in m.
Htray = Vertical distance betn. Trays in m.     Σ Mpipe = (MpipeL + MpipeR) x spcg. in kNm.
Σ Apipe = Sum of the projected area of pipe against wind.
Soil bearing pressure :   $p_{max}$    and    $p_{min}$    in kN/m2

## 12.4.5 Design of Sleeper

*Design Input*

| Sleeper mkd. | Spcg. of Sleeper m | Pipe | | | | Cable tray | | | | Sleeper | |
|---|---|---|---|---|---|---|---|---|---|---|---|
| | | Left of CL | | Right of CL | | Left of CL | | Right of CL | | LF (m) | LP (m) |
| | | nos. | dia mm | nos. | dia mm | a nos. | b nos. | c nos. | d nos. | BF (m) | BP (m) |
| | | | | | | | | | | HF (m) | HP (m) |
| PS1 | 4.5 | 1 | 100 | 1 | 500 | 4 | 4 | 2 | 2 | 3.000 | 2.700 |
| | | 1 | 100 | 1 | 150 | Cross beams-Left & Right (m x m) | | | | 1.200 | 0.200 |
| | | 1 | 100 | | | B | D | B | D | 0.250 | 0.950 |
| | | 1 | 100 | | | 0.25 | 0.3 | 0.25 | 0.3 | Hc(m) | 0.6 |
| | | $\Sigma$ Apipe = | | 1.05 m. | | | | | | | |

*Analysis and design results:*

*a) Load per sleeper*

| Fdn weight | Soil Fill | Surch-arge | Pipe | | | | | | Cable load | Wind load | Seismic load |
|---|---|---|---|---|---|---|---|---|---|---|---|
| | | | Left | | | Right | | Total | | | |
| Wc kN | Wsoil kN | Wsur kN | Self wt kN/m | Load & Moment | | Self wt kN/m | Load & Moment | | $\Sigma$Wpipe kN | Wcable kN | WLcable kN | SLcable kN |
| 52.20 | 41.62 | 30.60 | 0.22 | Wpl | 0.88 | 2.99 | Wpr | 3.42 | 19.35 | 64.8 | 5.31 | 3.89 |
| | | | 0.22 | Epl | 0.68 | 0.43 | Epr | 0.68 | $\Sigma$Mpipe | Mcable | WLpipe | SLpipe |
| | | | 0.22 | MpipeL | -0.59 | 0.00 | MpipeR | 2.31 | kNm | kNm | kN | kN |
| | | | 0.22 | | | 0.00 | | | 7.74 | -29.16 | 12.40 | 1.16 |

*b) Foundation design*

| Load case: | P (kN) | S (kN) | M (kNm) | $P_{max}$ | $P_{min}$ | FOSslide | FOS ovr |
|---|---|---|---|---|---|---|---|
| 1. DL + Pipe + Cable + G.Surch | 209 | 0 | 21.42 | 69.96 | 46.16 | | 6 |
| 2. DL + Pipe + Cable + Wind/Seismic | 178 | 18 | 48.25 | 76.24 | 22.63 | 3 | 4 |
| allowable soil bearing pressure = | | 150 kN/m2 | Observation | Safe. | No-uplift. | Safe. | |

*c) Strength design*

| Rebars: | Layer | dia mm | spcg mm | $Ast_{prov}$ mm$^2$ | $Ast_{rqd}$ mm$^2$ | $Ast_{min}$ mm$^2$ | Observation |
|---|---|---|---|---|---|---|---|
| Foundation | bottom | 12 | 200 | 566 | 468 | 300 | Ast prov > Ast rqd; Safe. |
| Pedestal | per side | 12 | 200 | 566 | 120 | 120 | Ast prov > Ast rqd; Safe. |

# Design of Industrial Structures

## Break up calculations for Pipe sleepers

The above tables are done in Excel sheet. The cell values are based on embedded formula and visual basic macros, which are given below in expanded form for clarity.

### Foundation weight, Wc in kN
*Visual basic function*
Function Wc(LF, BF, HF, LP, BP, HP, B, D, spcg, gconc)
    Wc = (LF * BF * HF + LP * BP * HP + 2 * B * D * spcg) * gconc
End Function

Wc = (3 x 1.2 x 0.25 + 2.7 x 0.2 x 0.95 + 2 x 0.25 x 0.3 x 4.5 ) x 25
   = 52.20 kN
Cell value, Wc =    | 52.20 | kN

### Soil fill, W soil in kN
*Visual basic function*
Function Wsoil(LF, BF, HF, LP, BP, HP, FGL, BOF, GWL, gsoil)
Wsoil = ((LF * BF - LP * BP) * (FGL - BOF - HF) * gsoil) - ((LF * BF - LP * BP) * (GWL - BOF - HF) * 10)
End Function

W soil = ( 3 x 1.2 -2.7 x 0.2) x (3.8 - 3.1 - 0.25) x 18
       -  (3 x 1.2 - 2.7 x 0.2) x (2.8 - 3.1 - 0.25) x 10
     = 41.616 kN
Cell value, Wc =    | 41.62 | kN

### Surcharge weight over foundation base, Wsur in kN
*Visual basic function*
Function Wsurch(LF, BF, LP, BP, gsurch)
    Wsurch = ((LF * BF - LP * BP) * gsurch)
End Function

Wsur = (3 x 1.2 - 2.7 x 0.2) x 10    =    30.6 kN

Cell value, Wc =    | 30.6 | kN

**Self weight of pipes - water fill**

| | nos | dia | wt (kN/m) | |
|---|---|---|---|---|
| | 1 | 100 | 0.22 | (ANSI B36.10) |
| | 1 | 150 | 0.43 | |
| | 1 | 500 | 2.99 | |

**Load & Moments for pipe on left side of pedestal centerline**
$W_{pl}$ = 0.22+0.22+0.22+0.22 =     0.88 kN         Cell value =    0.88
$E_{pl}$ =  2.7 x 0.25=                       0.68 m              Cell value =    0.68
$M_{pipeL}$ = -1 x 0.88 x 0.68=       -0.60 kNm         Cell value =    -0.59

**Load & Moments for pipe on right side of pedestal centerline**
$W_{pr}$ = 2.99+0.43 =                    3.42 kN           Cell value =    3.42
$E_{pr}$ = 2.7 x 0.25=                     0.68 m              Cell value =    0.68
$M_{pipeR}$ = 3.42 x 0.68=              2.33 kNm          Cell value =    2.31

$\Sigma W_{pipe}$ =     (0.88 + 3.42) x 4.5 =      19.35 kN       Cell value =   19.35
$\Sigma M_{pipe}$ =     (-0.59 + 2.31) x 4.5 =     7.74 kN        Cell value =   7.74

**Cable Load**
$W_{cable}$ =   (4 + 4 + 2 + 2) x 1.2 x 4.5 =           64.8 kN            Cell value =   64.8
$M_{cable}$ =   (-4 + -4 + 2 + 2) x 1.2 x 2.7/2 x 4.5 =   -29.16 kN       Cell value =   -29.16

**Wind load**
$WL_{cable}$ =   (4 - 1) x 0.3 x 1.28 x 2.05 x 0.5 x 4.5 =    5.31 kN     Cell value =   5.3
WL pipe =      1.05 x 2.05 x 1.28 x 4.5 =           12.40 kN           Cell value =   12.4

**Seismic load**
$SL_{cable}$ =   64.8 x 0.06 =      3.89 kN       Cell value =   3.89
SL pipe =      19.35 x 0.06 =      1.16 kN       Cell value =   1.16

**Foundation design**
**Load case 1**
P =   52.2 + 41.62 + 30.6 + 19.35 + 64.8 =         209 kN           Cell value =   209
M =      7.74  (+)    -29.16    =    -21.42 kNm    21.42 kNm        Cell value =   21.42
**Load case 2**
P =   52.2 + 41.62 + 19.35 + 64.8 =         178 kN                     Cell value =   178
S =   Max[ (5.31+12.4),(3.888+1.161) ] =          18 kN                Cell value =   18
M =   21.42+5.31 x 2.25 + 12.4 x 1.2 =            48.25 kNm          Cell value =   48.25

Cable:  Horz load =      5.31 kN     *Wind load governs.*
             Ht above base =        0.25+0.95+0.6+((4-1)x0.3x0.5) =         2.25 m
Pipe :   Horz load =      12.40 kN    *Wind load governs.*
             Ht above base =        0.25 + 0.95 =       1.20 m

## Design of Industrial Structures

**Soil bearing pressure - pmax and pmin**
Load Case 1
Max.bearing pressure, pmax
Function pmax(P, M, LF, BF)
    $A = P / (LF * BF)$                      A =    209 / (3 x 1.2) =          58.06
    $B = M * 6 / (BF * LF \wedge 2)$       B =    21.42 x 6 / (1.2 x 3^2) =        11.9
    $pmax = A + B$                       pmax = 58.06 + 11.9=      69.96 kN/m2
End Function                             Cell value =    | 69.96 |
Min.bearing pressure, pmax
Function pmin(P, M, LF, BF)
    $A = P / (LF * BF)$                      A =    209 / (3 x 1.2) =          58.06
    $B = M * 6 / (BF * LF \wedge 2)$       B =    21.42 x 6 / (1.2 x 3^2) =        11.9
    $pmin = A - B$                       pmax = 58.06 - 11.9=     46.16 kN/m2
End Function                             Cell value =    | 46.16 |

Factor of safety against sliding, FOSslide          Not applicable as shear force is nil.

Factor of safety against overturning, FOSovr

Function FOSovr(M, Scable, Spipe, HF, HP, HC, Wc, Ws, Wpipe, LF, LP, WcableL, WcableR)
    MOVR = M + Scable * (HF + HP + HC) + Spipe * (HF + HP)
    MRESIST = (Wc + Ws + Wpipe) * LF * 0.5 + WcableL * 0.5 * (LF + LP) + WcableR * 0.5 * (LF - LP)
    FOSovr = MRESIST / MOVR
End Function

Scable = wind force on cable rack = WLcable=     5.31 kN
Spipe = wind force on pipe = WLpipe =          12.40 kN

Movr = 21.42 + 5.31 x (0.25 + 0.95 + 0.6) + 12.4 x (0.25 + 0.95) =          46
Mresist =     0.9 x (52.2+41.62+19.35)x3x0.5 + 43.2x0.5(3+2.7) + 21.6x0.5x(3-2.7)
       =       279
FOSovr =     279 / 46 =       6                   Cell value =    | 6 |

Load Case 2
Max.bearing pressure, pmax
Function pmax(P, M, LF, BF)
    $A = P / (LF * BF)$                      A =    177.97 / (3 x 1.2) =       49.44
    $B = M * 6 / (BF * LF \wedge 2)$       B =    48.25 x 6 / (1.2 x 3^2) =        26.81
    $pmax = A + B$                       pmax = 49.44 + 26.81=    76.25 kN/m2
End Function                             Cell value =    | 76.24 |

Min.bearing pressure, pmax
*Function pmin(P, M, LF, BF)*

$A = P / (LF * BF)$           $A =$   $177.97 / (3 \times 1.2) =$        49.44
$B = M * 6 / (BF * LF \wedge 2)$   $B =$   $48.25 \times 6 / (1.2 \times 3^{\wedge}2) =$        26.81
$pmin = A - B$                pmax $= 49.44 - 26.81 =$   22.63 kN/m2
*End Function*                Cell value =   | 22.63 |

Factor of safety against sliding, FOSslide

*Function FOSslide(P, MEWsoil, cohesion, LF, BF, S)*
 $FOSslide = (P * MEWsoil + cohesion * LF * BF) / S$
*End Function*

MEWsoil =   μ      0.3 granular      cohesion =     coh     0.0 kg /cm2

FOS slide =   $(177.97 \times 0.3 + 0 \times 3 \times 1.2) / 17.71 =$       3
                                                  Cell value =    | 3 |

Factor of safety against overturning, FOSovr

*Function FOSovr(M, Scable, Spipe, HF, HP, HC, Wc, Ws, Wpipe, LF, LP, WcableL, WcableR)*
 $MOVR = M + Scable * (HF + HP + HC) + Spipe * (HF + HP)$
 $MRESIST = (Wc + Ws + Wpipe) * LF * 0.5 + WcableL * 0.5 * (LF + LP) + WcableR * 0.5 * (LF - LP)$
 $FOSovr = MRESIST / MOVR$
*End Function*

Scable = wind force on cable rack = WLcable=   5.31 kN
Spipe = wind force on pipe = WLpipe =          12.40 kN

Movr =  $48.25 + 5.31 \times (0.25 + 0.95 + 0.6) + 12.4 \times (0.25 + 0.95) =$         73
Mresist =    $0.9 \times (52.2 + 41.62 + 19.35) \times 3 \times 0.5 + 43.2 \times 0.5 (3+2.7) + 21.6 \times 0.5 \times (3-2.7)$
         =   279
FOSovr =   $279 / 73 = 4$                            Cell value =     | 4 |

**Strength design**

*Function Ast(BF, BP, HF, Pressure, cover, fdia)*
 $Proj = (BF - BP) * 0.5$
 $MOM = (Pressure * Proj \wedge 2) / 2$
 $de = HF * 1000 - 0.5 * fdia - cover$
 $Ast = MOM * 10 \wedge 6 / (230 * 0.87 * de)$
 *If* $Ast < 0.12 * HF * 10000$ *Then*
  $Ast = 0.12 * HF * 10000$
 *Else*
  $Ast = MOM * 10 \wedge 6 / (230 * 0.87 * de)$
 *End If*
*End Function*

# Design of Industrial Structures

## Foundation design
### Base slab
Projection of slab beyond pedestal face, Proj =    (1.2 - 0.2) / 2 =    0.5 m.
Net upward pressure =    150 -25 x 0.25 =         143.75 kN/m2
Moment at face of pedestal, M =         143.75 x 0.5^2 / 2 =    17.97 kNm /m width
Max. dia of bar =                16 mm    Clear cover =    50 mm
Effective depth, de =    0.25 x 1000 -0.5 x 16 - 50 =       192 mm
Permissible stree in steel reinforcement. σst =         230 Mpa         ( IS 456 2000 - Table 22)

Area of steel, Ast =    17968750 / (230 x 0.87 x 192) =         468 mm2    Cell value =    468
Min steel, Ast min =    0.12 x 0.25 x 10000=    300 mm2         ( IS 456 2000 - 26.5.2.1)

### Pedestal
                                                                         ( IS 456 2000 - 32.5)
Min steel, Ast min per face =    0.12 x 0.2 x 10000 / 2 =    120 mm2    Cell value =    120

## 12.4.6 Cross Beams

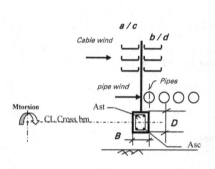
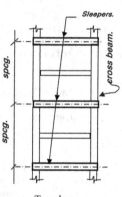

Top plan

### Design of cross beam

| Sleeper mkd. | Spcg. of Sleeper m | Load case: DL+LL+Pipe+Cable+Wind/Seismic | | | | | Design forces (as per IS 456: 2000 - Annex B-6) | | | |
|---|---|---|---|---|---|---|---|---|---|---|
| | | UDL vertical kN/m | UDL Horz. kN/m | Moment Vert kNm | Moment Torsion kNm | Shear vert kN | Shear horz kN | Ve kN | M kNm | Mt kNm | Me2 kNm |
| PS1 | 4.5 | 11.2 | 3.9 | 28.4 | 2.8 | 7.3 | 2.6 | 25.2 | 28.4 | 3.6 | 0.0 |
| B (m) | 0.25 | d reqd | 0.35 m | | | | $t_{ve}$ | | 0.3 N/mm2 | | |
| D (m) | 0.30 | d provi | 0.242 m | | Doubly Reinforced | | $t_c$ | | 0.40 N/mm2 | Safe. | |
| Reinf. design | Main Reinforcement : | | | | | nos | dia | + | nos | dia | |
| | Ast reqd = | | 567 mm2 | Provide (Ast) | | 2 | 20 | + | 1 | 16 | |
| | Ast provided = | | 830 mm2 | > Ast reqd; OK. | at Top and at Bottom layers (Ast=Asc). | | | | | | |
| | Transverse reinforcement, Asv : | | | dia | spcg | | | | | | |
| | Provide 2 legged closed Stirrups | | | | 8 | dia @ | | 150 mm c/c. | | | |
| | Asv reqd= | | 81 mm2 | Asv provided | | 101 mm2 | | > Asv reqd; OK. | | | |

Max dia of pipe =        500 mm
CG of pipe above CG of cross beam =    0.5 x 0.5 + 0.3/2 =        0.4 m
CG of cable trays above CG cross beam =        0.5 x 0.3 + 0.6 + ( 3 x 0.3) /2 =    1.2 m

## Back up calculation for Cross beam

Structural design by Working stress method           ( IS 456 2000 - Annex B)

Conc **M20**

| | | | | | |
|---|---|---|---|---|---|
| $\sigma cbc$ | 7 | Mpa | m | 13.33 | [ 280/3 $\sigma cbc$ ] |
| $\sigma st$ | 230 | Mpa | k | 0.29 | [m . $\sigma cbc$ / ( m. $\sigma cbc$ + $\sigma st$)] |
| $\sigma sv$ | 230 | Mpa | j | 0.90 | [1 - k/3] |
| | | | R | 0.91 | [1/2 . $\sigma cbc$ . k . j ] |

### Loading/beam

| | | |
|---|---|---|
| UDL Vertical = | 0.25x0.3x25 + (19.35+64.8) x0.5/4.5+1   = | 11.2 kN/m |
| UDL Horz = | Max(5.31+12.4)or(3.888+1.161) /4.5 = | 3.9 kN/m |
| M vert = | 11.2 x 4.5^2 / 8 = | 28.4 kNm |
| Moment torsion  = | (12.4 x 0.4 + 5.31x1.2)/2/2= | 2.8 kNm |
| Shear vert = | 0.65 x 11.2 = | 7.28 kN |
| Shear horz = | 0.65 x 3.9 = | 2.6 kN |

Ve =   7.3 + 1.6 x 2.8 / 0.25 =      25.2 kN
Mt =   2.8 x (1 + 0.3 / 0.25) / 1.7 =           3.6 kNm

d reqd =        (28.4 x 10^6 / (0.91 x 0.25 x 1000))^0.5)/1000 =        0.35 m
d prov =        300 - 50 -0.5x16 =         242 mm =    0.242 m

Ast reqd =      28400000 / (230 x 0.9 x 249) =     567 mm2

Asv =    T. sv / (b1. d1 .$\sigma sv$)       +      V. sv / ( 2.5 . d1, $\sigma sv$)              ( IS 456 2000 - Annex B)

here,                         T = Mt =      2.8 kNm
                                  sv =      150 mm c/c
        b1 = B  - 2 x (cover -1/2 $\phi$ bar) =      166 mm.
        d1 = D  - 2 x (cover -1/2 $\phi$ bar) =      216 mm.
                                $\sigma sv$ =      230 Mpa
                                V =      25.2 kN

        Asv =    51 + 30 =  81 mm$^2$

End

Design of Industrial Structures 485

## 12.5 DESIGN OF RIVERBED PROTECTION WORKS FOR AN UNDERWATER DISCHARGE PIPELINE

### 12.5.1 Introduction

This document consists of design of bedding and riprap protection for plant water discharge pipe laid into a river bed. Stone size computation has been done for flow conditions that produce maximum velocity at exit point in river bed at riprapped enclosure. The computations are being done following Hydraulic design guide lines furnished in EM 1110-2-1601 US Army Corps of Engineers (Engineering and Design) and Indian standard codes of practice.

### 12.5.2 Reference Documents

a) US army corps of engineers hydraulic design of flood control channels EM_1110-2-1601

b) IS 14262 - Planning and Design of Revetment guidelines

### 12.5.3 Design Parameters

| | | |
|---|---|---|
| Finish Grade level, FGL = | 3.15 M | at river bank |
| Highest water level of River = | 2.15 M | |
| Low water level = | -2.30 M | |
| River bed level = | -8.00 M | |
| Diameter of discharge pipes = | 4 M | |
| Total Nos of Outlet pipe = | 4 | |
| CL of Outlet pipes of 4M Dia = | -12.0 M | 15.2 M below FGL |
| River bed level (dredged) at pipe outlet = | -15.0 M | |
| Max Flow Rate through disch channel, Q = | 52 cumec | including storm water |
| Water level at Discharge Channel = | 2.8 M | |

### 12.5.4 Material

Well graded angular shaped stone of specific weight = 155 pcf (2.48 T/m3).
$D_{100}$ = 9 in (max)
$D_{50}$ = 10 in (max)   7 in (min)

[ Here, Stone size D50 means 50% of stone boulders are smaller than 10 inch.]

## 12.5.5 Conceptual Sketch

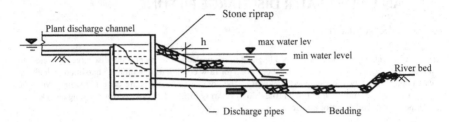

**Typical Section through discharge pipe**

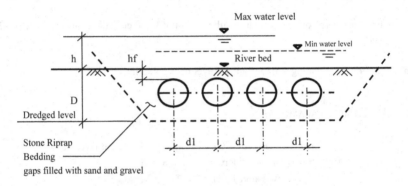

**Typical section through Pipe bedding on River bed**

# Design of Industrial Structures

## 12.5.6 Riprap and Bedding [Indian Standard Code Guideline - IS 14262]

| | | |
|---:|---:|:---|
| Max River water level = | 2.15 | M Elev |
| Min River water level = | -2.30 | M Elev |
| River bed level = | -8.00 | M Elev |
| PIPE ID = | 4 | m |
| Bottom of bed/Dredged level = | -15.0 | m |
| Dist between pipes, d1 = | 6 | m |
| CL of Pipe = | -12.0 | M Elev |
| Fill height above pipe, hf = | 2 | m |
| Fill density, $\gamma$ fill = | 25 | kN/m3 |
| River bed soil density, $\gamma$ soil = | 18 | kN/m3 |

**Estimation of Scour depth, $d_{sm}$:**

$$d_{sm} = 1.34 (D_b^2 / Ksf)^{\wedge} 1/3 = \text{depth below HFL}$$

| | | |
|---:|:---|:---|
| Load case, LC = | 1 | |
| Db, Dischage in Cumec /m wide = Vm. h = | 1.52 | Cumec |
| Ksf, Silt factor = | 0.5 | ( for fine silt ) |
| Mean Velocity of River flow, Vm = | 0.15 | M/sec |
| h = (2.15 (-) -8) = | 10.15 | m. |
| $d_{sm}$ = | 2.23 | m |

*Estimated value of $d_{sm}$ for different mean velocity, Vm :*

| LC | Vm | $d_{sm}$ |
|---:|---:|---:|
| 1 | 0.15 | 2.23 |
| 2 | 0.3 | 3.55 |
| 3 | 0.6 | 5.63 |
| 4 | 1.2 | 8.94 |

For determination of shallow foundation in river bed, scour value considered = 1.27 $d_{sm}$

| | | | |
|---|---|---|---|
| $d_{sm}$ | = | 8.94 m | |
| 1.27 $d_{sm}$ | = | 11.35 m | below HFL |
| $D_{provided}$ | = | 7.00 m | below river bed (15 - 8 = 7) |
| | and | 17.15 m | below HFL |

*Depth of pipe support bed is below 1.27 times of scour depth, Hence Okay.*

### Additonal surcharge of river bed:

| | | |
|---|---|---|
| Effective width of foundation bed = | 18 m | (3 x 6) |
| Unit weight of fill material = | 22 kN/m2 | |
| Submerged weight of granular fill less vol of pipes = | 909 kN/m | (18x7x(22-10)-4x(3.14x4^2/4)x(22-1 |
| Weight of pipes = | 99 kN/m | (4 x 3.14 x 4x2.5 x 0.79) |
| | 1048 kN/m | |
| Pressure on bed = | 58 kN/m2 | (1048 / 18) |
| Weight of soil replaced by dredging = | 56 kN/m2 | ((18 - 10) x 7) |

Effective surcharge = 2 kN/m2.   (58 - 56)

### Check for erosion of soil above pipes by riverbed flow

| | |
|---|---|
| River bed level = | -8 RL |
| Top of pipe = | -10 RL |
| Bot of pipe = | -14 RL |
| Max scour depth , 1.27 $d_{sm}$= | 11.35 m |
| Depth of fill above pipe, hf = | 2 m |
| Depth of pipe bottom below bed = | 6 m |
| Fill above pipe = | 2 m |

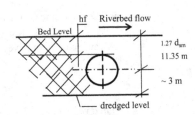

### Horizontal forces due to water currents at top of pipe level, P

$P = KV^2$   kg/cm2

K =   **1.5** for square ended obstruction
V =   The velocity of the current at the point, where
pressure intensity is being calculated in Meter/sec.

H =   11.35 M
X =   9.35 M

$U^2 = 2 V^2 \cdot X/H$

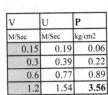

| V | U | P |
|---|---|---|
| M/Sec | M/Sec | kg/cm2 |
| 0.15 | 0.19 | 0.06 |
| 0.3 | 0.39 | 0.22 |
| 0.6 | 0.77 | 0.89 |
| 1.2 | 1.54 | 3.56 |

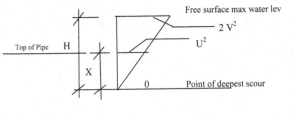

Frictional resistance of fill above pipe =   = fill ht x density x 0.2  = 2 x 25 x 0.2
                                                                        =  10 kN/m2
                                                                        =  10 kg/cm2   **Safe.**

> P = Horz. Forces due to water current.

## Design of Industrial Structures

### Velocity at discharge pipe outlet point :

Q dischrage = 52 Cumec.
No of outflow pipe = 4
flow velocity/pipe = **1.04 m/sec**

Corresponding P = 0.89 kg/cm2
< 10 kg/cm2; Fill material resistance.

### Stone revetment for River bed protection at discharge pipe outlet  [ IS 14262]

Weight of stone on horizontal bed = $\quad W = [0.02323\, S_s\, /(S_s - 1)^3]\, V^6$

Weight of stone = W in kg
Ss = specific gravity of stones
Density of stone = 2500 kg/m3
V = mean velocity of water near scour face.

Size of Stone in m, $d_{st} = 0.124*(W/S_s)^{\wedge}3$

Thickness of Pitching, $T = V^2 / 2\,g\,(S_s - 1)$ $\quad$ OR $\quad$ 2 Layers of dst, whichone is worse.

**T = Thickness of Riprap / pitching in m**
V = velocity in m/s
g = Acceleration due to gravity in m/s2
Ss = specific gravity of stones

| V | Ss | W in kg | $d_{st}$ in m | FOS | $d_{st}$ in mm | T in mm |
|---|---|---|---|---|---|---|
| 0.25 | 2.5 | 0.0000 | 0.0023 | 5 | 11 | 22 |
| 0.5 | 2.5 | 0.0003 | 0.0080 | 5 | 40 | 80 |
| 1.04 | 2.5 | 0.0212 | 0.0296 | 5 | 148 | 296 |
| 1.2 | 2.5 | 0.0514 | 0.0387 | 5 | 193 | 386 |

Thickness of bedding provided = $\quad$ 1.0 m
Thickness of riprap at bank = $\quad$ 450 mm

## 12.5.7 Riprap/Bedding Stone Sizing [US Army Corps of Engrs Hydraulic Design EM _ 1110-2-1601]

**Velocity estimation.**

| | | |
|---|---|---|
| Max Flow Rate through disch channel, Q = | 52 cumec | including storm water |
| Total No of Outlet pipe = | 4 | |
| Dia of pipe = | 4 m. | |
| Flow rate / Pipe = | 13 cumec | ( = 52 / 4) |
| Exit velocity = Q/Area = | 1.04 m/sec | |

**Stone size relations**

$$D_{30} = S_f\, C_s\, C_v\, C_t\, d\, [\,\{\gamma_w / (\gamma_s - \gamma_w)\}^{0.5} * \{V / (K_1\, g\, d)^{0.5}\}\,]^{2.5}$$

| | | | |
|---|---|---|---|
| $D_{30}$ = | riprap size of which 30 % is finer by weight, length | | |
| $S_f$ = | Safety factor = | | 3 |
| $C_s$ = | stability coeffecient for incipient failure = | | 0.3 |
| $C_v$ = | vertical velocity distribution coefficient = | | 1 |
| $C_t$ = | thickness coefficient = | | 1.5 |
| d = | local depth of flow (same location as V) = | 9.7 m | 31.82 ft at LWL |
| | (12 - 2.3 = 9.7) | | |
| $\gamma_w$ = | | 1.0 T/m3 | 62.43 pcf |
| $\gamma_s$ = | | | 155 pcf |
| V = | | 1.04 m/sec | 3.39 ft/sec |
| $K_1$ = | | | 0.6284 |
| | θ = 30 deg | φ = 40 deg | |
| | (angle of side slope) | (angle of repose) | |
| g = | | 9.81 m/sec2 | 32.2 ft/sec2 |

| | | | |
|---|---|---|---|
| $\{\gamma_w / (\gamma_s - \gamma_w)\}^{0.5}$ = | 0.8212 | | |
| $\{V / (K_1\, g\, d)^{0.5}\}$ = | 0.1338 | $D_{30}$ = | 0.17 ft |
| | | | 2.06 inch |
| $D_{50} = D_{30}\, (D_{85}/D_{15})^{1/3}$ | $D_{85}/D_{15}$ = | 1.7  to | 5.2 |
| | $D_{50}$ = | 2.46 to | 3.57 inch |
| Thickness of riprap = 2 times $D_{50}$ = | | 4.93 | 7.15 inch |

Provide riprap at sides of bank with 9 (D _100_ ) inch size stone in layers upto 18 inch (450 mm) [ D50 = 5 to 7 inch].
Bedding thickness provided =     15 - 12 - 2 =     1.0 m =     39 incnes

EM 1110-2-1601
Change 1
30 Jun 94

**Table 3-1**
Gradations for Riprap Placement in the Dry, Low-Turbulence Zones

| $D_{100}$(max) in. | Limits of Stone Weight, lb[1], for Percent Lighter by Weight | | | | | | $D_{30}$(min) ft | $D_{90}$(min) ft |
|---|---|---|---|---|---|---|---|---|
| | 100 | | 50 | | 15 | | | |
| | Max | Min | Max[2] | Min | Max[2] | Min | | |
| Specific Weight = 155 pcf | | | | | | | | |
| 9  | 34    | 14    | 10    | 7     | 5     | 2   | 0.37 | 0.53 |
| 12 | 81    | 32    | 24    | 16    | 12    | 5   | 0.48 | 0.70 |
| 15 | 159   | 63    | 47    | 32    | 23    | 10  | 0.61 | 0.88 |
| 18 | 274   | 110   | 81    | 55    | 41    | 17  | 0.73 | 1.06 |
| 21 | 435   | 174   | 129   | 87    | 64    | 27  | 0.85 | 1.23 |
| 24 | 649   | 260   | 192   | 130   | 96    | 41  | 0.97 | 1.40 |
| 27 | 924   | 370   | 274   | 185   | 137   | 58  | 1.10 | 1.59 |
| 30 | 1,268 | 507   | 376   | 254   | 188   | 79  | 1.22 | 1.77 |
| 33 | 1,688 | 675   | 500   | 338   | 250   | 105 | 1.34 | 1.94 |
| 36 | 2,191 | 877   | 649   | 438   | 325   | 137 | 1.46 | 2.11 |
| 42 | 3,480 | 1,392 | 1,031 | 696   | 516   | 217 | 1.70 | 2.47 |
| 48 | 5,194 | 2,078 | 1,539 | 1,039 | 769   | 325 | 1.95 | 2.82 |
| 54 | 7,396 | 2,958 | 2,191 | 1,479 | 1,096 | 462 | 2.19 | 3.17 |

# Index

Admixtures for concrete, 34
Aggregates, 33
As-built drawing, 97

Batching plants, 41
Beams and girders, 101, 125, 155, 188
Blinding concrete, 43, 45
Bolts and weld, 56, 108

Cement
    sulphate-resisting cement, 32
    types of cements, 32
Cement Deep Mixing (CDM) Pile, 28
Civil input data, 89
Coal bunkers, 118–124
Cold weather concreting, 42
Column, 132, 163
Column base plate and anchor bolts, 105, 193, 199, 205
Column bracing and tie member, 104, 137, 142, 145, 174, 181
Compaction piles, 28
Conceptual design, 75
Concrete mixes
    design mix concrete, 39
    nominal mix concrete, 39
    ready mix concrete, 41
Concreting underwater, 43
Connections
    bolted connections, 109
    welded connection, 110
Construction joints, 48
Contraction joint, 53
Contracts
    BOT contract, 2, 3
    EPC contract, 2, 3
    general conditions of contract, 4
    item-rate contract, 2
    supplementary conditions of contracts, 5
    technical specifications, 5, 7
    turn-key contract, 2, 3
Cost optimization, 80
Crack control joint, 54
Crane girder, 102, 210
Curing, 46

Design basis report, 57
Drawings
    bar bending drawing, 94
    design drawings, 94, 95
    excavation drawings, 94
    fabrication drawing, 95
    formwork, 95
Dynamic compaction
    falling drop hammers, 25
    vibro-compaction process, 27
    vibro-replacement method, 27

Embankment lining, 261
Embedded steels, 54
Enabling work, 16
Expansion joints, 51

Fly ash concrete, 44
Formwork, 36–38
    metal deck forms, 37
    slip form, 37
    steel sheet piles, 37
Foundation resting on pile or caisson
    Caisson or well foundation, 249
    pile foundations, 239, 245
Foundations
    combined footing and strip footing, 237
    isolated footing, 233
    mat or raft foundation, 238

Gable end structure, 106
GGBS (Ground granulated blast furnace slag), 35
Grades of concrete, 38
Ground improvement, 23
Grouting, 105

High strength steel, 55
Hot weather concreting, 43
Hybrid structures, 76, 79

Lattice girders, 101
Liquefaction potential, 18
Load
    combination of loads, 60, 63, 70
    dead loads (DL), 58, 62, 68
    equipment load (DLE), 59
    live loads (LL), 59, 63, 69
    seismic load (SL), 59, 63, 69
    temperature load (TL), 59, 63, 69
    wind (WL), 59, 63, 69

Machine foundation
    block foundations, 253
    frame foundation, 254

inertia block and vibration isolator type
  foundations, 255
pile foundation under machine foundation
  block, 255
rubber pad mounted equipment foundation, 255
spring mounted equipment foundations, 255
Mass concrete, 44
Materials and tests for quality control, 48
Method of analysis, 93
Method of construction document, 44
Methods of strength design
  limit state method, 223
  ultimate load method, 223
  working stress method, 223

Pipe and cable rack, 106
Pipe and cable trenches, 258
Plant area drainage, 318
Plant roads, 316–318
Prefabricated vertical drain (PVD), 24
Pre-stressed beam, 307
Price schedule, 7
Procurement list
  cement, 86
  reinforcement bars, 87–88
  structural steel work, 81–86
Protection against corrosion, 33, 60, 65, 72
Purlins, 99, 148

Quality control of concrete structure, 46
Quantity estimation, 96

RCC culverts at road crossing, 259
RCC reservoir, 257
Reconnaissance survey, 15
Reference codes of practice and standards
  American Standards, 11
  British Standards, 12
  Indian Standards, 12
Reinforced concrete members
  beam, 226, 279
  columns, 226, 294
  concrete wall, 225
  floor slabs, 224
  ground floor slab, 225
  roof slabs, 224, 288, 291
Reinforcement steel, 35
Retaining wall, 259
Road and pavements
  flexible pavement, 315
  interlocking paver blocks, 316
  rigid pavement, 315

Road curb, 318
Roof bracings, 101, 140, 145, 174
Roof trusses, 100

Safe bearing capacity, 18, 20
Sand drains, 25
Schedule of rates, 5, 6
Self-compacting concrete, 44
Separation or isolation joints, 53
Settlement, 18, 20, 21, 232
Shear lugs, 106, 198
Shear strength, 18, 20, 22
Sheet pile and diaphragm wall, 260
Shore protection work, 262
Side girt, 99
Silos, 262–270
Soil investigation work, 17
Soil properties and test designation, 19
Stack, 270–278
Stairs and platforms, 106
Steel member connection
  connection to RCC member, 117
  moment connection joint, 115
  shear connection, 114
  splice joints, 116
Steel tank foundations
  RCC ring wall foundation, 251
  RCC mat resting on pile, 251
  sand pad, 250
Structural framing system, 91
Structural steel materials, 55
Surge girder, 221

Tender document, 4
Tender evaluation, 5
Timeline planning, 89
Topographical survey, 15

Ultimate bearing capacity, 20

Water, 33
Welding
  butt weld, 112
  fillet weld or lap weld, 111, 112
  manual arc welding, 110
  metal inert gas welding, 111
  stud welding, 113
  submerged arc welding, 111
Wind load
  basic wind speed, 63
  design wind pressure, 63, 69
  wind speed, 63

Printed in the United States
by Baker & Taylor Publisher Services